CISM COURSES AND LECTURES

The series presents lecture notes, monographs, edited works and proceedings in the field of Mechanics, Engineering, Computer Science and Applied Mathematics.
Purpose of the series in to make known in the international scientific and technical community results obtained in some of the activities organized by CISM, the International Centre for Mechanical Sciences.

INTERNATIONAL CENTRE FOR MECHANICAL SCIENCES

COURSES AND LECTURES - No. 348

ENGINEERING MECHANICS OF FIBRE REINFORCED POLYMERS AND COMPOSITE STRUCTURES

EDITED BY

J. HULT
CHALMERS UNIVERSITY OF TECHNOLOGY

AND

F.G. RAMMERSTORFER
VIENNA UNIVERSITY OF TECHNOLOGY

SPRINGER-VERLAG WIEN GMBH

Le spese di stampa di questo volume sono in parte coperte da
contributi del Consiglio Nazionale delle Ricerche.

This volume contains 131 illustrations

Originally published by Springer-Verlag Wien New York in 1994

In order to make this volume available as economically and as rapidly as possible the authors' typescripts have been reproduced in their original forms. This method unfortunately has its typographical limitations but it is hoped that they in no way distract the reader.

ISBN 978-3-211-82652-2 ISBN 978-3-7091-2702-5 (eBook)
DOI 10.1007/978-3-7091-2702-5

PREFACE

This monograph is based on the lecture notes of the 4th International IUTAM Summer School on "Engineering Mechanics of Fibre Reinforced Polymers and Composite Structures" held at the International Center for Mechanical Sciences (CISM) in Udine, Italy, from July 5 to 9, 1993 by (in alphabetic order) Professors M. CHRZANOWSKI - Politechnika Krakowska, Krakow, Poland; P. GUDMUNDSON - Royal Institute of Technology, Stockholm, Sweden; J. HULT - Chalmers University of Technology, Gothenburg, Sweden; G. NONHOFF - Flugzeug- und Triebwerksbau, FH-Aachen, Germany; F.G. RAMMERSTORFER - Vienna University of Technology, Vienna, Austria and - last but not least - G.S. SPRINGER- Stanford University, Stanford, CA, USA.

The book aims at giving an overview of current methods in engineering mechanics of FRP composites as well as composite or hybrid components and structures. Main emphasis is laid on basic micro and macro mechanics of laminates. Continuous as well as short fibre composites are studied, and criteria for different kinds of failure are treated. Micromechanical considerations for material characterization and mechanisms of static ductile and brittle rupture are studied, as well as FRP structures under thermal and dynamic loading programs. Optimum design, manufacturing process simulations and environmental effects are described as well.

These contents should make designers familiar with the opportunities and limitations of modern high quality fibre composites. Practical engineering applications of the described analytical and numerical methods are also presented.

We, as the organizers of the Summer School and as the editors of this monograph, would like to thank all our colleagues who contributed by their lectures and by preparing chapters of this book. We also thank the IUTAM General Assembly, especially the Secretary

General, Professor Franz Ziegler, for selecting our course to be held as the 4th IUTAM International Summer School. We thank CISM for organizing this course and for the hospitality which the lecturers were provided with during their stay at Udine. The lecturers and authors owe special thanks to Professor Sandor Kaliszky, resident Rector of CISM, for his efforts in supporting this course, and to Professor Carlo Tasso for encouraging the lecturers to write this monograph.

J. Hult

F.G. Rammerstorfer

CONTENTS

CHAPTER 1

INTRODUCTION

J. Hult
Chalmers University of Technology, Göteborg, Sweden

ABSTRACT

The specific properties of fibre reinforced materials, both natural and man made, are described. Early examples are mud bricks reinforced with straw, known already in antiquity. Modern examples are glass fibre reinforced polymers or ceramic fibre reinforced metals. The need to combine low weight with high stiffness has prompted the use of fibre composites in aircraft design, both military and civil.

Rational methods have been developed to analyse the mechanical behaviour of fibre composites under various kinds of loading. Material properties are derived by methods of micro mechanics, whereas structural properties are derived by methods of macro mechanics. A key concept in these analyses is the anisotropy occurring in fibre composites with parallel fibres.

Strength and stiffness data are briefly presented for various types of fibre composite materials. Examples of recent standard literature on fibre composites are given.

Natural and artificial fibre composites

Fibrous materials have been used by man since prehistoric times. Woven cloth as well as ropes are examples of materials entirely made up of fibres in different configurations. Wood consists of cellulose fibres cemented together by lignin; it is a *composite material*. The material in which the fibres are embedded, such as the lignin, is called the *matrix*. This term also applies to the substance between the cells in a living substance. It derives from the latin word *mater*, meaning also the female *uterus*. Since the cellulose fibres in wood are essentially parallel, and since their strength and stiffness are very different from those of lignin, the mechanical properties of wood are highly anisotropic. The longitudinal strength is much greater than the transverse strength.

A great number of fibrous composite materials have been developed, which utilize such properties. Fibre reinforced polymers (FRP), notably with glass fibres, were among the first artifical composites to reach large scale industrial production. More advanced - and more expensive - fibres have subsequently come into use, such as carbon or boron fibres. Carbon fibre reinforced polymers (CFRP) found their first market in the sports fields: vaulting poles, golf clubs, tennis rackets, etc. Gradually the cost has gone down, but is still prohibitively high for large scale use (ca 25 times the price of steel). Advanced fibre composites of this kind are however finding increasing use in the aircraft and automobile industries.

A recent development in the composites field is metal matrix composites (MMC). Common matrix metals are aluminium, magnesium and titanium. Fibres may be silicon carbide, alumina and boron. The cheapest types of MMC are particulate composites, often with silicon carbides as reinforcing particles. MMC compare favourably with polymer matrix composites (PMC) with respect to wear resistance and thermal performance (resistance of the material to fire). The improved properties at elevated temperatures have made it possible to use MMC in such applications as gas turbine engines.Among other advantages of modern composite materials the following are important:

- Corrosion is usually much less of a problem than for metals
- Damage tolerance is usually very high for CFRP
- Injection moulding and highly cost effective production methods may often be used
- Tooling is usually cheaper than for metals.

Disadvantages, sometimes causing problems, with the use of polymer composites are:

- Popular misconceptions about the lack of durability ("plastic toy syndrome")
- Lack of tradition among designers
- Lack of openness in industry; protection of trade secrets not found in the old metal industry
- Cost of changing existing from metal production tools to composite production tools
- Health problems (skin diseases)
- Fire hazard, liberation of smoke dangerous to health

The properties of a fibre composite depend on the mechanical properties of the fibres and of the matrix, but also on the geometrical distribution of the fibres and on the adhesion between fibres and matrix. Fibres may be of different length, from continuous fibres extending all through the material to very short fibres only a few diameters in length. One typical characteristic of fibre composites is that the fibre length is of only secondary importance. The strength of a rope is rather independent of the individual fibre length.

The primary reason for developing fibre reinforced materials has been to achieve improved strength and stiffness properties. A very early example of fibre reinforcement is mentioned in the Bible (*Exodus*, Ch. II), where Pharaoh ordered the children of Israel to collect straw to make bricks. The purpose of the straw was not primarily to increase the strength of the sun dried mud bricks but to regulate the drying process so as to avoid excessive cracking. But the presence of straw also had a beneficial influence on the strength of the bricks.

Another kind of "fibre" reinforcement came into use in the 19th century, viz. steel reinforcement of concrete. Flower pots made of reinforced concrete were patented in France in 1867. Subsequently iron and steel reinforced concrete found increasing use in water cisterns, staircases, bridges, etc. As theoretical analysis of the interaction between concrete and reinforcement developed, the results could be used to predict the optimum distribution of reinforcement in various elements such as beams, plates, etc. Much of this work has been of importance in the subsequent development of the mechanics of fibre reinforced polymer composites.

Fibre composite aircraft

Simultaneous requirements of low weight and high stiffness in aircraft design have led to early and extensive use of fibre composites.

Severe shortage of aluminium during the second World War prompted the British aircraft company de Havilland to design a military plane largely out of plywood. The fighter plane *Mosquito* proved to be a most successful design, and a very great number were produced and used until the end of the war. Much new knowledge, in particular relating to design of an advanced fibre composite structure and to the use of adhesives in composite structural design, was gained with the *Mosquitos*.

Much less successful was the huge transport aircraft *Hercules H-4*, designed by Howard Hughes in the USA towards the end of the second World War. Its spruce framework was covered by birch plywood. With a wing span of 97.5 meters it weighed 180 tonnes. Its first and only flight took place in 1947. Even though useful new knowledge was gained also with this project, aircraft development took a different turn with the arrival of the jet engine, and no new large plywood aircraft were built.

The arrival, after the second World War, of CFRP and other advanced artificial fibre composites gave new possibilities in aircraft design, first for military and then for civil aircraft. The US Airforce fighter plane *F-111* was the first to use boron fibre composites. European fighter planes such as the Saab *Gripen*, the Dassault *Rafale* and the Eurofighter Consortium *EFA* are using carbon fibre composites in the wings, in the tail fin and also in the front fuselage. The first commercial aircraft to use Kevlar fibre panels instead of aluminium (thereby saving 400 kilograms in weight) was the Lockheed *L-1011-500 Tristar*. In the European *A320 Airbus* composite materials make up 15 percent of the total weight. In the early 1980's the *Boeing 767* used nearly two tons of composite materials (floor beams and all control surfaces). The French Aerospatiale Aeritalia *ATR72 70* was the first commercial aircraft with all composite wings.

A convincing demonstration of the potential of advanced fibre composites in aircraft design was the round the world flight, without refueling, of the all composite *Voyager* aircraft in 1988.

Composite material mechanics

Composite material mechanics is closely related to the mechanics of homogeneous materials, such as metals. Equilibrium and compatibility relations are the same, irrespective of material properties. The constitutive relations are, however, different for different types of material. Whereas metals may usually be described as isotropic, composite materials with parallel fibres are often strongly *anisotropic*.

The mechanical behaviour of fibre reinforced structures may be analyzed on two different levels of scale:

Micro mechanics deals with the resulting *material properties* in terms of the constituent materials (fibres, matrix; fibre distribution). The most important aspects here are local stiffness and basic failure mechanisms of the material;

Macro mechanics deals with the resulting *structural properties* in terms of the material properties and the structural configuration. The most important aspects here are the stiffness and strength of the entire composite structure.

In micromechanical analysis each phase (matrix, fibres, particles) is often assumed to be a homogeneous and isotropic material. The composite can then be considered as a heterogeneous material with piecewise constant material properties. But fibres or particles are also assumed to be of such small size in relation to the material volume under study, that the composite material itself may be considered to be homogeneous on a macroscopic scale. This homogeneization is the key to a mechanical analysis of structural components of composite materials.

In most applications composite materials are subjected to such loads that the deformation is *linearly elastic*. This implies that the constitutive behaviour can be described by a generalized Hooke's law. A compact notation, using matrix algebra, has been developed, which greatly simplifies the mechanical analysis. The nomenclature is rather standardized, but readers of handbooks etc. should be aware of certain discrepancies. Some symbols are given different meanings by different authors.

Stiffness and strength data

The advantages and disadvantages of advanced fibre composites appear in comparisons with metals, the traditional materials of industrial production.

The until now most common type of fibre composite material is glass fibre reinforced epoxy. Typical values of Young's modulus is 70-85 GPa for standard glass fibres and 4-5 GPa for epoxies. These data should be compared to 200 GPa for steel or 70 GPa for aluminium. Since the density is also an important parameter in comparing the potentials of different materials, the lower density of glass and epoxy improves the relative situation of the glass fibre composites. The specific modulus (Young's modulus divided by density) for a typical glass fibre composite may be about half that of steel or aluminium. The specific tensile strength (tensile strength divided by density) may however be 3 times that of steel or aluminium. Thus, on a weight basis, glass fibre composites may offer advantages in comparison to steel or aluminium. The cost of glass fibres epoxies is also relatively low.

Continuous carbon fibres with a tensile stiffness exceeding 700 GPa and a tensile strength exceeding 7 GPa have been produced for commercial purposes. Unidirectional aramid (Kevlar) and carbon fibre reinforced epoxies, available since the early 1970's, have a specific modulus equal to 3½ - 5 times that of steel or aluminium and a specific tensile strength equal to 4-6 times that of steel or aluminium. Their fatigue endurance limit is nearly 60 percent of the ultimate strength, as compared to ca. 50 percent (often much less) for steel or aluminium. The cost of carbon fibres is however still prohibitively high for widespread use.

The usually higher damping in FRP may be utilized in applications where the noise level has to be suppressed. Carbon fibre reinforced nylon bearings with very low friction require almost no lubrication, but must not be subjected to large loads.

Examples of literature on fibre composites

A great number of textbooks, monographs, handbooks and journals deal with the theory and application of fibre composites. More recent results are often available through dissertations and other research reports. An *ad hoc* selection is given below.

1. Introductory and popular accounts

Tsu-Wei, Chou, Roy L. McCullough & R. Byron Pipes: "Composites". *Scientific American*, October 1986, pp 167-177.

James Edward Gordon, *The New Science of Strong Materials*. Penguin Books, Harmondsworth, England 1968. 269 pages.

James Edward Gordon, *The Science of Structures and Materials*. W.H. Freeman & Co, New York, USA 1988. 217 pages.

Anthony Kelly: "The Nature of Composite Materials". In: *Materials*. W.H. Freeman & Co, London 1967, pp 96-110 (from *Scientific American*, September 1967).

J.W. Martin, *Strong Materials*. Wykeham Publications, London 1972. 113 pages.

2. Composite materials

Donald F. Adams, *Test Methods for Composite Materials*. Technomic Publishing Company, Lancaster, PA, USA 1990. 340 pages.

Lawrence J. Broutman & Richard H. Krock, *Modern Composite Materials*. Addison-Wesley, Reading, MA, USA 1967. 581 pages.

Derek Hull, *An Introduction to Composite Materials*. Cambridge University Press 1981. 246 pages.

Antony Kelly, *Strong Solids*. Clarendon Press, Oxford 1966. 212 pages.

Ramesh Talreja, *Fatigue of Composite Materials*. Technomic Publishing Company, Lancaster, PA, USA 1987. 191 pages.

3. Composite material mechanics; textbooks and monographs

J. Aboudi, *Mechanics of Composite Materials - A Unified Micromechanical Approach*. Elsevier Science Publishers B.V.

Bhagwan D. Agarwal & Lawrence J. Broutman, *Analysis and Performance of Fiber Composites*. Wiley-Interscience, New York 1990. 355 pages.

K.H.G. Ashbee, *Fundamental Principles of Fiber Reinforced Composites*. Technomic Publishing Company, Lancaster, PA, USA 1989. 390 pages.

Krishan K. Chawla, *Composite Materials*. Springer-Verlag, Berlin 1987.

Daniel Gay, *Matériaux Composites*. 3. ed. Hermes, Paris 1991. 569 pages.

Leonard Holloway, *Glass Reinforced Plastics in Construction*. Surrey University Press, Glasgow 1978. 228 pages.

M. Holmes & D.J. Just, *GRP in Structural Engineering*. Applied Science Publishers, London 1983. 298 pages.

R. Hussein, *Composite Panels/Plates: Analysis and Design*. Technomic Publishing Company, Lancaster, PA, USA 1986. 366 pages.

Robert M. Jones, *Mechanics of Composite Materials*. Mc Graw-Hill, New York 1975. 355 pages.

Golam M. Nevaz, *Delamination in Advanced Composites*. Technomic Publishing Company, Lancaster, PA, USA 1991. 497 pages.

A.M. Skudra *et al*, *Structural Analysis of Composite Beam Systems*. Technomic Publishing Company, Lancaster, PA, USA 1991. 312 pages.

Y.M. Tarnopolski *et al*, *Spatially Reinforced Composites*. Technomic Publishing Company, Lancaster, PA, USA 1991. 352 pages.

Stephen W. Tsai, *Composites Design*. 3. ed. Think Composites, Dayton, OH, USA 1987. 509 pages.

Stephen W. Tsai & H. Thomas Hahn, *Introduction to Composite Materials*. Technomic Publishing Company, Lancaster, PA, USA 1980. 475 pages.

Jack R. Vinson & Tsu-Wei Chou, *Composite Materials and their Use in Structures*. Applied Science Publishers, London 1990. 435 pages.

Jack R. Vinson & R.L. Sierakowski, *The Behavior of Structures Composed of Composite Materials*. Martinus Nijhoff Publishers, Dordrecht, Netherlands 1986. 323 pages.

4. Anisotropic Elasticity

R.F.S. Hearmon, *An Introduction to Applied Anisotropic Elasticity*. Oxford University Press 1961. 136 pages.

S.G. Lekhnitskii, *Theory of Elasticity of an Anisotropic Elastic Body*. Holden-Day, San Francisco 1963. 404 pages.

5. Handbooks

Leif A. Carlsson & John W. Gillespie (Eds.), *Delaware Composites Design Encyclopedia*. Technomic Publishing Company, Lancaster. PA, USA:

Vol. 1, *Mechanical Behavior and Properties of Composite Materials*, 1989, 160 pages
Vol. 2, *Micromechanical Materials Modeling*, 1990, 248 pages
Vol. 3, *Processing and Fabrication Technology*, 1990, 256 pages
Vol. 4, *Failure Analysis of Composite Materials*, 1990, 216 pages
Vol. 5, *Design Studies*, 1990, 256 pages
Vol. 6, *Test Methods*, 1990, 170 pages
Index for Volumes 1-6, 1991, 40 pages

Cyril A. Dostal (Ed.), *Engineered Materials Handbook*, Volume 1: *Composites*. ASM International, Metals Park, OH, USA 1987. 983 pages.

S.V. Hoa, *Analysis for Design of Fiber Reinforced Plastic Vessels and Piping*. Technomic Publishing Company, Lancaster, PA, USA 1991. 587 pages.

A. Kelly (Ed.), *Concise Encyclopedia of Composite Materials*, Pergamon Press, Oxford 1989.

Stuart M. Lee (Ed.), *Dictionary of Composite Materials Technology*. Vol. 1-2. Technomic Publishing Company, Lancaster, PA, USA 1989. 333+205 pages.

George S. Springer, *Composite Plates Impact Damage: An Atlas*. Technomic Publishing Company, Lancaster, PA, USA 1991. 425 pages.

6. Symposia, etc.

Proceedings of the American Society for Composites. Technomic Publishing Company, Lancaster, PA, USA:

First Technical Conference 1986, 502 pages
Second Technical Conference 1987, 600 pages
Third Technical Conference 1988, 761 pages
Fourth Technical Conference 1989, 1009 pages
Fifth Technical Conference 1991, 1049 pages.

Proceedings of the Institution of Mechanical Engineers Conference: Design in Composite Materials 1989. Mechanical Engineering Publications Ltd, Bury St Edmunds, Suffolk, England 1989.

Proceedings of the SPI Composites Institute Annual Conference. Technomic Publishing Company, Lancaster, PA, USA:

42nd Conference 1987, 790 pages
43rd Conference 1988, 680 pages
44th Conference 1989, 684 pages
45th Conference 1990, 666 pages
46th Conference 1991, 726 pages

Zvi Hashin & Carl T. Herakovic (Eds.), *Mechanics of Composite Materials. Recent Advances.* Proceedings of IUTAM Symposium, Blacksburg, VA, USA 1982. Pergamon Press, New York 1983. 499 pages.

H. Lilholt & R. Talreja (Eds.), *Fatigue and Creep of Composite Materials.* Proceedings of the 3rd Risø International Symposium on Metallurgy and Materials Science, Risø National Laboratory, Roskilde, Denmark 1982. 341 pages.

A.J.M. Spencer (Ed.), *Continuum Theory of the Mechanics of Fibre-Reinforced Composites.* Lectures at International Centre for Mechanical Sciences (CISM), Udine, Italy 1981. Springer-Verlag, Vienna 1984. 284 pages.

F.W. Wendt, H. Liebowitz & N. Perrone (Eds.), *Mechanics of Composite Materials.* Proceedings of the Fifth Symposium on Naval Structural Mechanics, Philadelphia, PA, USA 1967. Pergamon Press, Oxford 1970. 886 pages.

7. Journals, Research Reports, etc.

ASTM Journal of Composites Technology and Research (until 1985: *Composites Technology Review*).

Composite Structures. Elsevier Applied Science Publishers, Barking, Essex, England.

Composites. Butterworths-Heinemann, Oxford, England.

Composites Engineering. Pergamon Press, Oxford, England.

Composites Manufacturing. Butterworths-Heinemann, Oxford, England.

Composites Science and Technology. Elsevier Applied Science Publishers, Barking, Essex, England.

Journal of Composite Materials. Technomic Publishing Company, Lancaster, PA, USA.

Journal of Reinforced Plastics and Composites. Technomic Publishing Company, Lancaster, PA, USA.

Journal of Thermoplastic Composite Materials. Technomic Publishing Company, Lancaster, PA, USA.

Polymer Composites, Society for Plastics Engineers.

CHAPTER 2

MICROMECHANICS FOR MACROSCOPIC MATERIAL DESCRIPTION OF FRPs

F.G. Rammerstorfer and H.J. Böhm
Vienna Technical University, Vienna, Austria

ABSTRACT

Overall or effective material parameters of composites usually are obtained by experimental methods applied to composite specimens. If such overall properties are not available, however, but constituents' parameters can be obtained, analytical or numerical micromechanical techniques must be used. The same methods can also be applied to the design of composites or to characterize the composite by recalculating individual parameters from the measured overall behavior. Therefore, in this Chapter some simple and some more advanced methods for determining the overall, i.e. effective or smeared out, material properties of fiber reinforced composites from the material data of their constituents and their micro-geometrical arrangement are presented.

It should be mentioned that some idealizations must be made in this engineering approach. For example, material parameters of the constituents measured separately for matrix (as a bulk material) and fibers normally differ from the in-situ properties.

List of Variables:

The following notations (with or without sub- or superscripts, respectively) are used for describing the material behavior:

E: Young's modulus
G: Shear modulus
ν: Poisson's ratio
K: Bulk modulus
K_q: Transverse bulk modulus (plane strain bulk modulus)
α: Linear coefficient of thermal expansion (CTE)
β: Linear coefficient of moisture expansion (CME)
κ: Thermal conduction coefficient
k: General transport coefficient
c_p: Heat capacity
μ: Moisture content
ρ: Mass density
σ_u: (Tensile) strength
$(\sigma_l^{(m)})_{\varepsilon_u^{(f)}}$: Longitudinal matrix stress at the fiber fracture strain
σ_{lCu}: Ultimate uniaxial compressive stress in fiber direction
σ_{lTu}: Ultimate tensile stress in fiber direction
σ_{qCu}: Ultimate compressive stress normal to fiber direction
σ_{qTu}: Ultimate tensile stress normal to fiber direction
τ_{lqu}: Ultimate shear stress
σ_Y: Yield stress
E_T: Hardening (tangential) modulus
$\Delta T_{cr,Y}$: Temperature difference giving rise to plastification due to CTE mismatch
E_{ij}: Elasticity tensor/matrix
C_{ij}: Compliance tensor/matrix; only in Section 1
V: Volume fraction
ξ: Fiber/particle volume fraction, for composite free of voids
ξ_{crit}: Critical fiber volume fraction
d: Fiber diameter
l: Fiber length
l_{crit}: Critical fiber length

Sub- and Superscripts:

$_l$: Longitudinal
$_q$: Transverse, mostly used in-plane
$_t$: Transverse, mostly used out-of-plane
$_p$: Fully yielded

$^{(i)}$: Inclusion
$^{(f)}$: Fibers
$^{(p)}$: Particles
$^{(m)}$: Matrix
: Overall effective – no superscript

Introduction

The overall material properties of the composites depend on the micro-scale behavior. The step from the micro- to the macro-scale behavior is described by homogenization techniques. As Table 1 shows, there are two main groups of micromechanics models for composites, mean field theories and unit-cell based methods. In this Chapter the main emphasis is put on the first group, i.e. the micro-stresses are not considered in detail by their local spatial distribution but by their mean values.

SOME MICRO-MECHANICAL METHODS

MEAN FIELD THEORIES:

- ⊙ **Rules of Mixture (bar models) - e,(ep)**

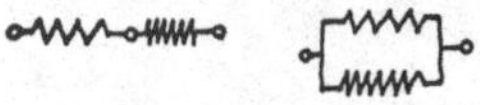

- ⊙ **Vanishing Fiber Diameter Model (VFD) - e, ep**
  (Dvorak, Bahei-el-Din, Krempl, Svobodnik ....)
- • **Composite Cylinder Assemblage - e**
  (Hashin ....)

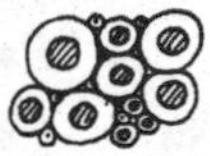

- ⊙ **Mori-Tanaka Type Models (MTM) - e, ep**
  (Wakashima, Pedersen, Weng, Benveniste,Dvorak, Böhm, Pettermann, Plankensteiner ....)
- • **Self Consistent Schemes (SCS) - e, ep**
  (Christensen, Berveiller ....)

UNIT CELL BASED METHODS

- • **Method of Cells - e, ep**
  (Aboudi ....)
- • **Semianalytical Methods based on Fourier-,Green's Functions - e, ep**
  (Nemat-Nasser, Walker ....)
- ⊙ **Numerical Unit Cell Approaches - e, ep**
  (Adams, Needleman, Tvergaard, Böhm, Weissenbek ....)

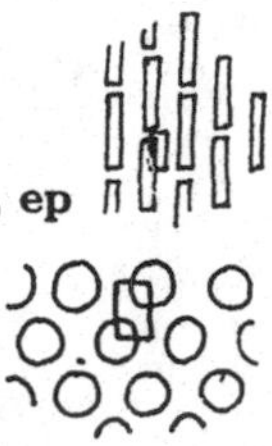

Table 1 Micromechanical methods

For convenience, simple engineering relations are covered in Sections 1 and 2, and more sophisticated mean field models are briefly discussed in Section 3. More detailed descriptions of mean field theories are given in another Chapter of this book, see [1]. It should be noted that the simple approaches do not necessarily lead to results which are consistent, i.e. the derived elastic constants are not assured to fulfill the required relations between each other; compare the "auxiliary relations" in Section 3.

With the notations $V^{(f)}$ for the volume fraction of the fibers, $V^{(m)}$ for the volume fraction of the matrix and the corresponding mass densities $\rho^{(f)}$ and $\rho^{(m)}$, the mass density of the composite is (for a composite free of voids)

$$\rho = \rho^{(f)} V^{(f)} + \rho^{(m)} V^{(m)} \qquad (1).$$

By eqn (1) the theoretical mass density is given. Assuming correct values for $V^{(j)}$ and $\rho^{(j)}$, $j = f, m$ as the in-situ values, the difference between the theoretical mass density ρ and the measured one ρ_c allows the determination of the void content in terms of void volume fraction

$$V^{(v)} = \frac{\rho - \rho_c}{\rho} \qquad (2).$$

Despite the fact that $V^{(v)}$ typically is smaller than 5% such small void contents can significantly affect the quality of composites. Especially the fatigue resistance and strength parameters are reduced by increased concentration of voids.

1 Some Simple Models for Unidirectional Continuous Fiber Composites

1.1 Stiffness Parameters

Assuming a perfect composite, i.e. no voids, perfect bonding between exactly aligned equally distributed fibers and a homogeneous matrix, no residual micro-stresses which normally appear as a consequence of the fabrication process, ..., the simplest description for unidirectional continuously reinforced composites are based on Voigt or Reuss models, i.e. models based on springs connected in series or in parallel, often called "rules of mixtures" (ROM).

Most of the following expressions in the present Section assume isotropic, linear elastic behavior for both fibers and matrix. Since the composite is assumed to be free of voids, the following relation holds: $V^{(f)} + V^{(m)} = 1$ and, hence, $V^{(m)} = 1 - V^{(f)}$. Therefore, the composite is characterized by the volume fraction of one component of the constituents. In the following considerations the fiber volume fraction is chosen to be this parameter and, in order to emphasize this fact, $V^{(f)}$ is replaced by ξ. Hence, $(1-\xi)$ stands for $V^{(m)}$.

For the description of the behavior in fiber direction, i.e. the longitudinal or l-direction, the following "strain coupling" assumptions

$$\varepsilon_l^{(f)} = \varepsilon_l^{(m)} = \varepsilon_l \quad , \qquad \sigma_l = \sigma_l^{(f)}\xi + \sigma_l^{(m)}(1-\xi)$$

lead to

$$E_l = E^{(f)}\xi + E^{(m)}(1-\xi) \tag{3}$$

known as "linear rule of mixtures".

Typical unidirectional composites exhibit transversely isotropic behavior. The simplest model for describing the behavior transverse to fiber direction, i.e. in any q-direction, assumes equal stresses together with the addition of strains:

$$\sigma_q^{(f)} = \sigma_q^{(m)} = \sigma_q \quad , \qquad \varepsilon_q = \varepsilon_q^{(f)}\xi + \varepsilon_q^{(m)}(1-\xi)$$

leading to

$$E_q = \frac{E^{(f)}E^{(m)}}{E^{(m)}\xi + E^{(f)}(1-\xi)} \tag{4}$$

A plot of E_l and E_q in terms of $E^{(m)}$ as a function of the fiber volume fraction ξ is shown in Fig. 1 for a unidirectional composite with $E^{(f)}/E^{(m)} \approx 25$ as calculated by eqns. (3) and (4). This figure shows clearly that reinforcing a matrix by stiff fibers mainly influences the stiffness in fiber direction, and rather high fiber volume fractions are necessary to obtain a significant stiffness increase in transverse direction. The fiber volume fraction cannot surpass a certain maximum value ξ_{max} which depends on the packing of the fibers. Theoretically, for fibers with constant circular cross sections $\xi_{max} = \pi/4 \approx 0.78$ for cubic packing and $\xi_{max} = \pi/(2\sqrt{3}) \approx 0.91$ for hexagonal packing. Hence, for ξ-values larger than ξ_{max} the curves in Fig. 1 are not physically meaningful.

Equation (4) is rather approximative. In practical engineering applications the semi-empirical Halpin–Tsai equations [2] are frequently used:

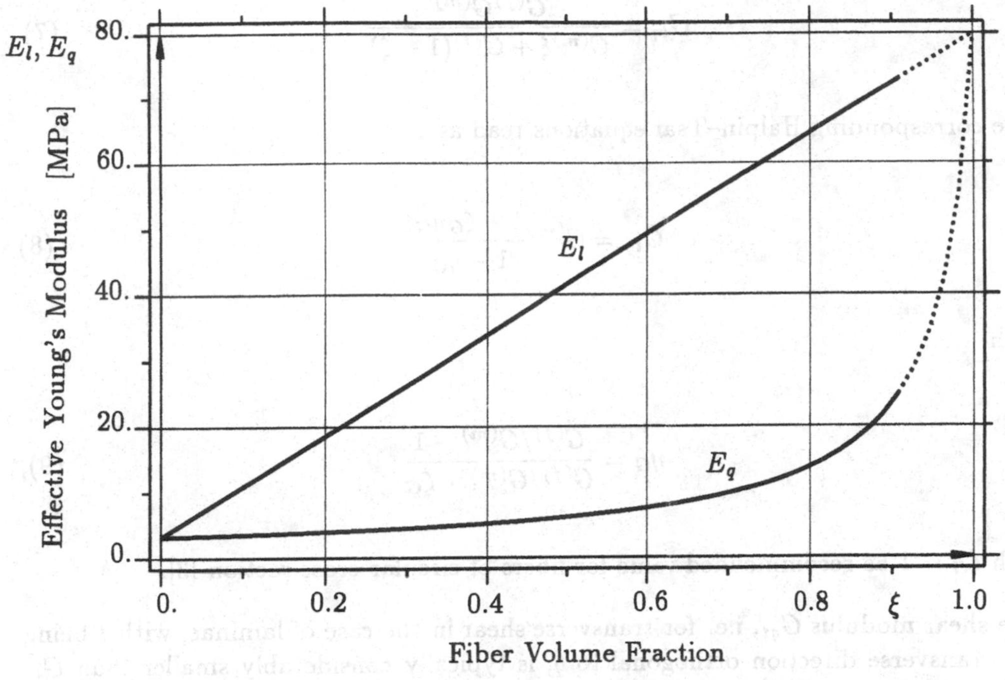

Fig. 1 Effective Young's moduli E_l and E_q (Voigt and Reuss models)

$$E_q = E^{(m)} \frac{1 + \zeta_E \eta_E \xi}{1 - \eta_E \xi} \tag{5}$$

with

$$\eta_E = \frac{E^{(f)}/E^{(m)} - 1}{E^{(f)}/E^{(m)} + \zeta_E} \tag{6},$$

In [2] a value of $\zeta_E = 2$ is recommended in the case of circular cross section of the fibers.

With respect to shear in l, q-planes, i.e. in-plane shear in the case of laminae, the simplest model assumes

$$\sigma_{lq} = \sigma_{lq}^{(f)} = \sigma_{lq}^{(m)} \quad , \quad \gamma_{lq} = \gamma_{lq}^{(f)} \xi + \gamma_{lq}^{(m)} (1 - \xi)$$

leading to

$$G_{lq} = \frac{G^{(f)}G^{(m)}}{G^{(m)}\xi + G^{(f)}(1-\xi)} \qquad (7).$$

The corresponding Halpin–Tsai equations read as

$$G_{lq} = G^{(m)}\frac{1+\zeta_G\eta_G\xi}{1-\eta_G\xi} \qquad (8)$$

with

$$\eta_G = \frac{G^{(f)}/G^{(m)}-1}{G^{(f)}/G^{(m)}+\zeta_G} \qquad (9),$$

with $\zeta_G = 1$ as recommended value for fibers of circular cross section [3].

The shear modulus G_{qt}, i.e. for transverse shear in the case of laminae, with t being any transverse direction orthogonal to q, is typically considerably smaller than G_{lq} and in [4] $G_{qt} \approx G^{(m)}$ is proposed. More precise values are derived by relations presented in Section 3.

Two Poisson's ratios, ν_{lq} and ν_{ql}, are required for the description of the behavior of unidirectional composites under in-plane loading. ν_{ij} is the Poisson's ratio for strain in direction j due to a uniaxial stress in direction i, i.e. $\nu_{lq} = -\frac{\varepsilon_q}{\varepsilon_l}$ in the case of uniaxial loading in l-direction. The Poisson's ratio ν_{tq} which relates transverse strains to orthogonal transverse ones is normally not needed when shell structures are considered (but can be computed easily as shown in Section 3).

The simple Voigt models assume

$$\nu_{lq}\varepsilon_l = \nu^{(f)}\varepsilon_l^{(f)}\xi + \nu^{(m)}\varepsilon_l^{(m)}(1-\xi) \quad , \qquad \varepsilon_l = \varepsilon_l^{(f)} = \varepsilon_l^{(m)} \quad ,$$

leading to

$$\nu_{lq} = \nu^{(f)}\xi + \nu^{(m)}(1-\xi) \qquad (10).$$

The Poisson's ratio ν_{ql} which typically is smaller than ν_{lq} can be obtained from the symmetry condition of the elasticity tensor, i.e. $E_{ij} = E_{ji}$, leading to

$$\nu_{ql} = \nu_{lq} \frac{E_q}{E_l} \tag{11}$$

with E_q and E_l from the above considerations.

The application of extremum principles of minimum potential and minimum complementary energy of the classical theory of elasticity leads to bounds on the elastic properties, see Section 3.

Very often fibers exhibit a transversely isotropic material behavior. For such cases simple alternative relations for the effective properties of unidirectional can be given, too, see e.g. [5]

$$E_l = \xi\, E_l^{(f)} + (1-\xi)\, E^{(m)}\,, \nu_{lq} = \xi\, \nu_{lq}^{(f)} + (1-\xi)\, \nu^{(m)} \qquad (12),$$

$$E_q = \frac{E^{(m)}}{1-\sqrt{\xi}(1-E^{(m)}/E_q^{(f)})} \qquad (13),$$

$$G_{lq} = \frac{G^{(m)}}{1-\sqrt{\xi}(1-G^{(m)}/G_{lq}^{(f)})} \qquad (14).$$

the above relations are specializations of the relations given in [6] where the influence of the void content, see eqn. (1), is also taken into account. In Section 3 additional expressions can be found for composites with anisotropic fibers.

Based on the so-called 'slice model' [7] in [4] the components of the elasticity tensor of a single unidirectional lamina can be derived directly from the properties of the transversely isotropic fibers, the isotropic matrix and the volume fractions.

1.2 Hygro-Thermal Behavior

In order to consider hygro-thermal stresses and deformations the knowledge of the corresponding effective expansion coefficients (describing the overall expansion behavior of the composite material) is needed in addition to the distribution of temperature changes ϑ (in degrees) and moisture changes μ (in mass fractions of the absorbed liquid related to the mass of the dry material, see eqn. (26)) with respect to the stress free reference states. In the present Section the thermal (and moisture induced) micro-stresses and strains due to the mismatch in the corresponding coefficients of the constituents are not considered (for this subject see e.g. [17,27]). Furthermore, one should bear in mind that temperature and moisture influence not only volume and shape changes but also mechanical properties (e.g. stiffness and strength parameters) as well as the transport coefficients (e.g. thermal or mass diffusivity).

Due to the globally orthotropic or, more strictly, transversely isotropic behavior of the unidirectional composite, temperature or moisture changes can lead to (unconstrained) strains only in the fiber direction or transverse to it:

$$^{th}\varepsilon_l = \alpha_l \vartheta \ , \quad ^{th}\varepsilon_q = \alpha_q \vartheta \ , \quad ^{th}\varepsilon_{lq} = 0 \qquad (15),$$

and

$$^{m}\varepsilon_l = \beta_l \mu \ , \quad ^{m}\varepsilon_q = \beta_q \mu \ , \quad ^{m}\varepsilon_{lq} = 0 \qquad (16).$$

Hence, the knowledge of the coefficients of thermal expansion (CTEs) α_l and α_q as well as of the coefficients of moisture expansion (CMEs) β_l and β_q is sufficient.

Based on energy considerations, simple assumptions in analogy to the ones used for stiffness considerations lead to useful formulas for the determination of the effective longitudinal and transverse CTE. Schapery [8] presents the following expressions:

$$\alpha_l = \frac{1}{E_l}(\alpha^{(f)}E^{(f)}\xi + \alpha^{(m)}E^{(m)}(1-\xi)) \qquad (17),$$

$$\alpha_q = (1+\nu^{(f)})\alpha^{(f)}\xi + (1+\nu^{(m)})\alpha^{(m)}(1-\xi) - \alpha_l \nu_{lq} \qquad (18).$$

E_l in eqn. (17) can be taken from eqn. (3), α_l and ν_{lq} in eqn. (18) are determined by eqns. (17) and (10), respectively.

For composite layers, i.e. laminae in laminates, with transversely isotropic fibers the corresponding equations are taken from [9]. α_l can be determined by eqn. (17) if $\alpha^{(f)}E^{(f)}$ is replaced by $\alpha_l^{(f)}E_l^{(f)}$, and the transverse CTE is given by

$$\alpha_q = (\alpha_q^{(f)} - \nu_{lq}^{(f)}(\alpha_l - \alpha_l^{(f)}))\xi + (\alpha^{(m)} - \nu^{(m)}(\alpha_l - \alpha^{(m)}))(1-\xi) \qquad (19).$$

Because the fibers usually have smaller (longitudinal) coefficients of thermal expansion (in some cases even negative ones) than the matrix materials, the longitudinal CTE of the composite is typically considerably smaller than the transverse one. The latter can, for small fiber volume fractions, even be larger than the thermal expansion coefficient of the matrix. This is a consequence of the constrained thermal expansion in longitudinal direction in combination with the Poisson effect. Similar to the effect of the reinforcement on the stiffness in fiber direction, i.e. E_l, even small fiber volume fractions may lead to a strong reduction of the longitudinal CTE and, consequently, to an improvement with respect to "thermal shape stability"; compare Fig. 2.

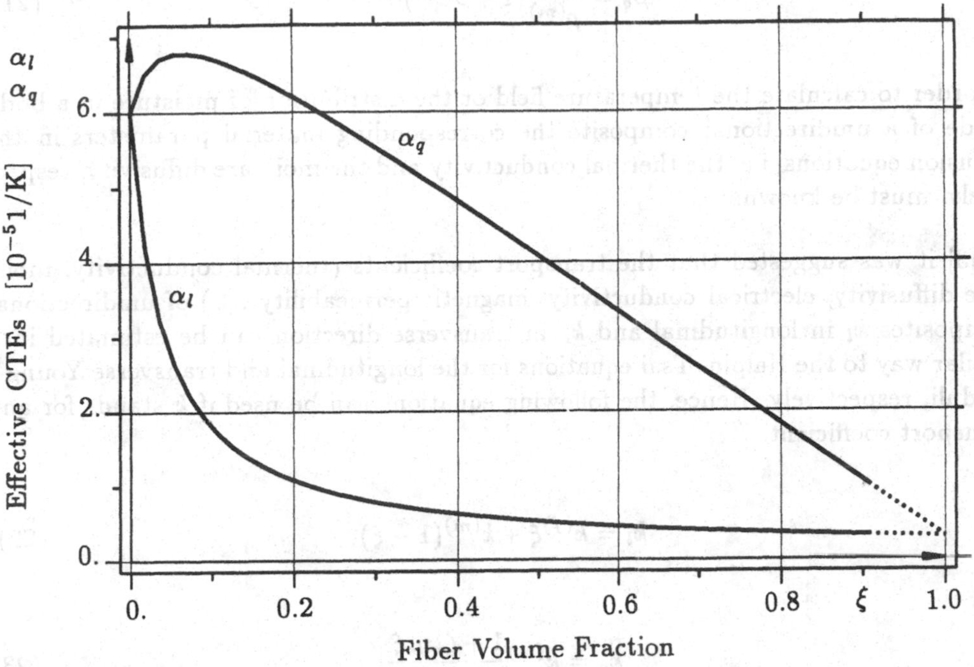

Fig. 2 Effective longitudinal and transverse coefficients of thermal expansion

If a material without voids absorbs moisture, as is the case for typical resin matrices, it shows volumetric changes (swelling). The linear coefficient of moisture expansion for an isotropic material is given by

$$\beta_i = \frac{1}{3}\frac{\rho_i}{\rho_{liq}} \tag{20}$$

where ρ_i and ρ_{liq} are the mass densities of the dry material and the absorbed liquid, respectively. If the material contains voids (even only a few percent) the coefficient of moisture expansion is significantly smaller than indicated by eqn. (20).

In typical fiber reinforced polymers the resin matrix absorbs moisture when exposed to a humid environment but the (inorganic) fibers do not. Hence, for relatively stiff fibers, i.e. large $E^{(f)}/E^{(m)}$ ratios, it is justified to take $\beta_l \approx 0$, compare [3]. The transverse CME is, however, influenced by this restriction of longitudinal swelling via the Poisson effect; according to [10] it is given by

$$\beta_q = \frac{\rho}{\rho^{(m)}}(1 + \nu^{(m)})\beta^{(m)} \qquad (21).$$

In order to calculate the temperature field or the distribution of moisture in a body made of a unidirectional composite the corresponding material parameters in the diffusion equations, i.e. the thermal conductivity and the moisture diffusivity, respectively, must be known.

In [3] it was suggested that the transport coefficients (thermal conductivity, moisture diffusivity, electrical conductivity, magnetic permeability ...) of unidirectional composites k_l in longitudinal and k_q in transverse direction can be estimated in a similar way to the Halpin–Tsai equations for the longitudinal and transverse Young's moduli, respectively. Hence, the following equations can be used if k stands for any transport coefficient:

$$k_l = k^{(f)}\xi + k^{(m)}(1 - \xi) \qquad (22),$$

$$k_q = k^{(m)}\frac{1 + \zeta_k \eta_k \xi}{1 - \eta_k \xi} \qquad (23)$$

with

$$\eta_k = \frac{k^{(f)}/k^{(m)} - 1}{k^{(f)}/k^{(m)} + \zeta_k} \qquad (24),$$

where $\zeta_k = 1$ in the case of fibers of circular cross section. The dependences of k_l and k_q on the fiber volume fraction ξ is similar to those of E_l and E_q, respectively, as shown in Fig. 1.

The calculation of the temperature fields follows the procedures well known for orthotropic monolithic materials if the thermal conductivity as described above and the specific heat, c_p, determined by the 'rule of mixtures'

$$c_p = c_p^{(f)}\xi + c_p^{(m)}(1 - \xi) \qquad (25)$$

are used.

Moisture distributions can be obtained by diffusive-type models in analogy to temperature fields, e.g. by finite element procedures. In many cases, as for example the determination of the moisture distribution over the thickness of a composite plate or shell with nearly constant conditions over the surfaces a one-dimensional consideration is sufficient. Simplified methods for analyzing the moisture diffusion process for these cases are given, for example, in [3,10].

The moisture content μ is defined by

$$\mu := \frac{W_{wet} - W_{dry}}{W_{dry}} \tag{26}$$

where W_{wet} and W_{dry} denote the weights of the moist and dry materials, respectively. The moisture content depends on the environmental conditions, the initial moisture distribution μ_0 in the body, the mass diffusivity of the composite material ${}^{m}k$ (which depends strongly on the temperature T in the body), the equilibrium moisture content μ_{eq} (which depends mainly on the moisture content of the environment) and the time t which the body is exposed to the environment under consideration.

The dependence of the equilibrium moisture content on the humidity of the environment, expressed by the relative humidity of the air Φ (in percent), can be approximated by [3]

$$\mu_{eq} = a\left(\frac{\Phi}{100}\right)^{b} \tag{27}$$

where a and b are experimentally determined material parameters of the composite.

1.3 Strength Predictions of Unidirectional Composites

Failure of fiber reinforced polymers is discussed in more detail in other Chapters of this book. In the present Section failure is only very briefly considered with respect to strength predictions under idealized conditions.

In typical fiber reinforced polymers the failure strain of the fibers is less than that of the matrix. Under this assumption two possible failure modes are considered for uniaxial tensile loading in fiber direction:

Mode a: If the fiber volume fraction is sufficiently large, i.e. $\xi > \xi_{min}$ (ξ_{min} will be determined later on), the matrix will not be able to support the entire load after failure of the fibers which – in this simple model – is assumed to take place if the composite is strained to the fiber fracture strain $\varepsilon_u^{(f)}$. Hence, under the assumptions stated in Subsection 1.1, the ultimate tensile strength can be estimated by

$$\sigma_{lTu} = \sigma_u^{(f)}\xi + (\sigma_l^{(m)})_{\varepsilon_u^{(f)}}(1-\xi) \quad \text{for} \;\; \xi > \xi_{min} \tag{28}$$

where $\sigma_u^{(f)}$ is the longitudinal tensile strength of the fibers, and $(\sigma_l^{(m)})_{\varepsilon_u^{(f)}}$ is the longitudinal matrix stress at the fiber fracture strain $\varepsilon_u^{(f)}$. It should be mentioned, however, that first local fiber breaks appear at much smaller stress levels (e.g. 50 % of the ultimate tensile strength) and up to complete rupture intermediate stages of the failure process can be observed such as fiber pullout, debonding of fibers, interface–matrix shear failure etc.

Mode b: For rather small fiber volume fractions, i.e. $\xi < \xi_{min}$, the matrix will be able to support the entire composite load when all the fibers are broken, i.e. for $\varepsilon_l > \varepsilon_u^{(f)}$, and the load can even be increased until the longitudinal matrix stress reaches the tensile strength of the matrix, i.e. $\sigma_l^{(m)} = \sigma_u^{(m)}$. If it is assumed that the fibers do not support any load for longitudinal composite strains larger than $\varepsilon_u^{(f)}$ the ultimate tensile strength can be estimated by

$$\sigma_{lTu} = \sigma_u^{(m)}(1-\xi) \quad \text{for} \;\; \xi < \xi_{min} \tag{29}$$

Figure 3 shows the variation of σ_{lTu} with ξ corresponding to eqns. (28) and (29), respectively.

From this figure it can be easily seen that ξ_{min} can be found by equating the right-hand sides of eqns. (28) and (29) which leads to

$$\xi_{min} = \frac{\sigma_u^{(m)} - (\sigma_l^{(m)})_{\varepsilon_u^{(f)}}}{\sigma_u^{(f)} + \sigma_u^{(m)} - (\sigma_l^{(m)})_{\varepsilon_u^{(f)}}} \tag{30}$$

As can also be seen from Fig. 3 the fiber volume fraction must exceed a certain critical value ξ_{crit} for achieving any strengthening effect at all. This value is obtained by equating $\sigma_u^{(m)}$ and the right-hand side of eqn. (28) leading to:

$$\xi_{crit} = \frac{\sigma_u^{(m)} - (\sigma_l^{(m)})_{\varepsilon_u^{(f)}}}{\sigma_u^{(f)} - (\sigma_l^{(m)})_{\varepsilon_u^{(f)}}} \tag{31}$$

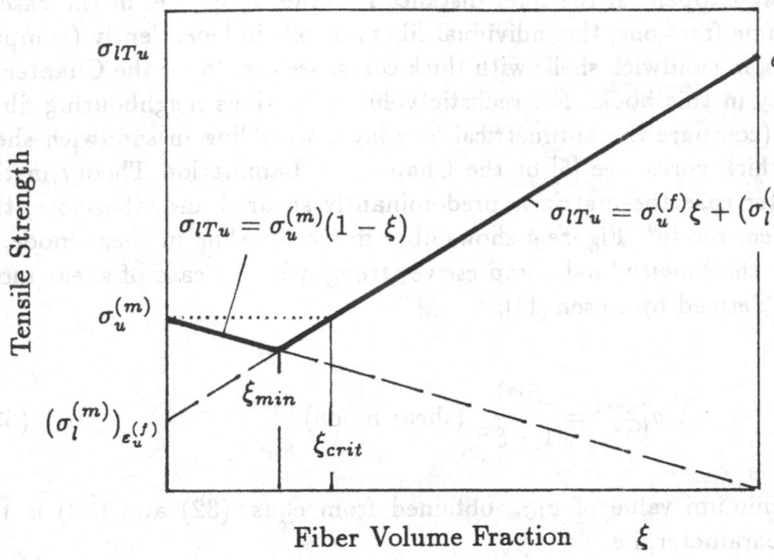

Fig. 3 Variation of longitudinal tensile strength with fiber volume fraction

For typical composites with polymeric matrices ξ_{crit} is a rather small value. However, when a strong matrix is to be reinforced by fibers of limited strength, like in the case of some metal matrix composites, large fiber volume fractions will be required in order to strengthen the matrix.

If a unidirectional composite is compressed in fiber direction several failure modes may appear:

Mode a: Due to the Poisson effect the global longitudinal compressive stress in the composite can result in transverse tensile stresses in the matrix, concentrated around the fibers, and fiber–matrix interface failure or transverse splitting may take place. This mode is called "transverse tensile mode". As outlined in [3] the corresponding longitudinal compressive strength can be estimated by

$$\sigma_{lCu}^{crack} = \frac{(E^{(f)}\xi + E^{(m)}(1-\xi))(1-\xi^{1/3})\varepsilon_u^{(m)}}{\nu^{(f)}\xi + \nu^{(m)}(1-\xi)} \quad \text{(transverse tensile mode)} \qquad (32),$$

where $\varepsilon_u^{(m)}$ is the matrix ultimate strain.

Mode b: The fibers in the unidirectional composite compressed in fiber direction act as long thin columns elastically supported by the surrounding matrix. Accordingly, micro-buckling may happen. If the fiber distance is rather large, i.e. in the case of very low fiber volume fractions, the individual fibers buckle independently (compare face layer wrinkling in sandwich shells with thick cores, see e.g. [5] or the Chapter on Lamination Theory in this book. For realistic volume fractions neighbouring fibers buckle in parallel (compare the antimetrical face layer wrinkling in sandwich shells with moderately thick cores, see [5] or the Chapter on Lamination Theory in this book). In the latter case the matrix is predominantly sheared and, therefore, this mode is called "shear mode". Figure 4 shows fiber micro-buckling in shear mode. A simple formula for the longitudinal compressive strength in the case of shear mode fiber buckling was derived by Rosen [11]:

$$\sigma_{lCu}^{buckl} = \frac{G^{(m)}}{1-\xi} \quad \text{(shear mode)} \qquad (33).$$

Of course, the minimum value of σ_{lCu} obtained from eqns. (32) and (33) is the relevant strength parameter, i.e.

$$\sigma_{lCu} = \min\,(\sigma_{lCu}^{crack}, \sigma_{lCu}^{buckl}) \qquad (34).$$

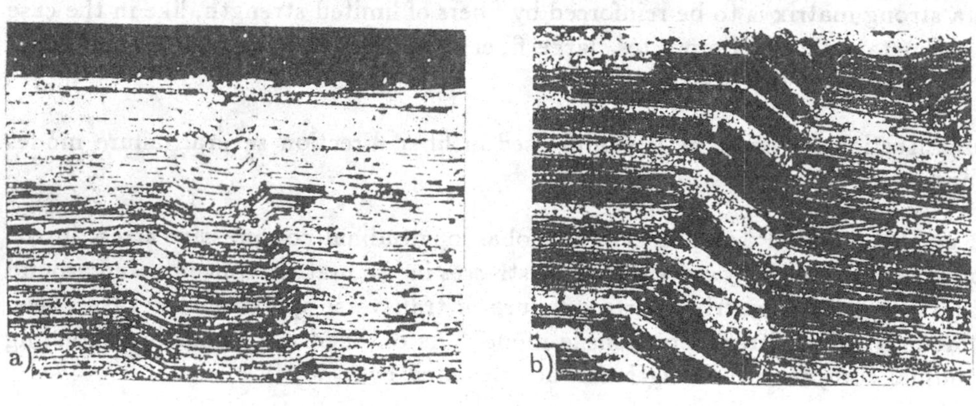

Fig. 4 Fiber buckling in shear mode; a) compression zone of a three-point bending specimen, b) detail: fiber breakage due to buckling

More advanced models for fiber buckling are presented e.g. in [12] and [13], where it is found that the compressive strength depends on the wavelength of initial imperfections. Without imperfections the results obtained in [13] correspond to Rosen's predictions, but taking imperfections into account, a better correlation with experimental results could be achieved.

These models do not take into account that, in the case of layered composites, the plies adjacent to the critical axial ply offer some support, see e.g. [14].

So far longitudinal loading has been considered. If a transverse tensile load is applied the fibers (typically having much higher Young's moduli than the matrix) act as constraints to the matrix deformations and lead to stress concentrations in the matrix. Hence, the fiber reinforced composite shows a much lower transverse ultimate tensile strain than the unrestrained matrix, leading to a reduction of the transverse tensile strength because of the fibers! In [15,16] micromechanical considerations are discussed based on analytically derived stress concentration factor matrices. More precise determinations of stress concentration factors, depending on the micro-geometry, can be obtained by numerical computations, see e.g. [17]. Micromechanical analyses which consider the stress field in detail, i.e. periodic micro-field approximations are used instead of mean field approximations, allow the application of damage mechanics on the micro level. This way individual failure modes (damage in the matrix, in the fibers or at the interface) and their interaction as well as micro-geometrical arrangement effects can be considered, see e.g. [18,19,20].

A rather rough estimate of the transverse tensile strength is based on the assumption of perfect bonding between the constituents, and on matrix failure if the maximum tensile stress reaches the matrix ultimate strength, $\sigma_u^{(m)}$. This leads, according to [3], to

$$\sigma_{qTu} = \frac{\sigma_u^{(m)}}{S} \quad (35),$$

with the stress concentration factor

$$S = \frac{1 - \xi(1 - E^{(m)}/E^{(f)})}{1 - (4\xi/\pi)^{1/2}(1 - E^{(m)}/E^{(f)})} \quad (36).$$

Failure under transverse compression normally takes place by matrix shear failure accompanied by debonding at the fiber–matrix interface resulting in a transverse compressive strength σ_{qCu} lower than the longitudinal compressive strength. However, if the deformations normal to the load–fiber plane are constrained the matrix shear failure is delayed and the mode may change to fiber shear failure leading to an increased transverse compressive strength especially for higher fiber volume fractions.

The mode of failure under in-plane shear loading depends strongly on the strength of the fiber–matrix interface. Strong interfaces combined with a weak matrix lead to matrix shear failure. However, typically matrix shear failure is accompanied by constituent debonding. For weak interfaces the in-plane shear strength τ_{lqu} is almost independent of the matrix and fiber strengths. It is recommended to determine τ_{lqu} and σ_{qCu} experimentally.

The above formulae for estimating strength parameters are derived under a number of idealizing assumptions. It should be mentioned that the following circumstances in real composites will lower the strength as compared to the values obtained from the above relations: misorientation of the fibers, fiber clustering, fibers of nonuniform strength, discontinuous fibers, imperfect bonding between fibers and matrix, residual micro-stresses resulting from fabrication of the composite due to thermal expansion mismatch (additional fabrication induced residual stresses in laminates are caused by the difference in the thermal contraction of the individual plies). Furthermore, it should be mentioned that hygro-thermal loading of composites leads to hygro-thermal stresses not only on the macro but also on the micro level. Hence, care has to be taken in assessing of the macroscopic hygro-thermal stresses by using strength data achieved from investigations under mechanical loading. For unidirectional metal matrix composites detailed descriptions of these phenomena can be found in [21].

From the above one can conclude that unidirectional continuous fiber composites show high stiffness and strength in fiber direction but less favorable properties for transverse loading. Hence, unidirectional composites are efficient for structures with pronounced unidirectional stress states. In most cases, however, engineering structures are exposed to loads leading to multiaxial stress states. Therefore, laminates consisting of unidirectional or woven fiber laminae are typically used with proper choice of the stacking sequence of the laminae with respect to fiber angle and other ply parameters. This topic is treated in the Chapter on Lamination Theory in this book. In situations where the stress state is not predictable to be more or less unidirectional or where normal stresses in all directions are approximately the same, the use of unidirectional composites may not be appropriate, and short-fiber composites with randomly oriented fibers might be much more cost effective. Hence, in the following Section short-fiber composites are considered.

2 Some Simple Models for Short-Fiber Composites

2.1 Stress Transfer Between Fiber and Matrix

In short-fiber composites the matrix transmits the applied load to the fibers, which are the principal load-bearing components. Figure 5 schematically shows the deformation pattern in a unit-cell of an aligned short-fiber composite (high-modulus fiber embedded in a low-modulus matrix) due to axial tensile loading.

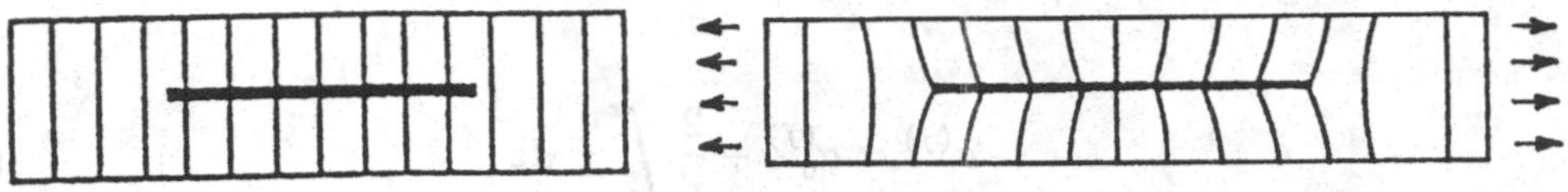

Fig. 5 Load transmission from matrix to fiber

In the situation shown the tensile load is transferred from the matrix to the fiber by tensile forces at the fiber end surfaces and by shear stresses along the cylinder surface of the fiber, especially near the fiber ends. These end effects are essential for characterizing discontinuous-fiber composites, i.e. the properties of the composite are functions of the fiber length. When the fiber lengths are much larger than the length along which the stress transfer takes place the composite can be treated as continuously reinforced, see the above Section.

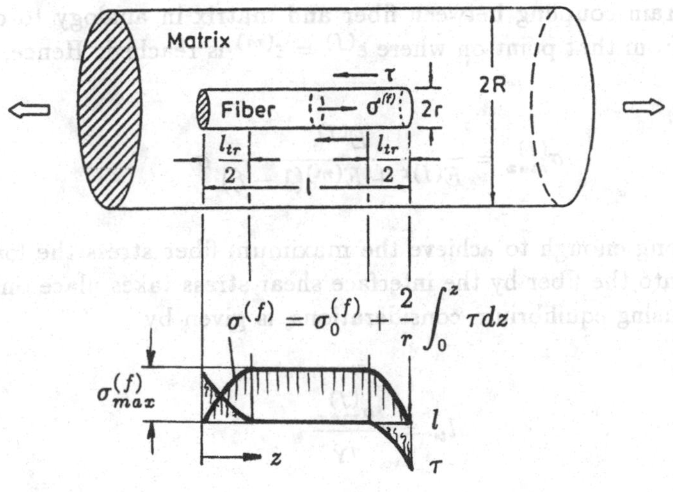

Fig. 6 Shear lag model

One of the simplest shear-lag analyses is based on Rosen's theory [11]. Assuming a simplified stress field as shown in Fig. 6, with τ being the longitudinal interface shear stress between fiber and matrix, equilibrium of the fiber element of length dz requires

$$\frac{d\sigma^{(f)}}{dz} = \frac{2\tau}{r} \qquad (37),$$

which, in integrated form, leads to

$$\sigma^{(f)} = \sigma_0^{(f)} + \frac{2}{r}\int_0^z \tau \, dz \qquad (38).$$

$\sigma_0^{(f)}$ is the fiber stress at the fiber end which in many analyses is neglected [3]. Doing this and assuming the interface shear stress to be constant with $\tau = \tau_Y$ along the load transfer length, which is a very rough approximation (justified by plastification of the matrix), eqn. (37) leads to

$$\sigma^{(f)} = \frac{2\,\tau_Y z}{r} \qquad (39),$$

as long as the maximum fiber stress accepted by the fiber embedded in the matrix is not reached. The limitation of the fiber stress by the maximum fiber stress $\sigma_{max}^{(f)}$ corresponds to strain coupling between fiber and matrix in analogy to continuous-fiber composites from that point on where $\varepsilon^{(f)} = \varepsilon^{(m)}$ is reached. Hence,

$$\sigma_{max}^{(f)} = \frac{E^{(f)}}{E^{(f)}\xi + E^{(m)}(1-\xi)}\,\sigma \qquad (40).$$

If the fibers are long enough to achieve the maximum fiber stress the load transfer from the matrix into the fiber by the interface shear stress takes place only over the length l_{tr} which, using equilibrium considerations, is given by

$$l_{tr} = r\frac{\sigma_{max}^{(f)}}{\tau_Y} \qquad (41).$$

l_{tr} is the sum of the lengths at both fiber ends at which shear stresses are acting between fiber and matrix. Since $\sigma_{max}^{(f)}$ is a function of the applied stress, σ, the "load-transfer length", l_{tr}, too, depends on σ. Figure 7 shows the result of this simplest model analysis.

Using this model a "critical fiber length", l_{crit}, can be defined as the minimum fiber length which allows the maximum allowable fiber stress, e.g. the fiber strength $\sigma_u^{(f)}$, to be reached:

$$l_{crit} = r\frac{\sigma_u^{(f)}}{\tau_Y} \quad \text{or} \quad \frac{l_{crit}}{d} = \frac{\sigma_u^{(f)}}{2\tau_Y} \qquad (42),$$

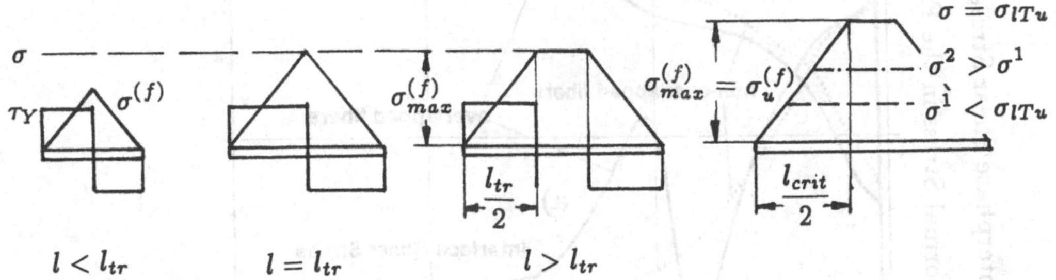

Fig. 7 Rosen's model

with $d = 2r$ being the fiber diameter. l_{crit} is an important parameter for short-fiber composites and affects the strength properties of the composite.

Furthermore, an "average fiber stress" can be defined, which – using the simple model – is given by

$$\bar{\sigma}^{(f)} = \frac{1}{l}\int_0^l \sigma^{(f)}\, dz \tag{43}$$

leading to

$$\bar{\sigma}^{(f)} = \frac{1}{2}\sigma^{(f)}_{max} \qquad \text{for} \quad l \le l_{tr} \tag{44}$$

$$\bar{\sigma}^{(f)} = \sigma^{(f)}_{max}\left(1 - \frac{l_{tr}}{2l}\right) \qquad \text{for} \quad l > l_{tr} \tag{45}$$

Even though this model with its simplifying assumptions is widely used, it should be mentioned that the load transfer mechanisms in reality are much more complicated. Hence, more advanced shear-lag models have been published, see for example [22,23]. In [20] numerical micromechanical analyses are presented which show that there is a pronounced effect of the arrangement of the short fibers on the micro-stress field. Using this kind of analysis, shear stress distributions along the fiber surface and normal stress distributions in the fiber can be computed as shown in Fig. 8. This figure shows that the assumption of a constant shear stress τ_Y along the load-transfer length is very approximative even for a fully yielded matrix.

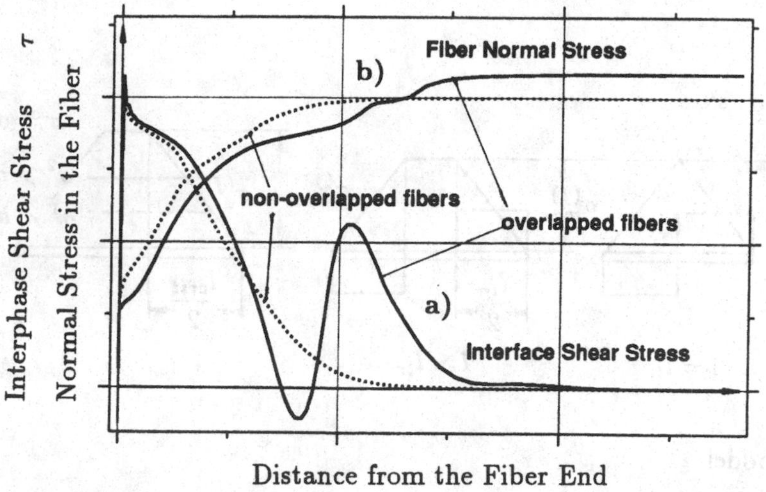

Fig. 8 a) Axial interphase shear stresses and b) fiber normal stresses near the fiber end, (for overlapping staggered and for non-overlapping fibers, respectively)

2.2 Stiffness Parameters

For engineering estimates the Halpin–Tsai equations presented in Section 1 for continuous fiber composites can also be used in modified form for short-fiber composites. In the case of aligned short-fiber composites, i.e. the fibers are all oriented in the same direction (which hardly is met in reality), transversely isotropic behavior can be assumed and the Young's modulus in fiber direction is estimated by

$$E_l = E^{(m)} \frac{1 + \eta_{E_l} \xi (2l/d)}{1 - \eta_{E_l} \xi} \qquad (46),$$

with

$$\eta_{E_l} = \frac{E^{(f)}/E^{(m)} - 1}{E^{(f)}/E^{(m)} + (2l/d)} \qquad (47),$$

and in transverse direction

$$E_q = E^{(m)} \frac{1 + 2\eta_{E_q} \xi}{1 - \eta_{E_q} \xi} \qquad (48),$$

with

$$\eta_{E_q} = \frac{E^{(f)}/E^{(m)} - 1}{E^{(f)}/E^{(m)} + 2} \quad (49).$$

Figure 9 shows the longitudinal stiffnesses of short-fiber composites as a function of fiber aspect ratio l/d. For comparison the longitudinal stiffnesses of the corresponding continuous-fiber composites with the same fiber volume fractions are also shown. It becomes obvious that for large fiber aspect ratios the behavior of short-fiber composites approaches that of continuous-fiber composites.

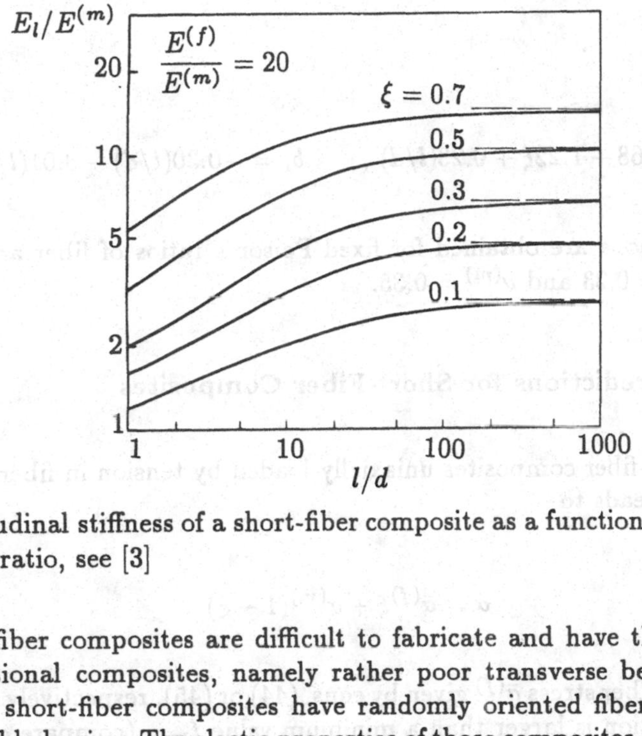

Fig. 9 Longitudinal stiffness of a short-fiber composite as a function of fiber aspect ratio, see [3]

Aligned short-fiber composites are difficult to fabricate and have the disadvantage of all unidirectional composites, namely rather poor transverse behavior. Hence, in many cases short-fiber composites have randomly oriented fibers leading to an isotropic overall behavior. The elastic properties of those composites can be predicted from the properties of aligned short-fiber composites by averaging over the angular distributions. Statistically distributed fiber orientations are treated for example in [24].

In [3] empirical formulae are given for predicting the elastic moduli of short-fiber composites with randomly distributed fibers:

$$E = \frac{3E_l + 5E_q}{8}, \qquad G = \frac{E_l + 2E_q}{8} \quad (50).$$

With eqn. (50) the Poisson's ratio of this isotropic composite is given, too, via $G = E/(2(1+\nu))$.

Based on a series of micromechanical finite element analyses, equations for the estimation of the effective Young's modulus of composites with short fibers which are randomly in-plane oriented are derived in [25]. As a starting point the general Halpin–Tsai equation – see eqns. (5) and (6) – are used, and the parameter $\zeta_E = \zeta(E^{(m)}, E^{(f)}, \xi, l/d)$ is found by regression analysis to be:

$$\zeta_E = b_0 + b_1 exp\left(-0.51\sqrt{E^{(f)}/E^{(m)}}\right) \qquad (51),$$

with

$$b_0 = 0.68 + 1.22\xi + 0.23(l/d)\ , \qquad b_1 = -0.30(l/d) - 0.01(l/d)^2 \qquad (52).$$

The above equations are obtained for fixed Poison's ratios of fiber and matrix, respectively: $\nu^{(f)} = 0.23$ and $\nu^{(m)} = 0.35$.

2.3 Strength Predictions for Short-Fiber Composites

For aligned short-fiber composites uniaxially loaded by tension in fiber direction the rule of mixtures leads to

$$\sigma = \bar{\sigma}^{(f)}\xi + \sigma^{(m)}(1-\xi) \qquad (53),$$

with the average fiber stress $\bar{\sigma}^{(f)}$ given by eqns. (44) or (45), respectively. Hence, if the fiber volume fraction is larger than a minimum value ξ_{min} (compare considerations in Subsection 1.2) and if the fiber length is greater than the critical fiber length, l_{crit}, the fiber stress can reach its maximum endurable value $\sigma_u^{(f)}$ and failure takes place due to fiber breaking at the longitudinal failure stress

$$\sigma_{lTu} = \sigma_u^{(f)}\left(1 - \frac{l_{crit}}{2l}\right)\xi + \left(\sigma^{(m)}\right)_{\epsilon_u^{(f)}}(1-\xi)\ , \quad \text{for} \quad l > l_{crit} \qquad (54)$$

and

$$\sigma_{lTu} = \sigma_u^{(f)}\xi + \left(\sigma^{(m)}\right)_{\epsilon_u^{(f)}}(1-\xi)\ , \quad \text{for} \quad l \gg l_{crit} \qquad (55),$$

where $(\sigma^{(m)})_{\varepsilon_u^{(f)}}$ is the matrix stress at the fiber fracture strain $\varepsilon_u^{(f)}$. For many practical applications $(\sigma^{(m)})_{\varepsilon_u^{(f)}}$ is replaced by the matrix tensile strength $\sigma_u^{(m)}$ and eqn. (55) reads

$$\sigma_{lTu} \approx \sigma_u^{(f)}\xi + \sigma_u^{(m)}(1-\xi)\,, \quad \text{for} \quad l \gg l_{crit} \tag{56}$$

In analogy to eqns. (30) and (31) the minimum and the critical fiber volume fractions, respectively, are given by

$$\xi_{min} = \frac{\sigma_u^{(m)} - (\sigma^{(m)})_{\varepsilon_u^{(f)}}}{\bar{\sigma}^{(f)} + \sigma_u^{(m)} - (\sigma^{(m)})_{\varepsilon_u^{(f)}}} \tag{57}$$

$$\xi_{crit} = \frac{\sigma_u^{(m)} - (\sigma^{(m)})_{\varepsilon_u^{(f)}}}{\bar{\sigma}^{(f)} - (\sigma^{(m)})_{\varepsilon_u^{(f)}}} \tag{58}$$

If the fiber length is smaller than the critical length, the fiber stress will not reach $\sigma_{max}^{(f)}$, i.e. the fiber will not be loaded up to its failure stress even for arbitrarily increased applied stress. Hence the aligned composite fails due to matrix or interface failure, and the longitudinal strength can be estimated by

$$\sigma_{lTu} = \frac{\tau_Y l}{d}\xi + \sigma_u^{(m)}(1-\xi)\,, \quad \text{for} \quad l < l_{crit} \tag{59}$$

Strength predictions on the basis of damage mechanics are presented for example in [19,20]. With this approach the interaction of matrix, interface and fiber failure can be studied.

For randomly oriented short-fiber composites predictions of strength may be based on an averaging of the angular dependence of the strength of the corresponding aligned short-fiber composite or, as proposed in [26], the strength of the randomly oriented composite is assumed equal to the strength of a quasi-isotropic laminate constructed from unidirectional plies. Construction of quasi-isotropic laminates and strength predictions for laminates are treated in the Chapter on Lamination Theory in this book.

3 Advanced Micromechanical Models

It was the aim of the previous Sections to describe the derivation of simple formulae for predicting stiffness and strength properties of fiber reinforced polymers based on some simplifying assumptions. Advanced micromechanical models can be based either on discretization methods, such as unit-cell approaches using the finite element method [19,27], or on analytical methods, essentially mean field theories. The advantage of the unit-cell approaches is that local effects can be treated with much more accuracy than in the case of mean field theories. However, the models must fulfill some symmetry requirements in order to retain the overall material symmetry of the composite under consideration. For example a simple square arrangement of fibers or a simple cubic arrangement of particles, respectively, would not lead to transversely isotropic or isotropic macroscopic behavior, respectively. To achieve transverse isotropy, hexagonal symmetry of the unit cell is required, compare the findings by Love for the elasticity of crystals [28].

In the present Section a summary of more advanced expressions for the analytical description of effective material parameters of composites is presented. This summary represents alternative approaches published in the literature. Unified notations are introduced. The inclusions, i.e. reinforcing fibers (continuous or short fibers) or particles, are assumed to be linear elastic (isotropic or transversely isotropic), and the linear elastic (or, in some expressions, elastic-plastic) matrix is assumed to behave isotropically. Perfect bonding between the constitutents and no voids are further assumptions. For the interested reader further expressions can be found, for example, in a review article by Hashin [29] and in [21]. For expressions for composites with imperfect interfaces see e.g. [30]. A review on and comparisons of different mean field theories can be found in [31]. The very recent review paper [32] is written from the point of view of a solid state physicist.

3.1 The Mori–Tanaka-Method

Advanced analytical models for the overall behavior of elastic short fiber reinforced composites typically make use of the so-called Eshelby tensor [33], from which expressions for the elastic fields in and around a single ellipsoidal inclusion in an infinite matrix can be obtained. Figure 10 shows a schematic sketch of Eshelby's idea.

Whereas Eshelby's original approach is applicable only to highly dilute composites ($\xi < 10\%$), improved mean field theories, such as the Mori–Tanaka methods [34–37] and many self consistent schemes [37], are capable of accounting for the interactions between neighbouring inclusions and thus can describe aligned composites throughout the physically meaningful range of volume fractions. Because the Eshelby tensor

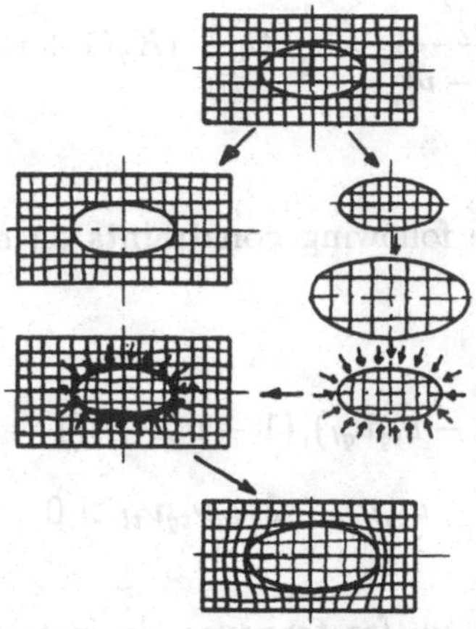

Fig. 10 Schematic scetch of Eshelby's model

depends on the inclusions' aspect ratio A, methods based on it cover aligned reinforcements ranging from continuous fibers ($A \to \infty$) to spheres ($A = 1$) to thin platelets ($A \to 0$). In another Chapter of this book, see [1], details of Mori–Tanaka-Approaches are given.

Generally, such approaches involve some amount of matrix algebra [31], which can be easily handled by simple computer codes. Analytical closed form expressions for the overall moduli can only be given for the limiting cases of continuous fibers and spherical inclusions; see following subsections.

3.2 Composites Unidirectionally Reinforced by Continuous Fibers

As already mentioned, unidirectional composites show a transversely isotropic behavior. For this kind of material symmetry only five independent material parameters, e.g. $E_l, E_q, G_{lq}, \nu_{lq}, \nu_{qt}$, are sufficient for describing the linear isothermal elastic behavior. Hence, the following relations are useful for expressing individual parameters by others:

Auxiliary Relations for Transversely Isotropic Media

$$\frac{4}{E_q} = \frac{1}{G_{qt}} + \frac{1}{K_q} + \frac{4\nu_{lq}^2}{E_l}, \quad K_q = \frac{E_l}{2[(1-\nu_{qt})(E_l/E_q) - 2\nu_{lq}^2]}$$

$$E_q = \frac{4K_qG_{qt}}{K_q + mG_{qt}}, \quad G_{qt} = \frac{E_q}{2(1+\nu_{qt})}, \quad \nu_{qt} = \frac{E_q - 2G_{qt}}{2G_{qt}}, \quad m = 1 + \frac{4K_q\nu_{lq}^2}{E_l} \quad (60).$$

$$K = \frac{E_l}{9}[1 + \frac{(1+\nu_{lq})^2}{\frac{(1-\nu_{qt})}{2}\frac{E_l}{E_q} - \nu_{lq}^2}] = \frac{1}{9}[E_l + 4K_q(1+\nu_{lq})^2]$$

In addition, Lempriere [38] stated the following constraints on elastic constants of orthotropic materials:

$$E_l, E_q, E_t, G_{lq}, G_{lt}, G_{qt} > 0, \quad (1-\nu_{lq}\nu_{ql}), (1-\nu_{lt}\nu_{tl}), (1-\nu_{tq}\nu_{qt}) > 0,$$
$$1 - \nu_{lq}\nu_{ql} - \nu_{lt}\nu_{tl} - \nu_{tq}\nu_{qt} - 2\nu_{lq}\nu_{tq}\nu_{tl} > 0 \qquad (61).$$

Because of $E_q = E_t, G_{lq} = G_{lt}, \nu_{lq} = \nu_{lt}$ for transversely isotropic materials, and $\nu_{lq}E_q = \nu_{ql}E_l$ for orthotropic layers under plane stress conditions, the above relations lead to the following constraints on the Poisson's ratios:

$$|\nu_{ij}| < \left(\frac{E_i}{E_j}\right)^{\frac{1}{2}} \quad \text{for} \quad i,j = l,q,t\ ; \quad i \neq j \qquad (62).$$

For further details on the constraints of elastic constants see e.g. [39,40].

a) Rules of Mixture

For the sake of completeness results of the simple ROM-models discussed in Subsection 1.1 are repeated here, and extensions, e.g. regarding transversely isotropic fibers, are presented, too.

The longitudinal behavior is approximated by the following expressions (Voigt models):

$$E_l = \hat{E}_l = \xi E_l^{(f)} + (1-\xi)E^{(m)},$$
$$\nu_{lq} = \hat{\nu}_{lq} = \xi \nu_{lq}^{(f)} + (1-\xi)\nu^{(m)},$$
$$\alpha_l = \frac{\xi E_l^{(f)}\alpha_l^{(f)} + (1-\xi)E^{(m)}\alpha^{(m)}}{\hat{E}_l},$$
$$\kappa_l = \hat{\kappa}_l = \xi \kappa_l^{(f)} + (1-\xi)\kappa^{(m)},$$
$$c_p = \hat{c}_p = \xi c_p^{(f)} + (1-\xi)c_p^{(m)}, \qquad (63).$$
$$\rho = \hat{\rho} = \xi \rho^{(f)} + (1-\xi)\rho^{(m)},$$

$$E_{l,T} = \xi E_l^{(f)} + (1-\xi)E_T^{(m)},$$

$$\alpha_{l,p} = \frac{\xi E_l^{(f)}\alpha_l^{(f)} + (1-\xi)E_T^{(m)}\alpha^{(m)}}{\xi E_l^{(f)} + (1-\xi)E_T^{(m)}},$$

$$\sigma_{l,Y} = \sigma_Y^{(m)}\frac{E_l}{E^{(m)}}, \quad |\Delta T_{cr,Y}| = \frac{(1-\xi)\sigma_Y^{(m)}}{\xi(\alpha_l - \alpha^{(f)})E_l^{(f)}}$$

The above expressions for c_p and ρ hold for all composites (in the absence of phase changes and chemical reactions). The formula for $\sigma_{l,Y}$ is valid only for purely uniaxial loading in fiber direction and does not hold for $\xi \to 1$, and the one for $|\Delta T_{cr,Y}|$ does not hold for $\xi \to 0$ and $\xi \to 1$.

The following expressions describe the transverse behavior (Reuss models):

$$E_q = \frac{E_q^{(f)}E^{(m)}}{\xi E^{(m)} + (1-\xi)E_q^{(f)}}, \quad G_{lq} = \frac{G_{lq}^{(f)}G^{(m)}}{\xi G^{(m)} + (1-\xi)G_{lq}^{(f)}}$$

$$\nu_{qt} = \frac{\nu_{qt}^{(f)}\nu^{(m)}}{\xi\nu^{(m)} + (1-\xi)\nu_{qt}^{(f)}}, \quad \kappa_q = \frac{\kappa_q^{(f)}\kappa^{(m)}}{\xi\kappa^{(m)} + (1-\xi)\kappa_q^{(f)}}, \qquad (64).$$

$$\sigma_{q,Y} = \sigma_Y^{(m)}$$

Schapery [8] presented improved expressions for the coefficient of thermal expansion (CTE) based on an energy approach:

$$\alpha_l = \frac{\xi E_l^{(f)}\alpha^{(f)} + (1-\xi)E^{(m)}\alpha^{(m)}}{\hat{E}_l},$$

$$\alpha_q = (1-\xi)(1+\nu^{(m)})\alpha^{(m)} + \xi(1+\nu_{lq}^{(f)})\alpha^{(f)} - \alpha\nu_{lq}, \qquad (65).$$

$$\alpha_{q,p} = \frac{3}{2}(1-\xi)\alpha^{(m)} + \xi(1+\nu_{lq}^{(f)})\alpha^{(f)} - \alpha_{l,p}(\frac{1}{2}(1-\xi) + \xi\nu_{lq}^{(f)})$$

For $\alpha^{(m)} > \alpha^{(f)}$, the above formula for the transverse CTE may give values of α_q larger than $\alpha^{(m)}$ at small fiber volume fractions ξ (the third term as given here is a ROM expression involving some approximations). The above expression for $\alpha_{q,p}$ is not valid for $\xi \to 0$.

b) CCA-Method - Direct Approach

Hashin [29,41] developed a direct analytical approach based on a model which considers cylinders of different cross sectional areas composed of central fibers surrounded by matrix material retaining the volume fractions of the composite in each cylinder. The transverse material cross section is completely filled with such cylinders, see Fig. 11. This model is called "Composite Cylinder Assemblage" (CCA), and it gives excellent predictions for the thermoelastic properties of unidirectional composites.

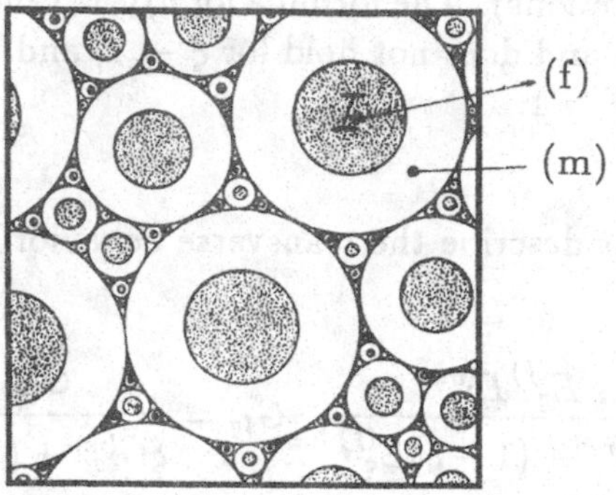

Fig. 11 Composite cylinder assembly – CCA-model

The following closed form solutions can be obtained from this model:

$$K_q = K_q^{(m)} + \frac{\xi}{1/(K_q^{(f)} - K_q^{(m)}) + (1-\xi)/(K_q^{(m)} + G^{(m)})},$$

$$G_{lq} = G^{(m)} + \frac{\xi}{1/(G_{lq}^{(f)} - G^{(m)}) + (1-\xi)/2G^{(m)}}, \qquad (66).$$

$$\kappa_l = \xi\kappa^{(f)} + (1-\xi)\kappa^{(m)}$$

The results for K_q and G_{lq} correspond to the lower bounds given in [42]. It should be noted that G_{qt} and κ_q cannot be obtained from the CCA-approach.

Rosen and Hashin [43] derived CTE expressions as follows:

$$
\begin{aligned}
\alpha_l = \hat{\alpha}_l + (C_{11} - \hat{C}_{11})[(\alpha_l^{(f)} - \alpha^{(m)})P_{11} + (\alpha_q^{(f)} - \alpha^{(m)})2P_{12}] + \\
(C_{12} - \hat{C}_{12})[(\alpha_l^{(f)} - \alpha^{(m)})2P_{12} + (\alpha_q^{(f)} - \alpha^{(m)})2(P_{22} + P_{23})], \\
\alpha_q = \hat{\alpha}_q + (C_{12} - \hat{C}_{12})[(\alpha_l^{(f)} - \alpha^{(m)})P_{11} + (\alpha_q^{(f)} - \alpha^{(m)})2P_{12}] + \\
(C_{22} - \hat{C}_{22})[(\alpha_l^{(f)} - \alpha^{(m)})P_{12} + (\alpha_q^{(f)} - \alpha^{(m)})(P_{22} + P_{23})] + \\
(C_{23} - \hat{C}_{23})[(\alpha_l^{(f)} - \alpha^{(m)})P_{12} + (\alpha_q^{(f)} - \alpha^{(m)})(P_{22} + P_{23})]
\end{aligned} \qquad (67).
$$

with

$$
P_{ij}(C_{jk}^{(f)} - C_{jk}^{(m)}) = I_{ik}
$$

and

$$
\begin{aligned}
\hat{\alpha}_l &= \xi\alpha_l^{(f)} + (1-\xi)\alpha^{(m)} \\
\hat{\alpha}_q &= \xi\alpha_q^{(f)} + (1-\xi)\alpha^{(m)} \\
\hat{C}_{ij} &= \xi C_{ij}^{(f)} + (1-\xi)C_{ij}^{(m)} \ .
\end{aligned}
$$

For isotropic fibers this reduces to

$$
\begin{aligned}
\alpha_l &= \hat{\alpha} + \frac{\alpha^{(f)} - \alpha^{(m)}}{1/K^{(f)} - 1/K^{(m)}}\left[\frac{3(1-2\nu_{lq})}{E_l} - \left(\frac{\xi}{K^{(f)}} + \frac{(1-\xi)}{K^{(m)}}\right)\right] \\
\alpha_q &= \hat{\alpha} + \frac{\alpha^{(f)} - \alpha^{(m)}}{1/K^{(f)} - 1/K^{(m)}}\left[\frac{3}{2K_q} - \frac{3(1-2\nu_{lq})\nu_{lq}}{E_l} - \left(\frac{\xi}{K^{(f)}} + \frac{(1-\xi)}{K^{(m)}}\right)\right]
\end{aligned} \qquad (68).
$$

The above formula is that given in [43], it does not fully agree with the expression in a more recent paper by Hashin [29], which uses $\frac{1}{K^{(m)}}$ instead of $\left(\frac{\hat{1}}{K}\right) = \frac{\xi}{K^{(f)}} + \frac{1-\xi}{K^{(m)}}$.

c) Variational Bounding

Based on extremum principles of minimum potential and minimum complementary energy, respectively, lower and upper bounds on the elastic properties of composites can be derived. The well known Hashin–Shtrikman bounds [44] for quasi-isotropic composites are classical representatives of such approaches. For unidirectional continuous fiber composites the following expressions are given in [29].

Here, variational bounding results are quoted for G_{qt} and κ_q which cannot be obtained from the CCA-method. For $G_{qt}^{(f)} > G^{(m)}$, $K_q^{(f)} > K_q^{(m)}$ the lower and upper bounds, L.B. and U.B., respectively, are given by

$$G_{qt}^{L.B.} = G^{(m)} + \frac{\xi}{1/(G_{qt}^{(f)} - G^{(m)}) + (1-\xi)(K_q^{(m)} + 2G^{(m)})/2G^{(m)}(K_q^{(m)} + G^{(m)})},$$
$$G_{qt}^{U.B.} = G^{(m)}\left(1 + \frac{\xi(1+\bar{\beta}^{(m)})}{\bar{\rho} - \xi[1 + 3(1-\xi)^2\bar{\beta}^{(m)^2}/(\lambda\xi^3 - \bar{\beta}^{(m)})]}\right),$$

(69),

$$\kappa_q^{L.B.} = \kappa^{(m)} + \frac{\xi}{1/(\kappa^{(f)} - \kappa^{(m)}) + (1-\xi)/2\kappa^{(m)}},$$
$$\kappa_q^{U.B.} = \kappa^{(f)} + \frac{1-\xi}{1/(\kappa^{(m)} - \kappa^{(f)}) + \xi/2\kappa^{(f)}}$$

with

$$\lambda = (\bar{\beta}^{(m)} - \gamma\bar{\beta}^{(f)})/(1+\gamma\bar{\beta}^{(f)}), \quad \bar{\rho} = (\gamma + \bar{\beta}^{(m)})/(\gamma - 1),$$
$$\bar{\beta}^{(f)} = K_q^{(f)}/(K_q^{(f)} + 2G_{qt}^{(f)}), \quad \bar{\beta}^{(m)} = 1/(3 - 4\nu^{(m)}), \quad \gamma = G_{qt}^{(f)}/G^{(m)}.$$

The above bounds for G_{qt} are valid if $G_{qt}^{(f)} \geq G^{(m)}$ and $K_q^{(f)} \geq K_q^{(m)}$; if these conditions are reversed, upper and lower bounds must be exchanged. These bounding expressions together with the CCA formulae correspond to Mori–Tanaka results [35] evaluated for an infinitely high fiber aspect ratio.

d) Halpin–Tsai Equations

Halpin–Tsai expressions are frequently used in Sections 1 and 2 to give engineering estimates for specific material parameters. The following generalized form [45]

$$p \approx p^{(m)}\left(\frac{1+\zeta\eta\xi}{1-\eta\xi}\right) \tag{70}$$

can be used for any thermoelastic modulus p. The parameter ζ depends on the micro-geometry, and η is defined as

$$\eta = \left(\frac{p^{(f)}}{p^{(m)}} - 1\right) \Big/ \left(\frac{p^{(f)}}{p^{(m)}} + \zeta\right) \qquad (71).$$

e) Hill's Relations

Hill [46] derived formulae for elastic material parameters which modify the above described ROM expressions by additional terms as shown for example for E_l and ν_{lq} as follows:

$$\begin{aligned} E_l &= \xi E_l^{(f)} + (1-\xi)E^{(m)} + \frac{4\xi(1-\xi)(\nu_{lq}^{(f)} - \nu^{(m)})^2}{(1-\xi)/K_q^{(f)} + \xi/K_q^{(m)} + 1/G^{(m)}}, \\ \nu_{lq} &= \xi\nu_{lq}^{(f)} + (1-\xi)\nu^{(m)} + \frac{\xi(1-\xi)(\nu_{lq}^{(f)} - \nu^{(m)})(1/K_q^{(m)} - 1/K_q^{(f)})}{(1-\xi)/K_q^{(f)} + \xi/K_q^{(m)} + 1/G^{(m)}} \end{aligned} \qquad (72).$$

There exist alternative improvements of the ROM models, as for example Poech's approach [47].

3.3 Short-Fiber Composites

In the present subsection short isotropic fibers embedded in an isotropic matrix are considered.

For composites reinforced by continuous fibers, Mori–Tanaka results for the thermoelastic moduli typically agree with the CCA-formulae listed in Subsection 3.2. Additional simple Mori–Tanaka expressions have been derived by Wakashima et al. [48] and Hatta and Taya [49] for aligned high aspect ratio fibers (for which the "auxiliary relations" for transversely isotropic media presented in Subsection 3.2 hold, too) and by Takahashi et al. [50] for randomly distributed high aspect ratio fibers (which give rise to an isotropic composite).

Since in Section 2 no CTE estimates are presented for aligned high aspect ratio short-fiber composites, the following expressions might be useful.

$$
\begin{aligned}
\alpha_l &= \xi\alpha^{(f)} + (1-\xi)\alpha^{(m)} + \xi(1-\xi)(\alpha^{(f)} - \alpha^{(m)})\left(\frac{A-B}{AC-BD}E^{(f)} - 1\right) \\
\alpha_q &= \xi\alpha^{(f)} + (1-\xi)\alpha^{(m)} + \xi(1-\xi)(\alpha^{(f)} - \alpha^{(m)})\left(\frac{C-D}{AC-BD}E^{(f)} - 1\right) \\
\kappa_l &= \xi\kappa^{(f)} + (1-\xi)\kappa^{(m)} \\
\kappa_q &= \kappa^{(m)} + \frac{2\xi\kappa^{(m)}(\kappa^{(f)} - \kappa^{(m)})}{(1-\xi)(\kappa^{(f)} - \kappa^{(m)}) + 2\kappa^{(m)}}
\end{aligned} \qquad (73),
$$

with

$$
\begin{aligned}
A &= (1-\xi)\frac{2\nu^{(m)}\nu^{(f)}G^{(f)} + 2(1-2\nu^{(f)})G^{(m)}}{1-\nu^{(m)}} + 2\xi(1-\nu^{(f)})G^{(f)} \\
B &= (1-\xi)\frac{\nu^{(m)}G^{(f)} + (1-2\nu^{(f)})\nu^{(m)}G^{(m)}}{1-\nu^{(m)}} + 2\xi\nu^{(f)}G^{(f)} \\
C &= (1-\xi)\frac{G^{(f)} + (1-2\nu^{(f)})G^{(m)}}{1-\nu^{(m)}} + 2\xi G^{(f)} \\
D &= (1-\xi)\frac{2\nu^{(f)}G^{(f)} + 2(1-2\nu^{(f)})\nu^{(m)}G^{(m)}}{1-\nu^{(m)}} + 4\xi\nu^{(f)}G^{(f)}
\end{aligned}
$$

For randomly distributed high aspect ratio fibers Takahashi et al. [50] derived the following expression for CTE:

$$
\alpha = \alpha^{(m)} + \frac{\xi K^{(f)}(G^{(f)} + 3K^{(m)} + 3G^{(m)})(\alpha^{(f)} - \alpha^{(m)})}{K^{(f)}[\xi(G^{(f)} + 3G^{(m)}) + 3K^{(m)}] + (1-\xi)K^{(m)}(G^{(f)} + 3G^{(m)})} \qquad (74).
$$

3.4 Particle Reinforced Composites

Despite the fact that this book mainly deals with fiber reinforced composites it might be worth the effort to complete the set of equations by presenting some expressions for particle reinforced composites, too. Here both the particles and the matrix are assumed to behave as isotropic, linear elastic materials. Spherical particles (as a model for particles which have similar extensions in all directions) randomly arranged in an isotropic matrix result in an isotropic composite. Hence, the linear elastic behavior is described by two elastic constants. The following expressions relate elastic parameters of isotropic materials to each other:

Auxiliary Relations for Isotropic Media

$$\frac{3}{E} = \frac{1}{G} + \frac{1}{3K}, \quad K = \frac{E}{3(1-2\nu)}, \quad G = \frac{E}{2(1+\nu)},$$

$$E = \frac{9KG}{3K+G}, \quad \nu = \frac{3K/2-G}{3K+G}, \quad K_q = \frac{E}{2(1-\nu-2\nu^2)} \qquad (75).$$

a) Dilute Spherical Particles, Eshelby Approach

As mentioned in Subsection 3.1, for a single ellipsoidal elastic inclusion in an infinite elastic matrix Eshelby derived in a classical paper [33] expressions for describing the thermoelastic stress field.

Based on this work a series of papers have been published for describing the thermoelastic behavior of particle reinforced composites, see e.g. [31]. The following expressions quoted in [29]

$$\begin{aligned} K &= K^{(m)} + \xi(K^{(p)} - K^{(m)})\frac{3K^{(m)} + 4G^{(m)}}{3K^{(p)} + 4G^{(m)}} \\ G &= G^{(m)} + \xi(G^{(p)} - G^{(m)})\frac{5(3K^{(m)} + 4G^{(m)})}{9K^{(m)} + 8G^{(m)} + 6(K^{(m)} + 2G^{(m)})G^{(p)}/G^{(m)}} \end{aligned} \qquad (76)$$

are useful for $\xi < 10\%$.

b) Spherical Particles, Direct Approaches

Levin [51] and Schapery [8] present expressions for the linear thermal expansion coefficient (CTE) of macrosopically isotropic particle reinforced composites:

$$\alpha = \alpha^{(m)} + \frac{\alpha^{(p)} - \alpha^{(m)}}{1/K^{(p)} - 1/K^{(m)}}\left(\frac{1}{K} - \frac{1}{K^{(m)}}\right) \qquad (77),$$

and Rosen and Hashin [29,41,43] used the model of "Composite Spheres Assemblage" (CSA), in analogy to the above mentioned CCA-method, to derive the following closed form solutions:

$$K = K^{(m)} + \frac{\xi}{1/(K^{(p)} - K^{(m)}) + 3(1-\xi)/(3K^{(m)} + 4G^{(m)})}$$
$$\alpha = \xi\alpha^{(p)} + (1-\xi)\alpha^{(m)} + \frac{4\xi(1-\xi)(K^{(p)} - K^{(m)})(\alpha^{(p)} - \alpha^{(m)})G^{(m)}}{3K^{(p)}K^{(m)} + 4G^{(m)}[\xi K^{(p)} + (1-\xi)K^{(m)}]} \qquad (78).$$
$$\kappa = \kappa^{(m)} + \frac{\xi}{1/(\kappa^{(p)} - \kappa^{(m)}) + (1-\xi)/3\kappa^{(m)}}$$

The above result for K corresponds to the Hashin–Shtrikman lower bound, see below.

c) Variational Bounding

The classical paper by Hashin and Shtrikman [44] uses extremum principles of the theory of elasticity, i.e. minimum potential and minimum complementary energy, with admissible linear displacement fields or with admissible constant stress fields, respectively, to obtain elementary bounds for the effective elastic moduli of macroscopically homogeneous composite materials. From [29] and [52] the following improved bounds for isotropic composites can be deduced (it is sufficient to present bounds on K and G):

$$K^{L.B.} = K^{(m)} + \frac{\xi}{\frac{1}{K^{(p)} - K^{(m)}} + \frac{3(1-\xi)}{3K^{(m)} + 4G^{(m)})}},$$
$$K^{U.B.} = K^{(p)} + \frac{1-\xi}{\frac{1}{K^{(m)} - K^{(p)}} + \frac{3(1-\xi)}{3K^{(p)} + 4G^{(p)})}},$$
$$G^{L.B.} = G^{(m)} + \frac{\xi}{\frac{1}{G^{(p)} - G^{(m)}} + \frac{6(1-\xi)(K^{(m)} + 2G^{(m)})}{5G^{(m)}(3K^{(m)} + 4G^{(m)})}}, \qquad (79).$$
$$G^{U.B.} = G^{(p)} + \frac{(1-\xi)}{\frac{1}{G^{(m)} - G^{(p)}} + \frac{6\xi(K^{(p)} + 2G^{(p)})}{5G^{(p)}(3K^{(p)} + 4G^{(p)})}}$$

The above bounds for K and G are valid if $K^{(f)} \geq K^{(m)}$ and $G^{(f)} \geq G^{(m)}$; if these conditions are reversed, upper and lower bounds must be exchanged.

In a recent paper Zimmermann [53] discusses the evaluation of Hashin–Shtrikman bounds on Poisson's ratio of isotropic composites. Since ν is an increasing function of K but a decreasing function of G, it is clear that the lower and upper bounds for ν must not be evaluated simply from the corresponding bounds for K and G using the above mentioned auxiliary relations between elastic parameters, but they have to be found as follows

$$\nu^{L.B.} = \frac{3K^{L.B.} - 2G^{U.B.}}{6K^{L.B.} + 2G^{U.B.}}, \qquad \nu^{U.B.} = \frac{3K^{U.B.} - 2G^{L.B.}}{6K^{U.B.} + 2G^{L.B.}} \tag{80}$$

d) Spherical Particles, Generalized Self Consistent Method

Christensen [37] derived effective elastic parameters by a generalized self consistent scheme. For example, the expressions for the effective bulk modulus

$$K = K^{(m)} + \xi(K^{(p)} - K^{(m)}) \frac{3K^{(m)} + 4G^{(m)}}{3K^{(m)} + 4G^{(m)} + 3(1-\xi)(K^{(p)} - K^{(m)})} \tag{81}$$

is exactly the same as obtained by a Mori–Tanaka approach, see below.

e) Spherical Particles, Mori–Tanaka Approach

Using a Mori–Tanaka type approach Benveniste [36] presents following relations:

$$\begin{aligned} K &= K^{(m)} + \frac{\xi(K^{(p)} - K^{(m)})(3K^{(m)} + 4G^{(m)})}{3K^{(m)} + 4G^{(m)} + 3(1-\xi)(K^{(p)} - K^{(m)})} \\ G &= G^{(m)} + \frac{\xi(G^{(p)} - G^{(m)})[5G^{(m)}(3K^{(m)} + 4G^{(m)})]}{5G^{(m)}(3K^{(m)} + 4G^{(m)}) + 6(1-\xi)(K^{(m)} + 2G^{(m)})(G^{(p)} - G^{(m)})} \\ \alpha &= \xi\alpha^{(p)} + (1-\xi)\alpha^{(m)} + \frac{\xi(1-\xi)(\alpha^{(p)} - \alpha^{(m)})(K^{(p)} - K^{(m)})}{3K^{(p)}K^{(m)}/4G^{(m)} + \xi K^{(p)} + (1-\xi)K^{(m)}} \\ \kappa &= \kappa^{(m)} + \frac{3\xi\kappa^{(m)}(\kappa^{(p)} - \kappa^{(m)})}{(1-\xi)(\kappa^{(p)} - \kappa^{(m)}) + 3\kappa^{(m)}} \end{aligned} \tag{82}$$

The above results agree with those obtained by the Hashin CSA approach and the Hashin–Shtrikman lower bounds.

References

1. Böhm H.J.: *Description of Thermoelastic Composites by a Mean Field Approach;* in this book,1994.

2. Halpin J.C., Tsai S.W.: *Effects of Environmental Factors on Composite Materials.* USAF Materials Laboratory Technical Report AFML–TR–67–423, Wright Patterson AFB, Dayton, OH, 1969.

3. Agarwal B.D., Broutman L.J.: *Analysis and Performance of Fiber Composites.* John Wiley & Sons, New York, NY, 1990.

4. Dorninger K.: *Entwicklung von nichtlinearen FE–Algorithmen zur Berechnung von Schalenkonstruktionen aus Faserverbundschalen.* VDI–Fortschrittsberichte **18/65**, VDI–Verlag, Düsseldorf, 1989.

5. Rammerstorfer F.G.: *Repetitorium Leichtbau.* Oldenbourg Verlag, Vienna, 1992.

6. Weeton J.W., Peters D.M., Thomas K.K.: *Engineer's Guide to Composite Materials;* American Society for Metals, Metal Park, OH, 1987.

7. Niederstadt G., Block J., Geier B., Rohwer K., Weiss R.: *Leichtbau mit kohlenstoffaserverstärkten Kunststoffen.* Expert–Verlag, Sindelfingen, 1985.

8. Schapery R.A.: *Thermal Expansion Coefficients of Composite Materials Based on Energy Principles;* J.Compos.Mater. **2**(3), 380–404, 1968.

9. Dorninger K., Rammerstorfer F.G.: *A Layered Composite Shell Element for Elastic and Thermoelastic Stress and Stability Analysis at Large Deformations;* Int.J.Num.Meth.Engng. **30**, 833–858, 1990.

10. Tsai S.W., Hahn H.T.: *Introduction to Composite Materials.* Technomic Publishing, Westport, CT, 1980.

11. Rosen B.W. (Ed.): *Fiber Composite Materials.* American Society for Metals, Metals Park, OH, 1965.

12. Waas A.M., Babcock C.D., Knauss W.G.: *A Mechanical Model for Elastic Fiber Microbuckling;* J.Appl.Mech. **57**, 138–149, 1990.

13. Lagoudas D.C., Tadjbakhsh I., Fares N.: *Approach to Microbuckling of Fibrous Composites;* J.Appl.Mech. **58**, 1–7, 1991.

14. Swanson S.R.: *A Micro-Mechanics Model for In-Situ Compression Strength of Fiber Composite Laminates;* J.Engng.Mater.Technol. **114**, 8–12, 1992.

15. Svobodnik A.J.: *Numerical Treatment of Elastic-Plastic Macromechanical Behavior of Longfiber-Reinforced Metal Matrix Composites.* VDI–Fortschrittsberichte **18/90**, VDI–Verlag, Düsseldorf, 1990.

16. Svobodnik A.J., Böhm H.J., Rammerstorfer F.G.: *A 3/D Finite Element Approach for Metal Matrix Composites Based on Micromechanical Models;* Int.J.Plast. **7**, 781–802, 1991.

17. Rammerstorfer F.G., Fischer F.D., Böhm H.J.: *Treatment of Micromechanical Phenomena by Finite Elements;* in *"Discretization Methods in Structural Mechanics"* (Eds. G.Kuhn, H.Mang), pp.393–404, Springer–Verlag, Berlin, 1990.

18. Mahishi J.M.: *An Integrated Micromechanical and Macromechanical Approach to Fracture Behavior of Fiber-Reinforced Composites;* Engng.Fract.Mech. **25**, 197–228, 1986.

19. Tvergaard V.: *Analysis of Tensile Properties for a Whisker-Reinforced Metal-Matrix Composite;* Acta metall.mater. **38**, 185–194, 1990.

20. Weissenbek E., Rammerstorfer F.G.: *Influence of the Fiber Arrangement on the Mechanical and Thermo-Mechanical Behavior of Short Fiber Reinforced MMCs;* Acta metall.mater. **43**, 1833–1844,1993.

21. Böhm H.J.: *Analytical Descriptions of the Material Behavior of Composites (A Summary of Formulas).* CDLμMW–Report **27–1991**, TU Wien, Vienna, 1991.

22. Favre J.P., Sigety P., Jaques D.: *Stress Transfer by Shear in Carbon Fibre Model Composites;* J.Mater.Sci. **26**, 189–195, 1991.

23. Hsueh C.H., Lu M.C.: *Elastic Stress Transfer from Fiber to Coating in a Fiber-Coating System;* Mater.Sci.Engng. **A117**, 115–123, 1989.

24. Staudinger G.: *Systematische numerische Untersuchung des Werkstoffverhaltens von kurzfaserverstärkten Kunststoffen.* Diploma Thesis, TU Wien, Vienna, 1988.

25. Courage W.M.G., Schreurs P.J.G.: *Effective Material Parameters for Composites with Randomly Oriented Short Fibers;* Comput.Struct. **44**, 1179–1185, 1992.

26. Kardos J.L.: *Structure-Property Relations for Short-Fiber Reinforced Plastics;* Division Technical Meeting, Engng.Prop.and Struct.Div., SPE, Akron, OH, 1975.

27. Böhm H.J.: *Computer Based Micromechanical Investigations of the Thermomechanical Behavior of Metal-Matrix Composites.* VDI–Fortschrittsberichte **18/101**, VDI–Verlag, Düsseldorf, 1991.

28. Love A.E.H.: *A Treatise of the Theory of Elasticity.* Cambridge University Press, Cambridge, 1927.

29. Hashin Z.: *Analysis of Composite Materials — A Survey;* J.Appl.Mech. **50**, 481–505, 1983.

30. Hashin Z.: *Extremum Principles for Elastic Heterogeneous Media with Imperfect Interfaces and Their Application to Bounding of Effective Moduli;* J.Mech.Phys.Sol. **40**(4), 767–782, 1992.

31. Böhm H.J.: *Notes on Some Mean Field Approaches for Composites.* CDLμMW-Report **57–1992**, TU Wien, Vienna, 1992.

32. Nan C.W.: *Physics of Inhomogeneous Inorganic Materials;* Progr.Mater.Sci. **37**, 1–116, 1993.

33. Eshelby J.D.: *The Determination of the Elastic Field of an Ellipsoidal Inclusion and Related Problems;* Proc.Roy.Soc.London **A241**, 376–391, 1957.

34. Mori T., Tanaka K.: *Average Stress in the Matrix and Average Elastic Energy of Materials with Misfitting Inclusions;* Acta Metall. **21**, 571–574, 1973.

35. Tandon G.P., Weng G.J.: *The Effect of Aspect Ratio of Inclusions on the Elastic Properties of Unidirectionally Aligned Composites;* Polym.Compos. **5**, 327–333, 1984.

36. Benveniste Y.: *A New Approach to the Application of Mori–Tanaka's Theory in Composite Materials;* Mech.Mater. **6**, 147–157, 1987.

37. Christensen R.M.: *A Critical Evaluation for a Class of Micromechanics Models;* J.Mech.Phys.Sol. **38**(3), 379–404, 1990.

38. Lempriere B.M.: *Poisson's Ratio in Orthotropic Materials;* AIAA J. **6**, 2226–2277, 1968.

39. Jones R.M.: *Mechanics of Composite Materials.* Scripta Book Company, Washington, DC, 1974.

40. Pedersen P.: *Bounds on Elasticity Energy in Solids of Orthotropic Materials;* Struct.Optimization **2**, 55–62, 1992.

41. Hashin Z.: *Analysis of Properties of Fiber Composites with Anisotropic Constituents;* J.Appl.Mech. **46**, 543–550, 1979.

42. Hashin Z.: *On Elastic Behavior of Fibre Reinforced Materials of Arbitrary Transverse Phase Geometry;* J.Mech.Phys.Sol. **13**, 119–134, 1965.

43. Rosen B.W., Hashin Z.: *Effective Thermal Expansion Coefficients and Specific Heats of Composite Materials;* Int.J.Engng.Sci. **8**, 157–173, 1970.

44. Hashin Z., Shtrikman S.: *A Variational Approach to the Theory of the Elastic Behavior of Multiphase Materials;* J.Mech.Phys.Sol. **11**, 127–140, 1963.

45. Halpin J.C., Kardos J.L.: *The Halpin–Tsai Equations: A Review;* Polym.Engng.Sci. **16**, 344–351, 1976.

46. Hill R.: *Elastic Properties of Reinforced Solids: Some Theoretical Principles;* J.Mech.Phys.Sol. **11**, 191–196, 1963.

47. Poech M.H.: *Deformation of Two-Phase Materials: Application of Analytical Elasticity Solutions to Plasticity;* Scripta metall.mater. **27**, 1027–1031, 1992.

48. Wakashima K., Otsuka M., Umekawa S.: *Thermal Expansion of Heterogeneous Solids Containing Aligned Ellipsoidal Inclusions;* J.Compos.Mater. **8**, 391–404, 1974.

49. Hatta H., Taya M.: *Effective Thermal Conductivity of a Misoriented Short Fiber Composite;* J.Appl.Phys. **58**, 2478–2486, 1985.

50. Takahashi K., Harakawa K., Sakai T.: *Analysis of the Thermal Expansion Coefficients of Particle Filled Polymers;* J.Compos.Mater. **14**, 144–159, 1980.

51. Levin V.M.: *On the Coefficients of Thermal Expansion of Heterogeneous Materials;* Mech.Sol. **2**, 58-61, 1967.

52. Kathrina T., Round R., Bridge B.: *An Investigation of the Composition Dependence of the Elasticity, Reaction Rate and Porosity of Orthophosphate Bonded Ceramics Using an Ultrasonic Double-Probe Method;* J.Phys.D **24**, 1673–1686, 1991.

53. Zimmerman R.W.: *Hashin–Shtrikman Bounds on the Poisson's Ratio of a Composite Material;* Mech.Res.Com. **16**, 563–569,1992.

CHAPTER 3

DESCRIPTION OF THERMOELASTIC COMPOSITES BY A MEAN FIELD APPROACH

H.J. Böhm
Vienna Technical University, Vienna, Austria

ABSTRACT

In this Chapter a Mori–Tanaka-type micromechanical method for modeling the thermoelastic behavior of composites with aligned reinforcements is described in some detail.

The basic assumptions underlying mean field approaches are discussed, and a number of general relations between the elastic tensors, the concentration tensors and the mean fields are given. Eshelby's solution and the equivalent inclusion method are used for obtaining expressions applicable to dilute composites. After introducing the concept of the average matrix stresses, Mori–Tanaka expressions for the concentration tensors for composites with non-dilute inclusion volume fractions are derived, from which the thermoelastic properties can be directly obtained. Finally, limitations as well as some extensions of the Mori–Tanaka approach are discussed.

List of Variables:

The following notations (with or without sub- or superscripts, respectively) are used for describing the material behavior:

$\langle\ \rangle$ Phase average
ε: Strain tensor
$\langle\varepsilon\rangle$: Tensor of effective strains (overall average)
ε_a: Tensor of the applied strain or strain response of the composite
$\varepsilon_a^{(m)}$: Tensor of the strain response of the unreinforced matrix
$\varepsilon_{tot}^{(i)}$: Tensor of total averaged strains in inclusion
$\varepsilon_{tot}^{(m)}$: Tensor of total averaged strains in matrix
ε_t: Tensor of total unconstrained eigenstrains (stress-free transformation strains) in inclusion
ε_c: Tensor of constrained strains in inclusion
ε_τ: Tensor of equivalent eigenstrains in inhomogeneous inclusion
σ: Stress tensor
$\langle\sigma\rangle$: Tensor of effective stresses (overall average)
σ_a: Tensor of applied stresses
$\sigma_{tot}^{(i)}$: Tensor of total averaged stresses in inclusion
$\sigma_{tot}^{(m)}$: Tensor of total averaged stresses in matrix
$\mathbf{A}(r)$: Position dependent strain concentration tensor
$\bar{\mathbf{A}}$: Phase averaged strain concentration tensor
$\mathbf{B}(r)$: Position dependent stress concentration tensor
$\bar{\mathbf{B}}$: Phase averaged stress concentration tensor
$\bar{\mathbf{a}}$: Phase averaged thermal strain concentration tensor
$\bar{\mathbf{b}}$: Phase averaged thermal stress concentration tensor
$\mathbf{C}$: Compliance tensor
$\mathbf{E}$: Elasticity tensor
$\mathbf{e}$: Specific thermal stress tensor
α: Tensor of coefficients of thermal expansion
$\mathbf{I}$: Identity tensor
$\mathbf{S}(r)$: Position dependent Eshelby tensor
$\mathbf{S}$: Eshelby tensor for ellipsoidal inclusions
A: Aspect ratio of the (ellipsoidal) inclusions
$\nu^{(m)}$: Poisson's ratio of the matrix
ΔT: Temperature difference with respect to a stress-free reference temperature

ξ: Inclusion volume fraction
Ω_s: Subregion of the composite
Ω_r: Reference volume for homogenization

Sub- and Superscripts:

$_{dil}$ Dilute
$_M$ Mori–Tanaka
$^{(p)}$ Phase p
$^{(i)}$ Inclusion
$^{(m)}$ Matrix
Overall effective – no superscript

Note:

Nye notation is employed throughout this chapter (compare e.g. [1]), i.e. 4th rank tensors are represented as 6×6 matrices and 2nd rank tensors are represented as 6-vectors.

1 Introduction

An important aim of theoretical descriptions of composite materials is the prediction of their overall properties — in the present case their thermoelastic response — from the material behavior of their constituents. Such models must account for at least two length scales, the macroscale, which is the length scale of the composite structure, component or sample, and the microscale, which for matrix–inclusion type composites is of the order of the diameters or the spacing of the reinforcing inclusions.

Most theoretical descriptions of composites are based on the assumption that the macroscale and the microscale are sufficiently different to allow the stress and strain fields to be split into contributions corresponding to each length scale, which can then be analyzed separately. Under such a scenario, the microscale variations of the stress and strain fields (the "fast variables") are assumed to influence the macroscale behavior via their averages, so that from the macroscale point of view the composite can be treated as a "material". At the microscale, where the composite's phases are distinguishable, gradients in the macroscale stress, strain and temperature distributions (the "slow variables", which may also include parameters describing the composite's microstructure) are taken to be sufficiently small for them to be approximated by constant far field loads.

Provided the above conditions are met and suitable algorithms for linking the microscale and macroscale responses of composites ("micromechanical models") are available, many engineering problems in the design of structures made of composite materials can be handled by using a micromechanical model as the "material model" in standard analytical or numerical engineering methods (e.g. the Finite Element Method) for predicting the response at the component level†. In addition, by inserting the macroscopic stresses and strains computed for a given location into the micromechanical model, estimates of the corresponding microscale stresses and strains in the composite's constituents can be obtained.

In accordance with the pivotal role played by micromechanical models in describing the thermoelastic behavior of composite materials, a large number of theories have been developed, encompassing different levels of sophistication and complexity. In the present chapter one such model, a Mori–Tanaka-type mean field approach, will be discussed. Models of this type have the capability of handling a wide range of composites with aligned inclusion–matrix topologies, and their relatively low computational requirements allow their use as material models in Finite Element programs.

† In the case of laminated composites, an intermediate length scale corresponding to the thickness of a single lamina may be introduced. The micromechanical model then serves to describe the response of each lamina, and lamination theory (compare [2]) is used to obtain the material behavior at the macroscopic level.

2 General Relations between Mean Fields in Thermoelastic Composites

For a given subregion (mesodomain) Ω_s of an elastic composite under constant far field loads, the microscale responses, $\varepsilon(r)$ and $\sigma(r)$ (where r denotes the position vector of the material point under consideration), and the corresponding macroscale responses, $\langle\varepsilon\rangle$ and $\langle\sigma\rangle$, can be formally linked by localization relations of the type

$$\begin{aligned} \varepsilon(r) &= \mathbf{A}(r)\langle\varepsilon\rangle \\ \sigma(r) &= \mathbf{B}(r)\langle\sigma\rangle \end{aligned} \quad (1)$$

and by homogenization relations of the form

$$\begin{aligned} \langle\varepsilon\rangle &= \frac{1}{\Omega_s}\int_{\Omega_s} \varepsilon(r)d\Omega \\ \langle\sigma\rangle &= \frac{1}{\Omega_s}\int_{\Omega_s} \sigma(r)d\Omega \end{aligned} \quad (2).$$

$\mathbf{A}(r)$ and $\mathbf{B}(r)$ are called the strain and stress concentration tensors [3] (or influence functions), respectively, and they contain all geometrical information necessary for describing the elastic response of the composite. Due to the high complexity of the microgeometry of actual composites, exact expressions for $\mathbf{A}(r)$, $\mathbf{B}(r)$, $\varepsilon(r)$ and $\sigma(r)$ cannot realistically be obtained and approximations have to be introduced.

For this purpose homogenization volumes (or reference volume elements) Ω_r may be considered, which are representative of the microgeometry of the composite (e.g. in a statistical way), are sufficiently large for meaningful sampling of the microfields, and are sufficiently small for the influence of macroscale gradients to be negligible. Various theories for describing the thermoelastic behavior of composites may be based on such homogenization volumes [4], an important group being mean field approaches, in which the microfields within each phase are approximated by their phase averages, $\varepsilon_{tot}^{(p)}$ and $\sigma_{tot}^{(p)}$ (i.e. the total averaged strains and stresses in each phase). The localization relations then take the form

$$\begin{aligned} \varepsilon_{tot}^{(p)} &= \bar{\mathbf{A}}^{(p)}\langle\varepsilon\rangle \\ \sigma_{tot}^{(p)} &= \bar{\mathbf{B}}^{(p)}\langle\sigma\rangle \end{aligned} \quad (3)$$

and the homogenization relations become

$$\varepsilon_{tot}^{(p)} = \frac{1}{\Omega^{(p)}} \int_{\Omega^{(p)}} \varepsilon(r) d\Omega \qquad \langle\varepsilon\rangle = \sum_p \frac{\Omega^{(p)}}{\Omega_r} \varepsilon_{tot}^{(p)}$$
$$\sigma_{tot}^{(p)} = \frac{1}{\Omega^{(p)}} \int_{\Omega^{(p)}} \sigma(r) d\Omega \qquad \langle\sigma\rangle = \sum_p \frac{\Omega^{(p)}}{\Omega_r} \sigma_{tot}^{(p)} \qquad (4),$$

where (p) stands for a phase of the composite and Ω_p is the corresponding phase volume. In contrast to the general expressions in eqn.(2), within the framework of mean field approaches the phase concentration tensors $\bar{\mathbf{A}}$ and $\bar{\mathbf{B}}$ are not functions of the spatial coordinates, but describe the average response of a given phase within the homogenization volume.

For the special case of void-free two-phase composites consisting of inclusions (i) embedded in a matrix phase (m), the total averaged strains and total averaged stresses in matrix and inclusions, respectively, take the form

$$\varepsilon_{tot}^{(m)} = \frac{1}{\Omega^{(m)}} \int_{\Omega^{(m)}} \varepsilon(r) d\Omega \qquad \sigma_{tot}^{(m)} = \frac{1}{\Omega^{(m)}} \int_{\Omega^{(m)}} \sigma(r) d\Omega$$
$$\varepsilon_{tot}^{(i)} = \frac{1}{\Omega^{(i)}} \int_{\Omega^{(i)}} \varepsilon(r) d\Omega \qquad \sigma_{tot}^{(i)} = \frac{1}{\Omega^{(i)}} \int_{\Omega^{(i)}} \sigma(r) d\Omega \qquad (5),$$

where $\Omega^{(m)}$ and $\Omega^{(i)}$ are the volumes taken up by matrix and inclusions, respectively, with $\Omega_r = \Omega^{(m)} + \Omega^{(i)}$. The inclusion volume fraction is given by $\xi = \Omega^{(i)}/\Omega_r$, and the matrix volume fraction is $(1-\xi)$. When both inclusions and matrix are assumed to display thermoelastic material behavior, the total averaged phase stresses and strains are connected by the material relations

$$\varepsilon_{tot}^{(m)} = \mathbf{C}^{(m)} \sigma_{tot}^{(m)} + \alpha^{(m)} \Delta T \qquad \sigma_{tot}^{(m)} = \mathbf{E}^{(m)} \varepsilon_{tot}^{(m)} + \mathbf{e}^{(m)} \Delta T$$
$$\varepsilon_{tot}^{(i)} = \mathbf{C}^{(i)} \sigma_{tot}^{(i)} + \alpha^{(i)} \Delta T \qquad \sigma_{tot}^{(i)} = \mathbf{E}^{(i)} \varepsilon_{tot}^{(i)} + \mathbf{e}^{(i)} \Delta T \qquad (6).$$

Here $\mathbf{E}^{(m)}$ and $\mathbf{E}^{(i)}$ are the elasticity tensors, $\mathbf{C}^{(m)}$ and $\mathbf{C}^{(i)}$ are the compliance tensors, $\mathbf{e}^{(m)}$ and $\mathbf{e}^{(i)}$ are the specific thermal stress tensors (i.e. the overall stress responses of the phases to a purely thermal unit load under fully constrained conditions), $\alpha^{(m)}$ and $\alpha^{(i)}$ are the tensors of the coefficients of thermal expansion (which will be called the "thermal expansion tensors" for brevity) of matrix and inclusions, respectively, and ΔT is the temperature difference with respect to some stress-free reference temperature. The relations $\mathbf{E}^{(p)} = [\mathbf{C}^{(p)}]^{-1}$ and $\mathbf{e}^{(p)} = -\mathbf{E}^{(p)} \alpha^{(p)}$ hold for both phases.

The overall stress–strain relations for thermoelastic composites take the form

$$\begin{aligned}\langle\varepsilon\rangle &= \mathbf{C}\langle\sigma\rangle + \alpha\Delta T\\ \langle\sigma\rangle &= \mathbf{E}\langle\varepsilon\rangle + \mathbf{e}\Delta T\end{aligned} \qquad (7),$$

where $\mathbf{E}$ and $\mathbf{C}$ are the overall elasticity and overall compliance tensors of the composite, respectively, α stands for the overall thermal expansion tensor and $\mathbf{e}$ is the overall specific thermal stress tensor. When both inclusions and matrix show isotropic material properties, the overall behavior of the composite is isotropic in the case of spherical inclusions and transversely isotropic for all types of aligned inclusions (continuous and discontinuous fibers, platelets) [5]. In the course of this chapter expressions for $\mathbf{E}$, $\mathbf{C}$ and α (from which the corresponding overall engineering moduli of the composite can be computed) will be derived in terms of the constituents' material properties and the composite's microgeometry.

The definition of volume averaging, eqn.(5), directly implies the following relations between the phase averaged fields

$$\begin{aligned}\langle\varepsilon\rangle &= \xi\varepsilon^{(i)}_{tot} + (1-\xi)\varepsilon^{(m)}_{tot} = \varepsilon_a + \alpha\Delta T\\ \langle\sigma\rangle &= \xi\sigma^{(i)}_{tot} + (1-\xi)\sigma^{(m)}_{tot} = \sigma_a\end{aligned} \qquad (8),$$

where $\varepsilon_a = \mathbf{C}\sigma_a$ is the overall strain response of the composite. The total phase averaged strains and stresses are related to the overall strains and stresses by the phase strain and stress concentration tensors $\bar{\mathbf{A}}$ and $\bar{\mathbf{B}}$, and to the temperature differences ΔT by the thermal strain and stress concentration tensors $\bar{\mathbf{a}}$ and $\bar{\mathbf{b}}$, respectively. For thermoelastic composites the concentration tensors are defined by the expressions

$$\begin{aligned}\varepsilon^{(m)}_{tot} &= \bar{\mathbf{A}}^{(m)}\langle\varepsilon\rangle + \bar{\mathbf{a}}^{(m)}\Delta T \qquad & \sigma^{(m)}_{tot} &= \bar{\mathbf{B}}^{(m)}\langle\sigma\rangle + \bar{\mathbf{b}}^{(m)}\Delta T\\ \varepsilon^{(i)}_{tot} &= \bar{\mathbf{A}}^{(i)}\langle\varepsilon\rangle + \bar{\mathbf{a}}^{(i)}\Delta T \qquad & \sigma^{(i)}_{tot} &= \bar{\mathbf{B}}^{(i)}\langle\sigma\rangle + \bar{\mathbf{b}}^{(i)}\Delta T\end{aligned} \qquad (9)$$

(compare eqn.(3) for the isothermal case). By combining eqns.(8) and (9) the phase averaged strain and stress concentration tensors (which for brevity will be called strain and stress concentration tensors, respectively, from now on) can be easily shown to fulfill the relations

$$\begin{aligned}\xi\bar{\mathbf{A}}^{(i)} + (1-\xi)\bar{\mathbf{A}}^{(m)} &= \mathbf{I} \qquad & \xi\bar{\mathbf{B}}^{(i)} + (1-\xi)\bar{\mathbf{B}}^{(m)} &= \mathbf{I}\\ \xi\bar{\mathbf{a}}^{(i)} + (1-\xi)\bar{\mathbf{a}}^{(m)} &= 0 \qquad & \xi\bar{\mathbf{b}}^{(i)} + (1-\xi)\bar{\mathbf{b}}^{(m)} &= 0\end{aligned} \qquad (10).$$

By inserting the expressions for the total averaged stresses and strains, eqns.(6), into eqns.(8) and then applying eqns.(9), the effective elasticity and compliance tensors of the composite can be obtained in terms of concentration tensors and of the material tensors of the constituents as

$$\begin{aligned}\mathbf{E} &= \xi\mathbf{E}^{(i)}\bar{\mathbf{A}}^{(i)} + (1-\xi)\mathbf{E}^{(m)}\bar{\mathbf{A}}^{(m)}\\ &= \mathbf{E}^{(m)} + \xi(\mathbf{E}^{(i)} - \mathbf{E}^{(m)})\bar{\mathbf{A}}^{(i)}\\ &= \mathbf{E}^{(i)} + (1-\xi)(\mathbf{E}^{(m)} - \mathbf{E}^{(i)})\bar{\mathbf{A}}^{(m)}\end{aligned} \tag{11}$$

$$\begin{aligned}\mathbf{C} &= \xi\mathbf{C}^{(i)}\bar{\mathbf{B}}^{(i)} + (1-\xi)\mathbf{C}^{(m)}\bar{\mathbf{B}}^{(m)}\\ &= \mathbf{C}^{(m)} + \xi(\mathbf{C}^{(i)} - \mathbf{C}^{(m)})\bar{\mathbf{B}}^{(i)}\\ &= \mathbf{C}^{(i)} + (1-\xi)(\mathbf{C}^{(m)} - \mathbf{C}^{(i)})\bar{\mathbf{B}}^{(m)}\end{aligned} \tag{12),}$$

compare e.g. [6]. In an analogous way, the overall specific thermal stress tensor and the overall thermal expansion tensor can be expressed as

$$\mathbf{e} = \xi(\mathbf{E}^{(i)}\bar{\mathbf{a}}^{(i)} + \mathbf{e}^{(i)}) + (1-\xi)(\mathbf{E}^{(m)}\bar{\mathbf{a}}^{(m)} + \mathbf{e}^{(m)}) \tag{13}$$

$$\alpha = \xi(\mathbf{C}^{(i)}\bar{\mathbf{b}}^{(i)} + \alpha^{(i)}) + (1-\xi)(\mathbf{C}^{(m)}\bar{\mathbf{b}}^{(m)} + \alpha^{(m)}) \tag{14),}$$

respectively.

By invoking the principle of virtual work and specifying appropriate loading conditions, Benveniste et al. [7,8] derived relations which link the thermal strain concentration tensors with the strain concentration tensors and the thermal stress concentration tensors with the stress concentration tensors, respectively

$$\begin{aligned}\bar{\mathbf{a}}^{(m)} &= (\mathbf{I} - \bar{\mathbf{A}}^{(m)})\big[\mathbf{E}^{(i)} - \mathbf{E}^{(m)}\big]^{-1}(\mathbf{e}^{(m)} - \mathbf{e}^{(i)})\\ \bar{\mathbf{a}}^{(i)} &= (\mathbf{I} - \bar{\mathbf{A}}^{(i)})\big[\mathbf{E}^{(m)} - \mathbf{E}^{(i)}\big]^{-1}(\mathbf{e}^{(i)} - \mathbf{e}^{(m)})\\ \bar{\mathbf{b}}^{(m)} &= (\mathbf{I} - \bar{\mathbf{B}}^{(m)})\big[\mathbf{C}^{(i)} - \mathbf{C}^{(m)}\big]^{-1}(\alpha^{(m)} - \alpha^{(i)})\\ \bar{\mathbf{b}}^{(i)} &= (\mathbf{I} - \bar{\mathbf{B}}^{(i)})\big[\mathbf{C}^{(m)} - \mathbf{C}^{(i)}\big]^{-1}(\alpha^{(i)} - \alpha^{(m)})\end{aligned} \tag{15).}$$

In addition, the overall specific thermal stress tensor and the overall thermal expansion tensor can be expressed in terms of the overall and component elasticity and compliance tensors as follows [7,8]

$$\begin{aligned}\mathbf{e} &= \mathbf{e}^{(m)} + (\mathbf{E}^{(m)} - \mathbf{E})\big[\mathbf{E}^{(i)} - \mathbf{E}^{(m)}\big]^{-1}(\mathbf{e}^{(m)} - \mathbf{e}^{(i)}) \\ &= (\mathbf{E} - \mathbf{E}^{(m)})\big[\mathbf{E}^{(i)} - \mathbf{E}^{(m)}\big]^{-1}\mathbf{e}^{(i)} + (\mathbf{E} - \mathbf{E}^{(i)})\big[\mathbf{E}^{(m)} - \mathbf{E}^{(i)}\big]^{-1}\mathbf{e}^{(m)} \qquad (16)\end{aligned}$$

$$\begin{aligned}\alpha &= \alpha^{(m)} + (\mathbf{C}^{(m)} - \mathbf{C})\big[\mathbf{C}^{(i)} - \mathbf{C}^{(m)}\big]^{-1}(\alpha^{(m)} - \alpha^{(i)}) \\ &= (\mathbf{C} - \mathbf{C}^{(m)})\big[\mathbf{C}^{(i)} - \mathbf{C}^{(m)}\big]^{-1}\alpha^{(i)} + (\mathbf{C} - \mathbf{C}^{(i)})\big[\mathbf{C}^{(m)} - \mathbf{C}^{(i)}\big]^{-1}\alpha^{(m)} \qquad (17).\end{aligned}$$

Accordingly, within the mean field framework the overall thermoelastic properties of two-phase composites are uniquely determined once the overall elastic response is known [8].

It should be noted that eqns.(10)–(17) are general relations which hold for all mean field theories. Thus, mean field methods often follow a strategy of deriving expressions for the phase stress and strain concentration tensors, from which the thermal phase concentration tensors as well as the overall thermoelastic material tensors of the composite, $\mathbf{E}$, $\mathbf{C}$, $\mathbf{e}$ and α, can be readily computed. Such an approach will also be used in the following sections.

3 Eshelby-Type Expressions for Dilute Composites

3.1 Misfit Stresses — Eshelby's Solution

The majority of mean field descriptions of composites are based on the work of Eshelby [9,10], who investigated the stress and strain distributions in infinite homogeneous media containing a subregion (the "inclusion") which spontaneously undergoes a shape change so that it no longer fits into its previous domain in the rest of the medium (the "matrix"). According to Eshelby's results, a homogeneous elastic inclusion in an infinite elastic matrix (i.e. an inclusion having the same material properties as the matrix) subjected to an unconstrained homogeneous strain ε_t (i.e. the inclusion would deform by ε_t if it were not constrained by the surrounding matrix), shows a constrained strain ε_c which can be expressed as [9]

$$\varepsilon_c(r) = \mathbf{S}(r)\varepsilon_t \qquad (18).$$

The tensor $\mathbf{S}(r)$ is called the Eshelby tensor, and it serves to describe the unknown constrained strain $\varepsilon_c(r)$ in terms of the unconstrained strain ε_t (stress-free strain, eigenstrain, "transformation strain") of the inclusion, which is typically known. For eqn.(18) to hold, ε_t may be any kind of eigenstrain which is uniform over the inclusion (e.g. a thermal strain or a strain due to some transformation which involves no changes in the elastic constants of the inclusion; note that ε_t may be anisotropic).

For general shapes of the inclusion, the Eshelby tensor is position dependent, but for the special case of ellipsoidal inclusions, **S** is constant and, consequently, the stress and strain states in the constrained inclusion are uniform, so that eqn.(18) takes the form

$$\varepsilon_c = \mathbf{S}\varepsilon_t \qquad (19).$$

All expressions given in the remainder of this chapter will correspond to this case.

For homogeneous ellipsoidal inclusions in an isotropic matrix, **S** can be evaluated analytically and depends only on the Poisson's ratio of the homogeneous material, $\nu^{(m)}$, and on the aspect ratio A of the inclusions†. If, in addition, the inclusions are ellipsoids of rotation (i.e. spheroids), the nonzero components of the Eshelby tensor can be expressed as [13,14]

$$S(1,1) = \frac{1}{2(1-\nu^{(m)})}\left\{\frac{4A^2-2}{A^2-1} - 2\nu^{(m)} + \left[\frac{4A^2-1}{1-A^2} + 2\nu^{(m)}\right] g(A)\right\}$$

$$S(2,2) = S(3,3) = \frac{1}{4(1-\nu^{(m)})}\left\{\frac{3A^2}{2(A^2-1)} + \left[\frac{4A^2-13}{4(A^2-1)} - 2\nu^{(m)}\right] g(A)\right\}$$

$$S(1,2) = S(1,3) = \frac{1}{2(1-\nu^{(m)})}\left\{\frac{A^2}{1-A^2} + 2\nu^{(m)} + \left[\frac{2A^2+1}{2(A^2-1)} - 2\nu^{(m)}\right] g(A)\right\}$$

$$S(2,1) = S(3,1) = \frac{1}{4(1-\nu^{(m)})}\left\{\frac{2A^2}{1-A^2} + \left[\frac{2A^2+1}{A^2-1} + 2\nu^{(m)}\right] g(A)\right\}$$

$$S(2,3) = S(3,2) = \frac{1}{4(1-\nu^{(m)})}\left\{\frac{A^2}{2(A^2-1)} + \left[\frac{4A^2-1}{4(1-A^2)} + 2\nu^{(m)}\right] g(A)\right\}$$

$$S(4,4) = S(5,5) = \frac{1}{2(1-\nu^{(m)})}\left\{\frac{2}{1-A^2} - 2\nu^{(m)} + \frac{1}{2}\left[\frac{2A^2+4)}{A^2-1} + 2\nu^{(m)}\right] g(A)\right\}$$

$$S(6,6) = \frac{1}{2(1-\nu^{(m)})}\left\{\frac{A^2}{2(A^2-1)} + \left[\frac{4A^2-7}{4(A^2-1)} - 2\nu^{(m)}\right] g(A)\right\} \qquad (20).$$

Here the 1-direction is understood to be the axis of rotation of the spheroid, $\nu^{(m)}$ is the Poisson's ratio of the matrix, A stands for the aspect ratio of the inclusions

† Analytical expressions for the Eshelby tensor can also be found for cuboidal inclusions [11,12], but in such cases the Eshelby tensors are position dependent and thus not directly applicable to for mean field methods.

(i.e. A is given by the length of the spheroids divided by their diameter, so that for continuous cylindrical fibers $A \to \infty$, for spherical inclusions $A = 1$, and for infinitely thin circular discs or platelets $A \to 0$), and the function $g(A)$ is given by the expressions

$$g = \frac{A}{(A^2-1)^{3/2}}\left[A(A^2-1)^{1/2} - \operatorname{arcosh} A\right]$$

for prolate (fiber-like) inclusions ($A \geq 1$) and

$$g = \frac{A}{(1-A^2)^{3/2}}\left[\arccos A - A(1-A^2)^{1/2}\right]$$

for oblate (disc-like) inclusions ($A \leq 1$).

For the special case of spherical inclusions ($A = 1$), the only nonzero components of the Eshelby tensor are

$$\begin{aligned} S(1,1) = S(2,2) = S(3,3) &= \frac{7-5\nu^{(m)}}{15(1-\nu^{(m)})} \\ S(1,2) = S(2,3) = S(1,3) &= \frac{5\nu^{(m)}-1}{15(1-\nu^{(m)})} \\ S(4,4) = S(5,5) = S(6,6) &= \frac{2(4-5\nu^{(m)})}{15(1-\nu^{(m)})} \end{aligned} \qquad (21),$$

for inclusions in the form of continuous fibers of circular cross section ($A \to \infty$) the Eshelby tensor takes the form

$$\begin{aligned} S(2,2) = S(3,3) &= \frac{5-4\nu^{(m)}}{8(1-\nu^{(m)})} \\ S(2,3) = S(3,2) &= \frac{4\nu^{(m)}-1}{8(1-\nu^{(m)})} \\ S(2,1) = S(3,1) &= \frac{\nu^{(m)}}{2(1-\nu^{(m)})} \\ S(4,4) = S(5,5) &= \frac{1}{2} \\ S(6,6) &= \frac{3-4\nu^{(m)}}{4(1-\nu^{(m)})} \end{aligned} \qquad (22),$$

and for infinitely thin circular discs the nonzero components can be evaluated as

$$S(1,1) = 1$$
$$S(1,2) = S(1,3) = \frac{\nu^{(m)}}{(1-\nu^{(m)})}$$
$$S(4,4) = S(5,5) = \frac{1}{2} \qquad (23).$$

Equations (20)–(23) follow the conventions of Nye notation, i.e. the 4th rank tensor **S** is represented as a 6×6 matrix. In addition, the shear angles, $\gamma_{ij} = 2\varepsilon_{ij}$, are used for the shear terms in the strain "vector". This notation corresponds to the one used by Pedersen [15], but a number of standard works dealing with misfitting inclusions (e.g. [11,13]) use other conventions for the shear strains and thus give expressions for $S(4,4)$, $S(5,5)$ and $S(6,6)$ differing from eqns.(20)–(23) by a factor of $\frac{1}{2}$.

Extensions of the above equations to general ellipsoids (which do not show rotational symmetry) and expressions for some anisotropic homogeneous matrix–inclusion systems are given in [11]. In cases where no analytical solutions are available, the Eshelby tensor can be evaluated numerically, compare [16]. In addition, it should be noted that in the case of inhomogeneous inclusions, which will be investigated in the following subchapters, eqns.(20)–(23) are valid for both isotropic and anisotropic inclusions in an isotropic matrix [11].

The stress and strain fields outside a transformed inclusion in an infinite matrix are not uniform in the general case [10]. Within the framework of mean field theories, which aim to link the average fields in matrix and inclusions with the overall response of inhomogeneous materials, however, it is only the average matrix stresses and strains that are of interest.

3.2 Dilute Inhomogeneous Inclusions Subjected to an Eigenstrain

Mean field methods for dilute (but inhomogeneous) matrix–inclusion composites typically make use of Eshelby's expressions for the fields within a homogeneous inclusion in combination with an "equivalent inclusion" (or "cutting and welding", compare [17]) approach. This strategy involves replacing the inhomogeneous inclusion (which has different material properties than the matrix) subjected to a given unconstrained eigenstrain with an "equivalent" homogeneous inclusion (to which eqn.(19) applies) subjected to an appropriate "equivalent" eigenstrain in such a way that the same constrained stress and strain fields are obtained in both cases.

In the constrained homogeneous inclusion the constrained strain ε_c can be computed from eqn.(19), and Hooke's law takes the form

$$\sigma^{(i)} = \mathbf{E}^{(i)}(\varepsilon_c - \varepsilon_t) \qquad (24),$$

where $\varepsilon_c - \varepsilon_t$ is the elastic strain in the inclusion (caused by the interaction with the matrix) which gives rise to local elastic stresses. Following the strategy outlined above, the case of inhomogeneous inclusions is handled by investigating a homogeneous equivalent inclusion (or "reference inclusion") subjected to an equivalent eigenstrain ε_τ, which is chosen in such a way that the inhomogeneous inclusion and the reference inclusion attain the same stress state $\sigma^{(i)}$ [9,17]. When $\sigma^{(i)}$ is expressed in terms of the elastic strains in the inclusion, the above condition translates into the equation

$$\sigma^{(i)} = \mathbf{E}^{(i)}(\varepsilon_c - \varepsilon_t) = \mathbf{E}^{(m)}(\varepsilon_c - \varepsilon_\tau) \qquad (25).$$

Because the homogeneous (equivalent) and the inhomogeneous (real) inclusions are required, on the one hand, to have the same shape and size in the strain-free state and, on the other hand, to show the same strains ε_c and stresses $\sigma^{(i)}$ in the constrained state, the stress-free strains will generally be different for the equivalent and the real inclusion, i.e. $\varepsilon_t \neq \varepsilon_\tau$ (compare e.g. the sketches in [17]).

Because eqn.(19) was derived for the homogeneous inclusion problem, it will also hold for the homogeneous reference inclusion, for which it takes the form

$$\varepsilon_c = \mathbf{S}\varepsilon_\tau \qquad (26).$$

By inserting this expression into eqn.(25) one obtains the relation

$$\sigma^{(i)} = \mathbf{E}^{(i)}(\mathbf{S}\varepsilon_\tau - \varepsilon_t) = \mathbf{E}^{(m)}(\mathbf{S} - \mathbf{I})\varepsilon_\tau \qquad (27)$$

which can be solved to give the equivalent eigenstrain ε_τ of the homogeneous inclusion in terms of the known stress-free eigenstrain ε_t of the inhomogeneous inclusion as

$$\varepsilon_\tau = \left[(\mathbf{E}^{(i)} - \mathbf{E}^{(m)})\mathbf{S} + \mathbf{E}^{(m)}\right]^{-1}\mathbf{E}^{(i)}\varepsilon_t \qquad (28).$$

Substituting this formula into the right hand side of eqn.(27), the stress in the inclusion, $\sigma^{(i)}$, is finally obtained as

$$\sigma^{(i)} = \mathbf{E}^{(m)}(\mathbf{S}-\mathbf{I})\left[(\mathbf{E}^{(i)}-\mathbf{E}^{(m)})\mathbf{S}+\mathbf{E}^{(m)}\right]^{-1}\mathbf{E}^{(i)}\varepsilon_t \qquad (29).$$

This expression holds, of course, for both the inhomogeneous inclusion and for the homogeneous reference inclusion.

3.3 Dilute Inclusions Subjected to an Eigenstrain and an Applied Stress

If a uniform external stress σ_a is applied to the elastic matrix–inclusion-system, the total stress in the inclusion, $\sigma_{tot}^{(i)}$, will be a superposition of this applied stress and of some additional stress caused by the constraining effect of the surrounding matrix on the inclusion. Following Withers et al. [17], such problems can be treated by an extension of the strategy used in Subsection 3.2. The approach takes the form of introducing a homogeneous reference inclusion subjected to both the external stress σ_a and a suitable equivalent eigenstrain ε_τ, the latter being chosen in such a way that the total average inclusion stress $\sigma_{tot}^{(i)}$ is the same in the inhomogeneous and in the equivalent inclusion. This equivalent eigenstrain, of course, depends both on the applied load and on the unconstrained eigenstrain of the inclusion.

By writing the total average inclusion stress in the inhomogeneous inclusion as $\mathbf{E}^{(i)}(\varepsilon_a^{(m)}+\varepsilon_c-\varepsilon_t)$ and that in the homogeneous reference inclusion as $\mathbf{E}^{(m)}(\varepsilon_a^{(m)}+\varepsilon_c-\varepsilon_\tau)$, the ansatz

$$\sigma_{tot}^{(i)} = \mathbf{E}^{(i)}(\varepsilon_a^{(m)}+\varepsilon_c-\varepsilon_t) = \mathbf{E}^{(m)}(\varepsilon_a^{(m)}+\varepsilon_c-\varepsilon_\tau) \qquad (30).$$

is obtained. Here $\varepsilon_a^{(m)} = \mathbf{C}^{(m)}\sigma_a$ is the strain response of the unreinforced matrix to the applied stress σ_a (which is used here for consistency with the concept of a homogeneous reference material). Equation (30) may be viewed as an extension of eqn.(25), where ε_c is replaced by the expression $\varepsilon_a^{(m)}+\varepsilon_c$. It should be noted, however, that even though σ_a can be directly identified with $\mathbf{E}^{(m)}\varepsilon_a^{(m)}$ in the expression for the homogeneous reference inclusion, no term in the expression for the inhomogeneous inclusion corresponds directly to σ_a.

By inserting Eshelby's relation in the form of eqn.(26) and solving for ε_τ the equivalent eigenstrain is obtained as

$$\varepsilon_\tau = \left[(\mathbf{E}^{(i)}-\mathbf{E}^{(m)})\mathbf{S}+\mathbf{E}^{(m)}\right]^{-1}\left[(\mathbf{E}^{(m)}-\mathbf{E}^{(i)})\varepsilon_a^{(m)}+\mathbf{E}^{(i)}\varepsilon_t\right] \qquad (31).$$

Substituting this relation into the right hand side of eqn.(30) results in the following expression for the total stress in the inclusion

$$\sigma_{tot}^{(i)} = \mathbf{E}^{(m)}\{\mathbf{I} + (\mathbf{S} - \mathbf{I})\left[(\mathbf{E}^{(i)} - \mathbf{E}^{(m)})\mathbf{S} + \mathbf{E}^{(m)}\right]^{-1}(\mathbf{E}^{(m)} - \mathbf{E}^{(i)})\}\varepsilon_a^{(m)} + \mathbf{E}^{(m)}(\mathbf{S} - \mathbf{I})\left[(\mathbf{E}^{(i)} - \mathbf{E}^{(m)})\mathbf{S} + \mathbf{E}^{(m)}\right]^{-1}\mathbf{E}^{(i)}\varepsilon_t \quad (32),$$

where the ε_t-term corresponds to eqn.(29).

By equating the effective stress in the composite $\langle\sigma\rangle$ with the applied stress σ_a, by setting the transformation strain $\varepsilon_t = 0$ (this essentially corresponds to the derivations used by Wakashima et al. [6], who consider only far field loads but no stress-free eigenstrains), and by comparing the definition of the inclusion stress concentration tensor $\bar{\mathbf{B}}^{(i)}$, eqns.(3) and (9), with eqn.(32), the following result for the stress concentration tensor of a dilute composite is obtained

$$\bar{\mathbf{B}}_{dil}^{(i)} = \mathbf{E}^{(m)}\{\mathbf{I} + (\mathbf{S} - \mathbf{I})\left[(\mathbf{E}^{(i)} - \mathbf{E}^{(m)})\mathbf{S} + \mathbf{E}^{(m)}\right]^{-1}(\mathbf{E}^{(m)} - \mathbf{E}^{(i)})\}\mathbf{C}^{(m)} \quad (33).$$

An analogous procedure, which consists of setting $\langle\varepsilon\rangle = 0$ and $\varepsilon_a^{(m)} = 0$ in eqns.(9) and (32), respectively, and inserting the thermal mismatch for the unconstrained eigenstrain, i.e. $\varepsilon_t = (\alpha^{(i)} - \alpha^{(m)})\Delta T$, leads to an equivalent expression for the thermal stress concentration tensor of the inclusions

$$\bar{\mathbf{b}}_{dil}^{(i)} = \mathbf{E}^{(m)}(\mathbf{S} - \mathbf{I})\left[(\mathbf{E}^{(i)} - \mathbf{E}^{(m)})\mathbf{S} + \mathbf{E}^{(m)}\right]^{-1}\mathbf{E}^{(i)}(\alpha^{(i)} - \alpha^{(m)}) \quad (34).$$

Matrix stress concentration tensors corresponding to eqn.(33) and eqn.(34) can be easily obtained by using eqns.(10).

An alternative approach, which results in simpler expressions for both the stress and strain concentration tensors of dilute composites was developed by Hill [19] and elaborated by Benveniste [18]. Here, the starting point is an inhomogeneous inclusion in the absence of stress-free transformation strains, i.e. $\varepsilon_t = 0$. Under this condition, the total strain in the inclusion can be described as

$$\varepsilon_{tot}^{(i)} = \varepsilon_a + \varepsilon_c \quad (35),$$

where ε_a again is the strain response of the composite to the applied stress σ_a. An equivalent expression to eqn.(30) then takes the form

$$\sigma_{tot}^{(i)} = \mathbf{E}^{(i)}(\varepsilon_a + \varepsilon_c) = \mathbf{E}^{(m)}(\varepsilon_a + \varepsilon_c - \varepsilon_\tau)$$

which, by using eqn.(35), can be rewritten as

$$\mathbf{E}^{(i)}\varepsilon^{(i)}_{tot} = \mathbf{E}^{(m)}(\varepsilon^{(i)}_{tot} - \varepsilon_\tau) \tag{36}$$

Solving eqn.(36) for ε_τ and substituting the result into eqn.(35) gives a relation between the total strain in the inclusion and the applied strain

$$\varepsilon^{(i)}_{tot} = \left[\mathbf{I} + \mathbf{S}\mathbf{C}^{(m)}(\mathbf{E}^{(i)} - \mathbf{E}^{(m)})\right]^{-1}\varepsilon_a \tag{37}$$

from which the strain concentration tensor for the inclusion follows by comparing with eqn.(9) as

$$\bar{\mathbf{A}}^{(i)}_{dil} = \left[\mathbf{I} + \mathbf{S}\mathbf{C}^{(m)}(\mathbf{E}^{(i)} - \mathbf{E}^{(m)})\right]^{-1} \tag{38}$$

Setting $\varepsilon^{(i)}_{tot} = \mathbf{C}^{(i)}\sigma^{(i)}_{tot}$ (this can be done due to the assumption that $\varepsilon_t = 0$) and using $\varepsilon_a = \mathbf{C}^{(m)}\sigma_a$ (this approximation to the exact relationship $\varepsilon_a = \mathbf{C}\sigma_a$ is valid at small inclusion volume fractions), the stress concentration tensor for the inclusion is finally found as

$$\bar{\mathbf{B}}^{(i)}_{dil} = \mathbf{E}^{(i)}\left[\mathbf{I} + \mathbf{S}\mathbf{C}^{(m)}(\mathbf{E}^{(i)} - \mathbf{E}^{(m)})\right]^{-1}\mathbf{C}^{(m)} \tag{39}$$

By using eqns.(15), thermal strain and stress concentration tensors corresponding to eqns.(38) and (39) can be easily obtained.

The expressions for stress and strain concentration tensors given in the present section were derived under the assumption that the inclusions are dilutely dispersed in the matrix and thus do not "feel" any effects due to their neighbors (i.e. each inclusion is assumed to be loaded only by the applied stress σ_a). Consequently, the concentration tensors for the inclusion, $\bar{\mathbf{A}}^{(i)}_{dil}$, $\bar{\mathbf{B}}^{(i)}_{dil}$ and $\bar{\mathbf{b}}^{(i)}_{dil}$ (eqns. (38), (39), (33) and (34), respectively) are independent of the inclusion volume fraction ξ. The inclusion volume fraction does, however, enter the corresponding expressions for the matrix concentration tensors via eqn.(10).

Equations (33), (34), (38) and (39) should only be used for the case of highly dilute inclusions, i.e. for inclusion volume fractions in the range $\xi \ll 0.1$.

4. A Mori–Tanaka Method for Non-Dilute Composites

Theoretical descriptions of the overall thermoelastic behavior of composites with inclusion volume fractions of more than a few percent must explicitly account for the effects of the surrounding inclusions on the stress and strain fields experienced by a given fiber or particle. One way for achieving this consists in approximating the stresses acting on an inclusion, which may be viewed as the perturbation stresses caused by the presence of other inclusions ("image stresses", "background stresses", "mean field stresses") superimposed on the applied far field stress, by an appropriate average matrix stress. The idea of combining such a concept of an average matrix stress with Eshelby-type equivalent inclusion approaches goes back to Brown and Stobbs [20] as well as Mori and Tanaka [21]. Theories of this type are generically called Mori–Tanaka methods or "Equivalent Inclusion — Average Stress" (EIAS) approaches.

It was shown by Benveniste [18] that in the isothermal case the central assumption involved in Mori–Tanaka approaches can be denoted as

$$\begin{aligned} \varepsilon_{tot}^{(i)} &= \bar{\mathbf{A}}_{dil}^{(i)} \varepsilon_{tot}^{(m)} \\ \sigma_{tot}^{(i)} &= \bar{\mathbf{B}}_{dil}^{(i)} \sigma_{tot}^{(m)} \end{aligned} \qquad (40).$$

These expressions can be directly interpreted as modifications of eqn.(9) in which the dilute inclusion strain and stress concentration tensors, $\bar{\mathbf{A}}_{dil}^{(i)}$ and $\bar{\mathbf{B}}_{dil}^{(i)}$, act on the non-dilute total averaged matrix strains or stresses, $\varepsilon_{tot}^{(m)}$ and $\sigma_{tot}^{(m)}$, instead of acting on the applied strains or stresses, ε_a and σ_a, respectively. This, of course, corresponds to the above concept of subjecting each inclusion to a mean matrix stress instead of the unperturbed applied stress.

With such an approach, the mean field methodology developed for dilute inclusions can be retained, the effects of the surrounding inclusions being accounted for by suitably modifying the loads applied to each inclusion. In contrast to the dilute case, however, where the problem of the thermoelastic response of the composite is solved as soon as $\bar{\mathbf{A}}_{dil}^{(i)}$ or $\bar{\mathbf{B}}_{dil}^{(i)}$ are known (ε_a and σ_a are typically input data), EIAS approaches use eqn.(40) as the starting point for obtaining expressions for the as-yet unknown total averaged matrix strains and stresses. In fact, the various Mori–Tanaka-type methods available from the literature (e.g. [6,13,14,15,18]) differ mainly in the way in which $\varepsilon_{tot}^{(m)}$ and $\sigma_{tot}^{(m)}$ are substituted for ε_a and σ_a, respectively, and in the algebra of solving for $\varepsilon_{tot}^{(m)}$ and $\sigma_{tot}^{(m)}$.

A simple and elegant EIAS approach was formulated by Benveniste [18], which is based on inserting eqns.(40) into eqns.(8) to generate relations between the applied (or overall) strains and stresses on the one hand and the total inclusion strains and stresses on the other hand

$$\begin{aligned} \langle \varepsilon \rangle = \varepsilon_a &= \xi \varepsilon_{tot}^{(i)} + (1-\xi)\varepsilon_{tot}^{(m)} = \xi \bar{\mathbf{A}}_{dil}^{(i)} \varepsilon_{tot}^{(m)} + (1-\xi)\varepsilon_{tot}^{(m)} \\ &= \left[(1-\xi)\mathbf{I} + \xi \bar{\mathbf{A}}_{dil}^{(i)}\right]\left[\bar{\mathbf{A}}_{dil}^{(i)}\right]^{-1} \varepsilon_{tot}^{(i)} = \left[\bar{\mathbf{A}}_{M}^{(i)}\right]^{-1} \varepsilon_{tot}^{(i)} \\ \langle \sigma \rangle = \sigma_a &= \xi \sigma_{tot}^{(i)} + (1-\xi)\sigma_{tot}^{(m)} = \xi \bar{\mathbf{B}}_{dil}^{(i)} \sigma_{tot}^{(m)} + (1-\xi)\sigma_{tot}^{(m)} \\ &= \left[(1-\xi)\mathbf{I} + \xi \bar{\mathbf{B}}_{dil}^{(i)}\right]\left[\bar{\mathbf{B}}_{dil}^{(i)}\right]^{-1} \sigma_{tot}^{(i)} = \left[\bar{\mathbf{B}}_{M}^{(i)}\right]^{-1} \sigma_{tot}^{(i)} \end{aligned} \qquad (41).$$

By rearranging these results, the following Mori–Tanaka expressions for the inclusion strain and stress concentration tensors for the non-dilute composite are obtained

$$\begin{aligned} \bar{\mathbf{A}}_{M}^{(i)} &= \bar{\mathbf{A}}_{dil}^{(i)}\left[(1-\xi)\mathbf{I} + \xi \bar{\mathbf{A}}_{dil}^{(i)}\right]^{-1} \\ \bar{\mathbf{B}}_{M}^{(i)} &= \bar{\mathbf{B}}_{dil}^{(i)}\left[(1-\xi)\mathbf{I} + \xi \bar{\mathbf{B}}_{dil}^{(i)}\right]^{-1} \end{aligned} \qquad (42).$$

The corresponding expressions for the non-dilute matrix strain and stress concentration tensors can be found by inserting eqns.(42) into eqns.(40)

$$\begin{aligned} \bar{\mathbf{A}}_{M}^{(m)} &= \left[(1-\xi)\mathbf{I} + \xi \bar{\mathbf{A}}_{dil}^{(i)}\right]^{-1} \\ \bar{\mathbf{B}}_{M}^{(m)} &= \left[(1-\xi)\mathbf{I} + \xi \bar{\mathbf{B}}_{dil}^{(i)}\right]^{-1} \end{aligned} \qquad (43).$$

In contrast to the dilute case, the Mori–Tanaka results for both the matrix and the inclusion concentration tensors are dependent on the inclusion volume fraction.

Equations (43) automatically fulfill eqns.(10) and may be evaluated with any strain and stress concentration tensors pertaining to dilute inclusions in a matrix (in fact, the $\bar{\mathbf{A}}_{dil}$ and $\bar{\mathbf{B}}_{dil}$ might come from any appropriate mean field theory and do not necessarily have to be based on Eshelby's formulae). If the Eshelby-type expressions eqns.(38) and (39) are employed, the matrix strain and stress concentration tensors for the non-dilute composite take the form

$$\begin{aligned} \bar{\mathbf{A}}_{M}^{(m)} &= \left\{(1-\xi)\mathbf{I} + \xi\left[\mathbf{I} + \mathbf{S}\mathbf{C}^{(m)}(\mathbf{E}^{(i)} - \mathbf{E}^{(m)})\right]^{-1}\right\}^{-1} \\ \bar{\mathbf{B}}_{M}^{(m)} &= \left\{(1-\xi)\mathbf{I} + \xi \mathbf{E}^{(i)}\left[\mathbf{I} + \mathbf{S}\mathbf{C}^{(m)}(\mathbf{E}^{(i)} - \mathbf{E}^{(m)})\right]^{-1}\mathbf{C}^{(m)}\right\}^{-1} \end{aligned} \qquad (44).$$

The expressions for the effective elasticity and compliance tensors of the composite quoted in [7,8,18] can be recovered by substituting $\bar{\mathbf{A}}_M^{(m)}$ and $\bar{\mathbf{B}}_M^{(m)}$ from eqns.(43) into eqns.(11) and (12) to give

$$\mathbf{E} = \mathbf{E}^{(m)} + \xi(\mathbf{E}^{(i)} - \mathbf{E}^{(m)})\bar{\mathbf{A}}_{dil}^{(i)}\left[(1-\xi)\mathbf{I} + \xi\bar{\mathbf{A}}_{dil}^{(i)}\right]^{-1}$$
$$\mathbf{C} = \mathbf{C}^{(m)} + \xi(\mathbf{C}^{(i)} - \mathbf{C}^{(m)})\bar{\mathbf{B}}_{dil}^{(i)}\left[(1-\xi)\mathbf{I} + \xi\bar{\mathbf{B}}_{dil}^{(i)}\right]^{-1} \qquad (45).$$

Finally, the specific thermal stress and thermal expansion tensors follow from eqns.(43), (16) and (17) as

$$\mathbf{e} = \mathbf{e}^{(m)} + \xi(\mathbf{E}^{(i)} - \mathbf{E}^{(m)})\bar{\mathbf{A}}_{dil}^{(i)}\left[(1-\xi)\mathbf{I} + \xi\bar{\mathbf{A}}_{dil}^{(i)}\right]^{-1}\left[\mathbf{E}^{(i)} - \mathbf{E}^{(m)}\right]^{-1}(\mathbf{e}^{(i)} - \mathbf{e}^{(m)})$$
$$\alpha = \alpha^{(m)} + \xi(\mathbf{C}^{(i)} - \mathbf{C}^{(m)})\bar{\mathbf{B}}_{dil}^{(i)}\left[(1-\xi)\mathbf{I} + \xi\bar{\mathbf{B}}_{dil}^{(i)}\right]^{-1}\left[\mathbf{C}^{(i)} - \mathbf{C}^{(m)}\right]^{-1}(\alpha^{(i)} - \alpha^{(m)})$$
$$(46).$$

For realistic inclusion–matrix stiffness ratios, Mori–Tanaka-type approaches such as Pedersen's Mean Field Theory [15], the Mori–Tanaka theory of Tandon and Weng [13] and Wakashima's method [6] (which introduce the mean matrix stress by modifying eqn.(30) or by replacing $\varepsilon_a^{(m)}$ with $\varepsilon_{tot}^{(m)}$ in eqn.(32), respectively, and subsequently arrive at different formulae for the concentration tensors) predict the same values for the overall thermoelastic moduli of two-phase composites as obtained by evaluating eqn.(45) [22]. For composites reinforced by particles or continuous aligned fibers, the results of the above Mori–Tanaka theories agree with the reference solutions obtained by Hashin's CSA and CCA methods [5] in combination with the Hashin–Shtrikman lower bounds [23], respectively. For detailed discussions of the range of validity of Mori–Tanaka methods and for comparisons with other mean field approaches see e.g. [24,25].

The Mori–Tanaka-type approach discussed in the present chapter is restricted to thermoelastic composites consisting of a matrix reinforced by aligned inclusions of a single type (i.e. all inclusions are assumed to have the same material properties and the same aspect ratio). By expanding such mean field methods to handle multiple inclusion phases, the formalism can be extended to cases involving nonaligned inclusions (described via orientation distributions) as well as inclusions of different shapes and material properties. The averaging over inclusions of different orientations, shapes and materials may be performed at the level of the elasticity tensors [26] or at the level of the applied, constrained and unconstrained strains in eqns.(25), (30) and (36), see e.g. [14,27]. EIAS approaches accounting for such complex composites are, however, limited by the fact that the resulting overall elastic tensors are not necessarily symmetrical [28]. In addition, it should be noted that in mean field methods based on the Eshelby tensor, inclusions of the same shape but of different size cannot be distinguished.

A number of procedures for extending Mori–Tanaka methods to inelastic composites have been reported in the literature. Incremental plasticity schemes may be based on a general approach, Dvorak's Transformation Field method [29], which can directly use the elastic phase concentration tensors given in eqns.(42) and (46) to obtain estimates for the thermo-elastic-plastic behavior of aligned composites at relatively low computational costs. In addition to their main field of application, the description of the thermoelastic behavior of composites, Mori–Tanaka methods can also be used to model porous materials, see e.g. [14].

References

1. Dvorak G.J.: *Plasticity Theories for Fibrous Composite Materials;* in *"Metal Matrix Composites: Mechanisms and Properties"* (Eds. R.K.Everett, R.J.Arsenault), pp. 1–77; Academic Press, Boston, MA, 1991.
2. Rammerstorfer F.G.: *Lamination Theory,* in this book, 1994.
3. Hill R.: *Elastic Properties of Reinforced Solids: Some Theoretical Principles;* J.Mech.Phys.Sol. **11**, 357–372, 1963.
4. Böhm H.J.: *Numerical Investigation of Microplasticity Effects in Unidirectional Longfiber Reinforced Metal Matrix Composites;* Modell.Simul.Mater.Sci.Engng. **1**, 649–671, 1993.
5. Hashin Z.: *Analysis of Composite Materials — A Survey;* J.Appl.Mech. **50**, 481–505, 1983.
6. Wakashima K., Tsukamoto H., Choi B.H.: *Elastic and Thermoelastic Properties of Metal Matrix Composites with Discontinuous Fibers or Particles: Theoretical Guidelines towards Materials Tailoring;* in *"The Korea–Japan Metals Symposium on Composite Materials"*, pp. 102–115; The Korean Institute of Metals, Seoul, Korea, 1988.
7. Benveniste Y., Dvorak G.J.: *On a Correspondence Between Mechanical and Thermal Effects in Two-Phase Composites;* in *"Micromechanics and Inhomogeneity"* (Eds. G.J.Weng, M.Taya, H.Abé), pp. 65–82; Springer–Verlag, New York, NY, 1990.
8. Benveniste Y., Dvorak G.J., Chen T.: *On Effective Properties of Composites with Coated Cylindrically Orthotropic Fibers;* Mech.Mater. **12**, 289–297, 1991.
9. Eshelby J.D.: *The Determination of the Elastic Field of an Ellipsoidal Inclusion and Related Problems;* Proc.Roy.Soc.London **A241**, 376–396, 1957.
10. Eshelby J.D.: *The Elastic Field Outside an Ellipsoidal Inclusion;* Proc.Roy.Soc. London **A252**, 561–569, 1959.
11. Mura T.: *Micromechanics of Defects in Solids.* Martinus Nijhoff, Dordrecht, NL, 1987.
12. Chiu Y.P.: *On the Stress Field due to Initial Strains in a Cuboid Surrounded by an Infinite Elastic Space;* J.Appl.Mech. **44**, 587–590, 1977.
13. Tandon G.P., Weng G.J.: *The Effect of Aspect Ratio of Inclusions on the Elastic Properties of Unidirectionally Aligned Composites;* Polym.Compos. **5**, 327–333, 1984.
14. Zhao Y.H., Tandon G.P., Weng G.J.: *Elastic Moduli for a Class of Porous Materials;* Acta Mech. **76**, 105–130, 1989.
15. Pedersen O.B.: *Thermoelasticity and Plasticity of Composites — I. Mean Field Theory;* Acta Metall. **31**, 1795–1808, 1983.

16. Gavazzi A.C., Lagoudas D.C.: *On the Numerical Evaluation of Eshelby's Tensor and its Application to Elastoplastic Fibrous Composites;* Comput.Mech. **7**, 12–19, 1990.

17. Withers P.J., Stobbs W.M., Pedersen O.B.: *The Application of the Eshelby Method of Internal Stress Determination to Short Fibre Metal Matrix Composites;* Acta Metall. **37**, 3061–3084, 1989.

18. Benveniste Y.: *A New Approach to the Application of Mori–Tanaka's Theory in Composite Materials;* Mech.Mater. **6**, 147–157, 1987.

19. Hill R.: *A Self Consistent Mechanics of Composite Materials;* J.Mech.Phys.Sol. **13**, 213–222, 1965.

20. Brown L.M., Stobbs W.M.: *The Work-Hardening of Copper–Silica. I. A Model Based on Internal Stresses, with no Plastic Relaxation;* Phil.Mag. **23**, 1185–1199, 1971.

21. Mori T., Tanaka K.: *Average Stress in the Matrix and Average Elastic Energy of Materials with Misfitting Inclusions;* Acta Metall. **21**, 571–574, 1973.

22. Böhm H.J.: *Notes on Some Mean Field Approaches for Thermoelastic Two-Phase Composites.* CDL–μMW Report **108–1994**, TU Wien, Vienna, 1994.

23. Hashin Z., Shtrikman S.: *A Variational Approach to the Theory of the Elastic Behavior of Multiphase Materials;* J.Mech.Phys.Sol. **11**, 127–140, 1963.

24. Christensen R.M.: *A Critical Evaluation for a Class of Micromechanics Models;* J.Mech.Phys.Sol. **38**, 379–404, 1990.

25. Christensen R.M., Schantz H., Shapiro J.: *On the Range of Validity of the Mori–Tanaka Method;* J.Mech.Phys.Sol. **40**, 69–73, 1992.

26. Allen D.H., Lee J.W.: *The Effective Thermoelastic Properties of Whisker-Reinforced Composites as Functions of Material Forming Parameters;* in "*Micromechanics and Inhomogeneity*" (Eds. G.J.Weng, M.Taya, H.Abé), pp. 17–40; Springer–Verlag, New York, NY, 1990.

27. Pettermann, H.: *Die Mori–Tanaka–Methode für Mehrphasenkomposite mit beliebigen Orientierungsverteilungen der Einschlüsse.* CDL–μMW Report **94–1993**, TU Wien, Vienna, 1993.

28. Benveniste Y., Dvorak G.J., Chen T.: *On Diagonal and Elastic Symmetry of the Approximate Effective Stiffness Tensor of Heterogeneous Media;* J.Mech.Phys.Sol. **39**, 927–946, 1991.

29. Dvorak G.J.: *Transformation Field Analysis of Inelastic Composite Materials;* Proc.Roy.Soc.London A **437**, 311–327, 1992.

CHAPTER 4

LAMINATION THEORY AND FAILURE MECHANISMS IN COMPOSITE SHELLS

F.G. Rammerstorfer
Vienna Technical University, Vienna, Austria
and
A. Starlinger
AIREX Composite Engineering, Sins, Switzerland

ABSTRACT

In this Section the classical lamination theory is described on the basis of Mindlin–Reissner's kinematics. Hygrothermal effects are included, and a formulation is achieved which can simply specified for specific laminates such as symmetric, quasi-orthotropic and quasi-isotropic ones. Furthermore, interlaminar stresses and edge effects as well as some failure criteria and the post-failure behavior with stiffness degradation are considered. Based on antiplane core conditions a sandwich theory is developed, and a procedure is presented for estimating local instability phenomena such as different modes of face layer wrinkling or intracell buckling and failure due to transverse normal stresses.

List of Variables:

The following notations (with or without sub- or superscripts, respectively) are used for describing the material behavior:

E: Young's modulus
G: Shear modulus
ν: Poisson's ratio
α: Linear coefficient of thermal expansion (CTE)
β: Linear coefficient of moisture expansion (CME)
σ_{lCu}: Ultimate uniaxial compressive stress in fiber direction
σ_{lTu}: Ultimate tensile stress in fiber direction
σ_{qCu}: Ultimate compressive stress normal to fiber direction
σ_{qTu}: Ultimate tensile stress normal to fiber direction
τ_{lqu}: Ultimate shear stress
$\sigma_{zz\,u}$: Ultimate transverse normal stress for interlaminar failure
$\tau_{xz\,u}, \tau_{yz\,u}$: Ultimate in-plane shear stress for interlaminar failure
$\underset{\approx}{\mathbf{E}}$: Elasticity matrix
$\underset{\approx}{\mathbf{A}}, \underset{\approx}{\mathbf{B}}, \underset{\approx}{\mathbf{D}}$: Laminate's stiffness submatrices
$\underset{\sim}{\mathbf{A}}, \underset{\sim}{\mathbf{B}}, \underset{\sim}{\mathbf{D}}$: Laminate's thermal or moisture expansion vectors
$\underset{\approx}{\mathbf{T}}(\Theta)$: Rotation transformation matrix
Θ: Ply-angle; with respect to x_1-axis
h: Thickness of layer or shell
t: Thickness of face layer (of sandwich shell)
c: Thickness of core (of sandwich shell)
C: Support stiffness of the core (of sandwich shell)
K: Bending stiffness (of face layer of sandwich shell)
R: Radius of shell curvature (of sandwich shell)
N, Q, M: Stress resultants
P: Membrane force per unit length (in face layer of sandwich shell)
σ, τ: Normal and shear stresses
σ', τ': Effective stresses with respect to interlaminar failure
u, v, w: Displacements
φ, ψ: Angles of rotation of shell's normal
ε, γ: Strains and shear angles
$\underset{\sim}{\chi}$: Derivatives of rotations of shell's normal
θ, μ: Temperature, moisture
α, β: Wavelength parameters (for wrinkling analysis)
ξ: Biaxiality parameter (for wrinkling analysis)
η: Bending parameter (for wrinkling analysis)

Operators:

$(\ )^T$: Transposed of matrix ( )
$(\ )^{-1}$: Inverse of matrix ( )

Sub- and Superscripts:

$_L$: Local, i.e. related to axes of orthotropy
$_l$: Longitudinal
$_q$: Transverse, mostly used in-plane
$_t$: Transverse, mostly used out-of-plane
$^{(f)}$: Fibers
$^{(m)}$: Matrix
: Overall effective – no superscript
$_f$: Face layers (of sandwich shell)
$_c$: Core (of sandwich shell)
^{c}: Critical (for wrinkling analysis)
^{n} : Layer index, $n = 1, \ldots, N$
th : Thermal effect
^{m} : Moisture effect
$\hat{\ }$: Combination of plane stress and transverse shear
$\bar{\ }$: At the reference surface
$\widehat{\ }$: Difference between upper and lower surface
$\tilde{\ }$: Related to specific face layer coordinate system

Introduction

Advanced composite components are in most cases composed of layered laminates consisting of a – sometimes high – number of layers, the so-called laminae or plies which are bonded together. A single lamina represents a unidirectional (or, in the case of a woven reinforcement or of a randomly oriented short-fiber composite layer, multidirectional) thin composite. The mechanical and the hygro-thermal behavior of a laminate depend on the corresponding behavior of the laminae and on their stacking sequence. The properties of the individual laminae and the stacking sequence (changing orientation of the material axes from lamina to lamina) are chosen by the designer to meet certain design requirements for the laminate. Hence, the knowledge of the behavior of the lamina is a prerequisite for understanding the behavior of the laminate. Therefore, in the next section the analysis of laminae is dicussed followed by sections dealing with the analysis of laminates.

1 Constitutive Relations of Laminae

In Section 1 the effective material behavior of composites is discussed on the basis of micro-mechanics. In this section the material laws are specialized for thin plies, assuming orthotropic behavior with the axes of orthotropy being denoted by l and q, as used in Section 1, and t being the out-of-plane axis.

As a starting point the material law of a thin unidirectional layer (UD-layer) under plane stress conditions is considered:

$$\underset{\sim}{\sigma}_L = \underset{\approx}{\mathrm{E}}_L \, \underset{\sim}{\varepsilon}_L \quad \Longleftrightarrow \quad \begin{pmatrix} \sigma_{ll} \\ \sigma_{qq} \\ \tau_{lq} \end{pmatrix} = \begin{pmatrix} (\mathrm{E}_{11})_L & (\mathrm{E}_{12})_L & 0 \\ & (\mathrm{E}_{22})_L & 0 \\ sym. & & (\mathrm{E}_{44})_L \end{pmatrix} \begin{pmatrix} \varepsilon_{ll} \\ \varepsilon_{qq} \\ \gamma_{lq} \end{pmatrix} \tag{1}$$

Since this material law is related to the *local* axes of orthotropy of the individual layer under consideration, the subscript L is used. The elements of the $\underset{\approx}{\mathrm{E}}_L$-matrix can be determined by methods described in Section 1 or by measuring the response of the UD-layer under uniaxial and shear loading in the direction of the orthotropy axes. The relations between the $(\mathrm{E}_{ij})_L$ and the elasticity parameters are given below, see e.g. [1]:

$$\begin{aligned} (\mathrm{E}_{11})_L &= \frac{\mathrm{E}_l}{1 - \nu_{lq}^2 \frac{\mathrm{E}_q}{\mathrm{E}_l}} \\ (\mathrm{E}_{22})_L &= \frac{\mathrm{E}_q}{1 - \nu_{lq}^2 \frac{\mathrm{E}_q}{\mathrm{E}_l}} \\ (\mathrm{E}_{12})_L &= \nu_{lq} \, (\mathrm{E}_{22})_L \\ (\mathrm{E}_{44})_L &= \mathrm{G}_{lq} \end{aligned} \tag{2}$$

Since the layer, and even more the laminate, shows a certain transverse stiffness, i.e. a stiffness under transverse shear in t-direction normal to the (l, q)-plane, this simple plane stress material law is modified, and transverse shear stresses are also included:

$$\underset{\sim}{\tau}_L = \underset{\approx}{\mathbf{E}}^t_L \, \underset{\sim}{\gamma}_L \quad \Longleftrightarrow \quad \begin{pmatrix} \tau_{lt} \\ \tau_{qt} \end{pmatrix} = \begin{pmatrix} (\mathrm{E}_{55})_L & 0 \\ 0 & (\mathrm{E}_{66})_L \end{pmatrix} \begin{pmatrix} \gamma_{lt} \\ \gamma_{qt} \end{pmatrix} \tag{3}$$

or in complete form:

$$\hat{\underset{\sim}{\sigma}}_L = \hat{\underset{\approx}{\mathbf{E}}}_L \, \hat{\underset{\sim}{\varepsilon}}_L \quad \Longleftrightarrow \quad \begin{pmatrix} \sigma_{ll} \\ \sigma_{qq} \\ \tau_{lq} \\ \tau_{lt} \\ \tau_{qt} \end{pmatrix} = \begin{pmatrix} (\mathrm{E}_{11})_L & (\mathrm{E}_{12})_L & 0 & 0 & 0 \\ & (\mathrm{E}_{22})_L & 0 & 0 & 0 \\ & & (\mathrm{E}_{44})_L & 0 & 0 \\ & sym. & & (\mathrm{E}_{55})_L & 0 \\ & & & & (\mathrm{E}_{66})_L \end{pmatrix} \begin{pmatrix} \varepsilon_{ll} \\ \varepsilon_{qq} \\ \gamma_{lq} \\ \gamma_{lt} \\ \gamma_{qt} \end{pmatrix} \tag{4}$$

The out-of-plane shear stiffnesses, $(\mathrm{E}_{55})_L = \mathrm{G}_{lt} \ldots$ for shear in the (l, t)-plane, $(\mathrm{E}_{66})_L = \mathrm{G}_{qt} \ldots$ for shear in the (q, t)-plane, are difficult to determine experimentally. In Section 3 of [2] some formulae are given for determining these parameters. In [3] the following estimates are proposed:

$$\begin{aligned} (\mathrm{E}_{55})_L &\approx (\mathrm{E}_{44})_L = \mathrm{G}_{lq} \\ (\mathrm{E}_{66})_L &\approx G^{(m)} < \mathrm{G}_{lq} \end{aligned} \tag{5}$$

For orthotropic materials the stress–strain relations are dependent on the angle of rotation Θ of the reference system (x_1, x_2, x_3) with respect to the material axes (l, q), see Fig. 1.

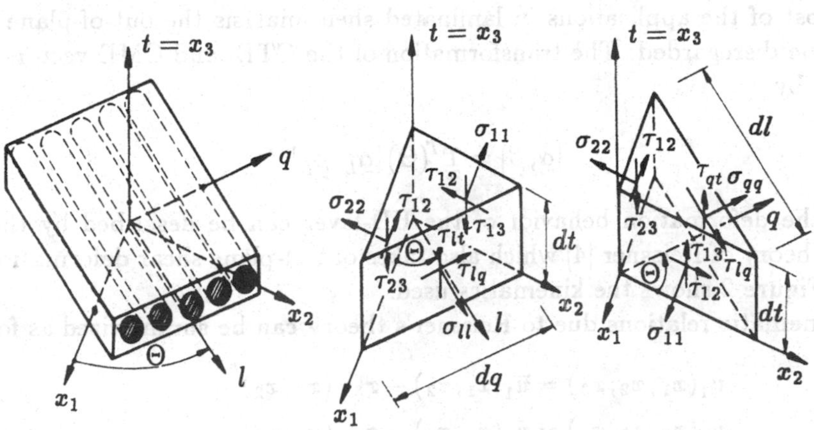

Fig. 1 Reference systems and stress components

Using the common transformation expressions for stress and strain tensor rotations, respectively, the following relation is obtained for the elasticity matrix relative to the (x_1, x_2, x_3)-system, compare [3]:

$$\hat{\underset{\approx}{E}} = \hat{\underset{\approx}{T}}^{-1}(\Theta)\,\hat{\underset{\approx}{E}}_L\left(\hat{\underset{\approx}{T}}^T(\Theta)\right)^{-1} = \hat{\bar{\underset{\approx}{T}}}^T(\Theta)\,\hat{\underset{\approx}{E}}_L\,\hat{\bar{\underset{\approx}{T}}}(\Theta) \qquad (6),$$

and, analogously,

$$\underset{\approx}{E} = \bar{\underset{\approx}{T}}^T(\Theta)\,\underset{\approx}{E}_L\,\bar{\underset{\approx}{T}}(\Theta) \quad , \qquad \underset{\approx}{E}^t = \bar{\underset{\approx}{T}}^{t\,T}(\Theta)\,\underset{\approx}{E}_L^t\,\bar{\underset{\approx}{T}}^t(\Theta) \qquad (7),$$

with the transformation matrices

$$\hat{\underset{\approx}{T}}(\Theta) = \begin{pmatrix} \underset{\approx}{T}(\Theta) & \underset{\approx}{0} \\ \underset{\approx}{0}^T & \underset{\approx}{T}^t(\Theta) \end{pmatrix} = \begin{pmatrix} c^2 & s^2 & 2sc & 0 & 0 \\ s^2 & c^2 & -2sc & 0 & 0 \\ -sc & sc & c^2 - s^2 & 0 & 0 \\ 0 & 0 & 0 & c & s \\ 0 & 0 & 0 & -s & c \end{pmatrix} \qquad (8).$$

$$\hat{\bar{\underset{\approx}{T}}}(\Theta) = \left(\hat{\underset{\approx}{T}}^T(\Theta)\right)^{-1}$$

Here c and s stand for $c := \cos\Theta$ and $s := \sin\Theta$, respectively.

Analogously, the vectors of the coefficients of in-plane thermal and moisture expansion, respectively, have, relative to the orthotropy axes, the form

$$[\underset{\sim}{\alpha}_L, \underset{\sim}{\beta}_L] = \begin{pmatrix} [\alpha_l\,, \beta_l\,] \\ [\alpha_q\,, \beta_q\,] \\ 0 \end{pmatrix} \qquad (9).$$

(The notation $[a, b]$ does not mean matrix notation but that the relations hold for each of the parameters, i.e. a and b, individually.)

For most of the applications in laminated shell analysis the out-of-plane expansions can be disregarded. The transformation of the CTE- and CME-vectors can be performed by

$$[\underset{\sim}{\alpha}, \underset{\sim}{\beta}] = \underset{\approx}{T}^T(\Theta)\,[\underset{\sim}{\alpha}_L, \underset{\sim}{\beta}_L] \qquad (10).$$

Now, the deformation behavior of the UD-layer can be described by the plate and shell theory of Reissner [4] which also takes out-of-plane shear deformations into account. Figure 2 shows the kinematics used.

The kinematic relations due to Reissner's theory can be summarized as follows:

$$\begin{aligned} u_1(x_1, x_2, x_3) &= \bar{u}_1(x_1, x_2) - x_3\psi(x_1, x_2) \\ u_2(x_1, x_2, x_3) &= \bar{u}_2(x_1, x_2) - x_3\varphi(x_1, x_2) \\ u_3(x_1, x_2, x_3) &= \bar{u}_3(x_1, x_2) \end{aligned} \qquad (11)$$

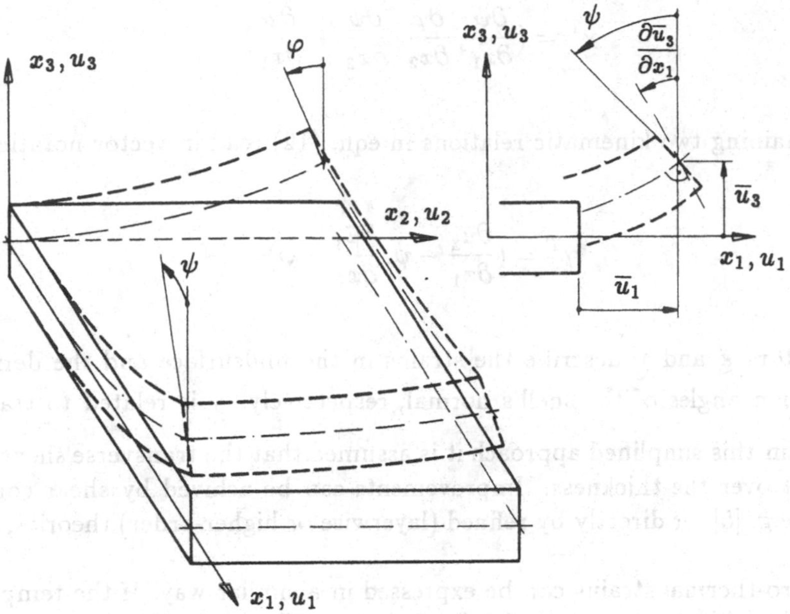

Fig. 2 Mindlin–Reissner kinematics

$$
\begin{aligned}
\varepsilon_{11} &= \frac{\partial u_1}{\partial x_1} &&= \frac{\partial \bar{u}_1}{\partial x_1} - x_3 \frac{\partial \psi}{\partial x_1} \\
\varepsilon_{22} &= \frac{\partial u_2}{\partial x_2} &&= \frac{\partial \bar{u}_2}{\partial x_2} - x_3 \frac{\partial \varphi}{\partial x_2} \\
\gamma_{12} &= \frac{\partial u_1}{\partial x_2} + \frac{\partial u_2}{\partial x_1} &&= \frac{\partial \bar{u}_1}{\partial x_2} + \frac{\partial \bar{u}_2}{\partial x_1} - x_3 \left(\frac{\partial \psi}{\partial x_2} + \frac{\partial \varphi}{\partial x_1}\right) \\
\gamma_{13} &= \frac{\partial u_3}{\partial x_1} + \frac{\partial u_1}{\partial x_3} &&= \frac{\partial \bar{u}_3}{\partial x_1} - \psi \\
\gamma_{23} &= \frac{\partial u_3}{\partial x_2} + \frac{\partial u_2}{\partial x_3} &&= \frac{\partial \bar{u}_3}{\partial x_2} - \varphi
\end{aligned}
\qquad (12),
$$

In vector notation the first three equations read

$$\underset{\sim}{\varepsilon}(x_1, x_2, x_3) = \bar{\underset{\sim}{\varepsilon}}(x_1, x_2) - x_3\, \underset{\sim}{\chi}(x_1, x_2) \qquad (13),$$

with

$$\bar{\underset{\sim}{\varepsilon}}^T = \left(\frac{\partial \bar{u}_1}{\partial x_1}, \frac{\partial \bar{u}_2}{\partial x_2}, \frac{\partial \bar{u}_1}{\partial x_2} + \frac{\partial \bar{u}_2}{\partial x_1}\right) \qquad (14)$$

and

$$\underset{\sim}{\chi}^T = (\frac{\partial \psi}{\partial x_1}, \frac{\partial \varphi}{\partial x_2}, \frac{\partial \psi}{\partial x_2} + \frac{\partial \varphi}{\partial x_1}) \qquad (15).$$

The remaining two kinematic relations in eqn. (12) read in vector notation

$$\underset{\sim}{\gamma}^T = (\frac{\partial \overline{u}_3}{\partial x_1} - \psi, \frac{\partial \overline{u}_3}{\partial x_2} - \varphi) \qquad (16).$$

The vectors $\underset{\sim}{\overline{\varepsilon}}$ and $\underset{\sim}{\chi}$ describe the strains in the midsurface and the derivatives of the rotation angles of the shell's normal, respectively; $\underset{\sim}{\gamma}$ is related to transverse shear. Within this simplified approach it is assumed that the transverse shear strains are constant over the thickness. Improvements can be achived by shear correction factors, see e.g. [5], or directly by refined (layerwise or higher-order) theories, see e.g. [6,7,8].

The hygro-thermal strains can be expressed in a similar way. If the temperature and the moisture fields, respectively, (both being differences with respect to stress free states) vary only linearly over the thickness the following relations hold:

$$[{}^{th}\underset{\sim}{\varepsilon}, {}^{m}\underset{\sim}{\varepsilon}](x_1, x_2, x_3) = [\underset{\sim}{\alpha}, \underset{\sim}{\beta}]\,[\overline{\vartheta}, \overline{\mu}](x_1, x_2) - \frac{x_3}{\mathrm{h}}[\underset{\sim}{\alpha}, \underset{\sim}{\beta}]\,[\widehat{\vartheta}, \widehat{\mu}](x_1, x_2) \qquad (17)$$

$[\overline{\vartheta}, \overline{\mu}]\ldots$ [temperature, moisture] field in the midsurface

$[\widehat{\vartheta}, \widehat{\mu}]\ldots$ field of [temperature, moisture] difference between surface at $x_3 = -\frac{\mathrm{h}}{2}$ and surface at $x_3 = +\frac{\mathrm{h}}{2}$, respectively,

with

$$[\vartheta, \mu](x_1, x_2, x_3) = [\overline{\vartheta}, \overline{\mu}](x_1, x_2) - \frac{x_3}{\mathrm{h}}[\widehat{\vartheta}, \widehat{\mu}](x_1, x_2) \qquad (18).$$

The following definitions of stress resultants are used, see Fig. 3:

$$\underset{\sim}{N} = \begin{pmatrix} N_{11} \\ N_{22} \\ N_{12} \end{pmatrix} \qquad \underset{\sim}{M} = \begin{pmatrix} M_{11} \\ M_{22} \\ M_{12} \end{pmatrix} \qquad \underset{\sim}{Q} = \begin{pmatrix} Q_{13} \\ Q_{23} \end{pmatrix} \qquad (19),$$

with

$$N_{11} = \int_{-h/2}^{+h/2} \sigma_{11}\, dx_3 \qquad M_{11} = \int_{-h/2}^{+h/2} \sigma_{11} x_3\, dx_3$$

$$N_{22} = \int_{-h/2}^{+h/2} \sigma_{22}\, dx_3 \qquad M_{22} = \int_{-h/2}^{+h/2} \sigma_{22} x_3\, dx_3$$

$$N_{12} = \int_{-h/2}^{+h/2} \tau_{12}\, dx_3 \qquad M_{12} = \int_{-h/2}^{+h/2} \tau_{12} x_3\, dx_3 \qquad (20).$$

$$Q_{13} = \int_{-h/2}^{+h/2} \tau_{13}\, dx_3$$

$$Q_{23} = \int_{-h/2}^{+h/2} \tau_{23}\, dx_3$$

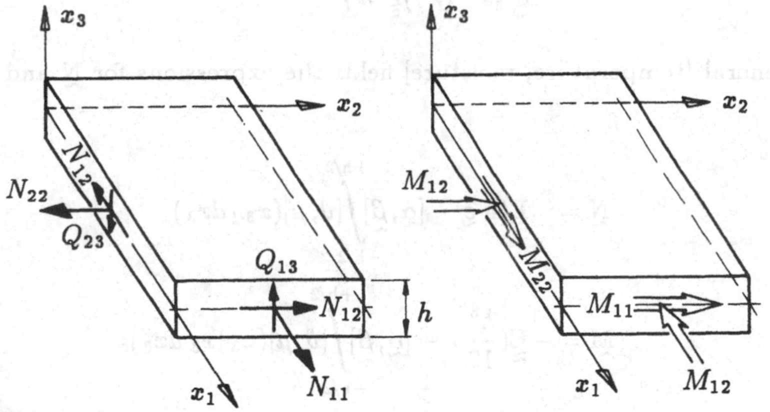

Fig. 3 Definition of stress resultants

Using the hygro-thermo-elastic material law

$$\underset{\sim}{\sigma} = \underset{\approx}{\mathbf{E}}(\underset{\sim}{\varepsilon} - {}^{th}\underset{\sim}{\varepsilon} - {}^{m}\underset{\sim}{\varepsilon}) \qquad (21)$$

and the above relations the following equations are obtained:

$$\begin{aligned}
\underset{\sim}{\mathrm{N}} &= \int_{-h/2}^{+h/2} \underset{\sim}{\sigma}\, dx_3 &&= \int_{-h/2}^{+h/2} \underset{\approx}{\mathbf{E}}\Big((\underset{\sim}{\varepsilon} - \underset{\sim}{\alpha}\overline{\vartheta} - \underset{\sim}{\beta}\overline{\mu}) - x_3(\underset{\sim}{\chi} - \underset{\sim}{\alpha}\frac{\widehat{\vartheta}}{\mathrm{h}} - \underset{\sim}{\beta}\frac{\widehat{\mu}}{\mathrm{h}})\Big)\, dx_3 \\
\underset{\sim}{\mathrm{M}} &= \int_{-h/2}^{+h/2} \underset{\sim}{\sigma} x_3\, dx_3 &&= \int_{-h/2}^{+h/2} \underset{\approx}{\mathbf{E}}\Big((\underset{\sim}{\varepsilon} - \underset{\sim}{\alpha}\overline{\vartheta} - \underset{\sim}{\beta}\overline{\mu}) - x_3(\underset{\sim}{\chi} - \underset{\sim}{\alpha}\frac{\widehat{\vartheta}}{\mathrm{h}} - \underset{\sim}{\beta}\frac{\widehat{\mu}}{\mathrm{h}})\Big) x_3\, dx_3 \qquad (22). \\
\underset{\sim}{\mathrm{Q}} &= \int_{-h/2}^{+h/2} \underset{\sim}{\tau}\, dx_3 &&= (k^*)\int_{-h/2}^{+h/2} \underset{\approx}{\mathbf{E}}^t \underset{\sim}{\gamma}\, dx_3 \qquad (k^* \ldots \text{ shear correction factor})
\end{aligned}$$

Neglecting for the time being a possible temperature dependence of the hygro-thermo-elastic material properties, it can be assumed that $\underset{\approx}{\mathbf{E}}$, $\underset{\approx}{\mathbf{E}}^t$, $\underset{\sim}{\alpha}$ and $\underset{\sim}{\beta}$ are constant within the UD-layer, and the above integrals render

$$\begin{aligned}
\underset{\sim}{\mathrm{N}} &= \underset{\approx}{\mathbf{E}}\,\mathrm{h}(\underset{\sim}{\varepsilon} - \underset{\sim}{\alpha}\overline{\vartheta} - \underset{\sim}{\beta}\,\overline{\mu}) \\
\underset{\sim}{\mathrm{M}} &= -\underset{\approx}{\mathbf{E}}\frac{h^3}{12}(\underset{\sim}{\chi} - \underset{\sim}{\alpha}\frac{\widehat{\vartheta}}{\mathrm{h}} - \underset{\sim}{\beta}\frac{\widehat{\mu}}{\mathrm{h}}) \qquad (23). \\
\underset{\sim}{\mathrm{Q}} &= (k^*)\underset{\approx}{\mathbf{E}}^t h \underset{\sim}{\gamma}
\end{aligned}$$

For general [temperature, moisture] fields the expressions for $\underset{\sim}{\mathrm{N}}$ and $\underset{\sim}{\mathrm{M}}$ are read as

$$\begin{aligned}
\underset{\sim}{\mathrm{N}} &= \underset{\approx}{\mathbf{E}}\Big(\mathrm{h}\,\underset{\sim}{\varepsilon} - [\underset{\sim}{\alpha}, \underset{\sim}{\beta}]\int_{-h/2}^{+h/2}[\vartheta, \mu](x_3)\, dx_3\Big) \\
\underset{\sim}{\mathrm{M}} &= -\underset{\approx}{\mathbf{E}}\Big(\frac{h^3}{12}\underset{\sim}{\chi} - [\underset{\sim}{\alpha}, \underset{\sim}{\beta}]\int_{-h/2}^{+h/2}[\vartheta, \mu](x_3) x_3\, dx_3\Big)
\end{aligned} \qquad (24).$$

2 Theory of Multilayered Laminates

2.1 Constitutive Laws for Laminates

As mentioned above laminates are constructed from laminae with usually varying orientations of the orthotropy axes of the individual laminae. Hence, in view of eqns. (7) and (10) the individual plies have different $\underset{\approx}{\mathbf{E}}$- and $\underset{\approx}{\mathbf{E}}^t$-matrices as well as different

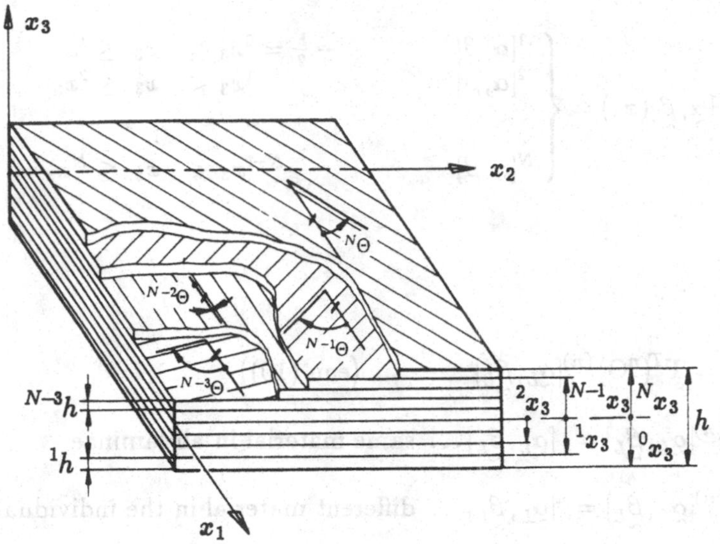

Fig. 4 Geometrical definitions of the laminate

$\underset{\sim}{\alpha}$- and $\underset{\sim}{\beta}$- vectors related to the common reference system x_1, x_2, x_3.

Using the definitions as shown in Fig. 4 the $\underset{\approx}{\mathbf{E}}$- and $\underset{\approx}{\mathbf{E}}^t$-matrices and the CTE- as well as CME-vectors can be derived as follows:

$$[\underset{\approx}{\mathbf{E}}, \underset{\approx}{\mathbf{E}}^t](x_3) = \begin{cases} {}^1[\underset{\approx}{\mathbf{E}}, \underset{\approx}{\mathbf{E}}^t] & -\frac{h}{2} = {}^0x_3 \leq \; x_3 \; \leq {}^1x_3 \\ {}^2[\underset{\approx}{\mathbf{E}}, \underset{\approx}{\mathbf{E}}^t] & {}^1x_3 < \; x_3 \; \leq {}^2x_3 \\ \vdots & \vdots \\ {}^N[\underset{\approx}{\mathbf{E}}, \underset{\approx}{\mathbf{E}}^t] & {}^{N-1}x_3 < \; x_3 \; \leq {}^Nx_3 = \frac{h}{2} \end{cases} \tag{25}$$

with

$${}^nx_3 = -\frac{h}{2} + \sum_{j=1}^{n} {}^jh \qquad n = 1, \ldots, N$$

$${}^n[\underset{\approx}{\mathbf{E}}, \underset{\approx}{\mathbf{E}}^t] = [\overline{\underset{\approx}{\mathbf{T}}}^T, \overline{\underset{\approx}{\mathbf{T}}}^{t^T}]({}^n\Theta)\,{}^{(n)}[\underset{\approx}{\mathbf{E}}_L, \underset{\approx}{\mathbf{E}}_L{}^t]\,[\overline{\underset{\approx}{\mathbf{T}}}, \overline{\underset{\approx}{\mathbf{T}}}^t]({}^n\Theta) \quad \ldots \quad \text{(eqn. (7))}$$

${}^{(n)}[\underset{\approx}{\mathbf{E}}_L, \underset{\approx}{\mathbf{E}}_L{}^t] = \; [\underset{\approx}{\mathbf{E}}_L, \underset{\approx}{\mathbf{E}}_L{}^t] \ldots$ equal material in all laminae

${}^{(n)}[\underset{\approx}{\mathbf{E}}_L, \underset{\approx}{\mathbf{E}}_L{}^t] = {}^n[\underset{\approx}{\mathbf{E}}_L, \underset{\approx}{\mathbf{E}}_L{}^t] \ldots$ different material in the individual laminae

$$[\underset{\sim}{\alpha},\underset{\sim}{\beta}](x_3) = \begin{cases} {}^1[\alpha,\beta] & -\frac{h}{2} = {}^0x_3 \le x_3 \le {}^1x_3 \\ {}^2[\alpha,\beta] & {}^1x_3 < x_3 \le {}^2x_3 \\ \vdots & \vdots \\ {}^N[\alpha,\beta] & {}^{N-1}x_3 < x_3 \le {}^Nx_3 = \frac{h}{2} \end{cases} \tag{26}$$

with

$${}^n[\underset{\sim}{\alpha},\underset{\sim}{\beta}] = \underset{\approx}{\mathbf{T}}^T({}^n\Theta)\,{}^{(n)}[\underset{\sim}{\alpha}_L,\underset{\sim}{\beta}_L] \qquad \text{(eqn (10))}$$

$${}^{(n)}[\underset{\sim}{\alpha}_L,\underset{\sim}{\beta}_L] = [\underset{\sim}{\alpha}_L,\underset{\sim}{\beta}_L] \ldots \text{ same material in all laminae}$$

$${}^{(n)}[\underset{\sim}{\alpha}_L,\underset{\sim}{\beta}_L] = {}^n[\underset{\sim}{\alpha}_L,\underset{\sim}{\beta}_L] \ldots \text{ different material in the individual laminae}$$

The eqns. (22) in combination with eqns. (25) and (26) after some algebraic manipulation lead to

$$\begin{aligned}
\underset{\sim}{N} =& \int_{-h/2}^{+h/2} \underset{\approx}{\mathbf{E}}(x_3)\,dx_3\,\underset{\sim}{\bar{\varepsilon}} - \int_{-h/2}^{+h/2} \underset{\approx}{\mathbf{E}}(x_3)\underset{\sim}{\alpha}(x_3)\,dx_3\bar{\vartheta} - \int_{-h/2}^{+h/2} \underset{\approx}{\mathbf{E}}(x_3)\underset{\sim}{\beta}(x_3)\,dx_3\bar{\mu} - \\
& \int_{-h/2}^{+h/2} \underset{\approx}{\mathbf{E}}(x_3)x_3\,dx_3\underset{\sim}{\chi} + \int_{-h/2}^{+h/2} \underset{\approx}{\mathbf{E}}(x_3)\underset{\sim}{\alpha}(x_3)x_3\,dx_3\widehat{\vartheta} + \int_{-h/2}^{+h/2} \underset{\approx}{\mathbf{E}}(x_3)\underset{\sim}{\beta}(x_3)x_3\,dx_3\widehat{\mu} \\
\underset{\sim}{M} =& \int_{-h/2}^{+h/2} \underset{\approx}{\mathbf{E}}(x_3)x_3\,dx_3\,\underset{\sim}{\bar{\varepsilon}} - \int_{-h/2}^{+h/2} \underset{\approx}{\mathbf{E}}(x_3)\underset{\sim}{\alpha}(x_3)x_3\,dx_3\bar{\vartheta} - \int_{-h/2}^{+h/2} \underset{\approx}{\mathbf{E}}(x_3)\underset{\sim}{\beta}(x_3)x_3\,dx_3\bar{\mu} - \\
& \int_{-h/2}^{+h/2} \underset{\approx}{\mathbf{E}}(x_3)x_3^2\,dx_3\underset{\sim}{\chi} + \int_{-h/2}^{+h/2} \underset{\approx}{\mathbf{E}}(x_3)\underset{\sim}{\alpha}(x_3)x_3^2\,dx_3\widehat{\vartheta} + \int_{-h/2}^{+h/2} \underset{\approx}{\mathbf{E}}(x_3)\underset{\sim}{\beta}(x_3)x_3^2\,dx_3\widehat{\mu} \\
\underset{\sim}{Q} =& (k^*)\int_{-h/2}^{+h/2} \underset{\approx}{\mathbf{E}}^t(x_3)\,dx_3\,\underset{\sim}{\gamma}
\end{aligned} \tag{27}$$

The above integrals can be split, which leads to the following expressions:

$$\underset{\approx}{A}=\int_{-h/2}^{+h/2}\underset{\approx}{E}(x_3)\,dx_3 =\sum_{n=1}^{N}\int_{^{n-1}x_3}^{^{n}x_3}{}^{n}\underset{\approx}{E}\,dx_3 =\sum_{n=1}^{N}{}^{n}\underset{\approx}{E}({}^{n}x_3-{}^{n-1}x_3)=\sum_{n=1}^{N}{}^{n}\underset{\approx}{E}\,{}^{n}h$$

$$\underset{\approx}{B}=\int_{-h/2}^{+h/2}\underset{\approx}{E}(x_3)x_3\,dx_3=\sum_{n=1}^{N}\int_{^{n-1}x_3}^{^{n}x_3}{}^{n}\underset{\approx}{E}\,x_3\,dx_3=\sum_{n=1}^{N}{}^{n}\underset{\approx}{E}\frac{{}^{n}x_3^2-{}^{n-1}x_3^2}{2} \qquad (28),$$

$$\underset{\approx}{D}=\int_{-h/2}^{+h/2}\underset{\approx}{E}(x_3)x_3^2\,dx_3=\sum_{n=1}^{N}\int_{^{n-1}x_3}^{^{n}x_3}{}^{n}\underset{\approx}{E}\,x_3^2\,dx_3=\sum_{n=1}^{N}{}^{n}\underset{\approx}{E}\frac{{}^{n}x_3^3-{}^{n-1}x_3^3}{3}$$

$$\underset{\approx}{A}=\int_{-h/2}^{+h/2}\underset{\approx}{E}^t(x_3)\,dx_3=\sum_{n=1}^{N}{}^{n}\underset{\approx}{E}^t\,{}^{n}h \qquad \textbf{(without shear correction factors)} \qquad (29),$$

$${}^{th}\underset{\sim}{A}=\int_{-h/2}^{+h/2}\underset{\approx}{E}(x_3)\underset{\sim}{\alpha}(x_3)\,dx_3 =\sum_{n=1}^{N}\int_{^{n-1}x_3}^{^{n}x_3}{}^{n}\underset{\approx}{E}\,{}^{n}\underset{\sim}{\alpha}\,dx_3 =\sum_{n=1}^{N}{}^{n}\underset{\approx}{E}\,{}^{n}\underset{\sim}{\alpha}({}^{n}x_3-{}^{n-1}x_3)$$

$${}^{th}\underset{\sim}{B}=\int_{-h/2}^{+h/2}\underset{\approx}{E}(x_3)\underset{\sim}{\alpha}(x_3)x_3\,dx_3=\sum_{n=1}^{N}\int_{^{n-1}x_3}^{^{n}x_3}{}^{n}\underset{\approx}{E}\,{}^{n}\underset{\sim}{\alpha}\,x_3\,dx_3=\sum_{n=1}^{N}{}^{n}\underset{\approx}{E}\,{}^{n}\underset{\sim}{\alpha}\frac{{}^{n}x_3^2-{}^{n-1}x_3^2}{2} \qquad (30)$$

$${}^{th}\underset{\sim}{D}=\int_{-h/2}^{+h/2}\underset{\approx}{E}(x_3)\underset{\sim}{\alpha}(x_3)x_3^2\,dx_3=\sum_{n=1}^{N}\int_{^{n-1}x_3}^{^{n}x_3}{}^{n}\underset{\approx}{E}\,{}^{n}\underset{\sim}{\alpha}\,x_3^2\,dx_3=\sum_{n=1}^{N}{}^{n}\underset{\approx}{E}\,{}^{n}\underset{\sim}{\alpha}\frac{{}^{n}x_3^3-{}^{n-1}x_3^3}{3}$$

and

$${}^{m}\underset{\sim}{A}=\int_{-h/2}^{+h/2}\underset{\approx}{E}(x_3)\underset{\sim}{\beta}(x_3)\,dx_3 =\sum_{n=1}^{N}\int_{^{n-1}x_3}^{^{n}x_3}{}^{n}\underset{\approx}{E}\,{}^{n}\underset{\sim}{\beta}\,dx_3 =\sum_{n=1}^{N}{}^{n}\underset{\approx}{E}\,{}^{n}\underset{\sim}{\beta}({}^{n}x_3-{}^{n-1}x_3)$$

$${}^{m}\underset{\sim}{B}=\int_{-h/2}^{+h/2}\underset{\approx}{E}(x_3)\underset{\sim}{\beta}(x_3)x_3\,dx_3=\sum_{n=1}^{N}\int_{^{n-1}x_3}^{^{n}x_3}{}^{n}\underset{\approx}{E}\,{}^{n}\underset{\sim}{\beta}\,x_3\,dx_3=\sum_{n=1}^{N}{}^{n}\underset{\approx}{E}\,{}^{n}\underset{\sim}{\beta}\frac{{}^{n}x_3^2-{}^{n-1}x_3^2}{2} \qquad (31).$$

$${}^{m}\underset{\sim}{D}=\int_{-h/2}^{+h/2}\underset{\approx}{E}(x_3)\underset{\sim}{\beta}(x_3)x_3^2\,dx_3=\sum_{n=1}^{N}\int_{^{n-1}x_3}^{^{n}x_3}{}^{n}\underset{\approx}{E}\,{}^{n}\underset{\sim}{\beta}\,x_3^2\,dx_3=\sum_{n=1}^{N}{}^{n}\underset{\approx}{E}\,{}^{n}\underset{\sim}{\beta}\frac{{}^{n}x_3^3-{}^{n-1}x_3^3}{3}$$

Using these expressions eqn. (27) leads the hygro-thermo-elastic constitutive equation of the laminate:

$$\begin{pmatrix} \underset{\sim}{N} \\ \underset{\sim}{M} \end{pmatrix} = \begin{pmatrix} \underset{\approx}{A} & \underset{\approx}{B} \\ \underset{\approx}{B} & \underset{\approx}{D} \end{pmatrix} \begin{pmatrix} \underset{\sim}{\bar{\varepsilon}} \\ -\underset{\sim}{\chi} \end{pmatrix} - \begin{pmatrix} {}^{th}\underset{\sim}{A} & {}^{th}\underset{\sim}{B} \\ {}^{th}\underset{\sim}{B} & {}^{th}\underset{\sim}{D} \end{pmatrix} \begin{pmatrix} \bar{\vartheta} \\ -\widehat{\vartheta} \end{pmatrix} - \begin{pmatrix} {}^{m}\underset{\sim}{A} & {}^{m}\underset{\sim}{B} \\ {}^{m}\underset{\sim}{B} & {}^{m}\underset{\sim}{D} \end{pmatrix} \begin{pmatrix} \bar{\mu} \\ -\widehat{\mu} \end{pmatrix} \quad (32).$$

$$\underset{\sim}{Q} = \underset{\approx}{\bar{A}}\, \underset{\sim}{\gamma}$$

The matrices $\underset{\approx}{A}$, $\underset{\approx}{B}$ and $\underset{\approx}{D}$ are called the extensional, coupling and bending stiffness matrices, respectively. It is essential to notice that due to a nonvanishing coupling matrix, $\underset{\approx}{B}$, effects appear which would not be expected from experience with homogeneous isotropic plates. For example, forces acting in the midplane, i.e. membrane forces, result in warping deformations, i.e. bending or twisting deformations – see Fig. 5 –, and pure bending loads may lead to stretching of the midsurface. These coupling effects can lead to undesirable stresses, in addition to the applied load stresses, if the deformations are restrained. Furthermore, temperature changes, even if constant over the thickness of the laminate, may lead to out-of-plane deformations – see Fig. 6 – and, hence, composites with nonvanishing $\underset{\approx}{B}$ will show warping when cooled down from the fabrication temperature to room temperature.

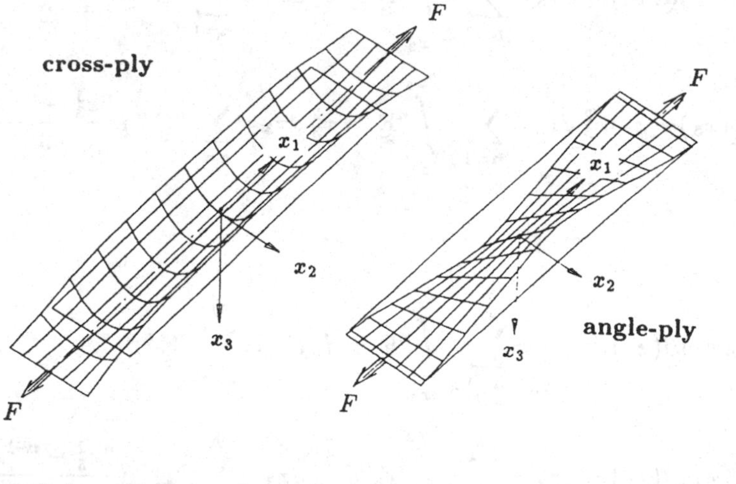

Fig. 5 Out-of-plane deformations of a two-layer laminate; a) cross-ply, b) angle-ply

Improvements for describing the out-of-plane shear behavior are for example discussed in [9,10,11].

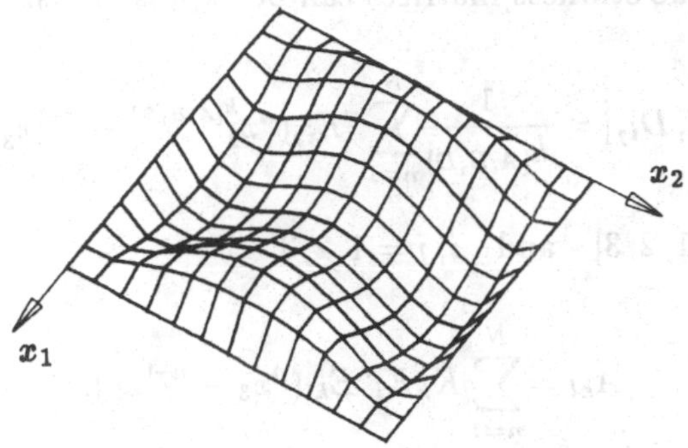

Fig. 6 Warping of a two-layer cross-ply laminate plate under constant temperature rise

2.2 Special Laminates

In summary, lamination theory is presented in a compact formulation for the isothermal case without moisture effects:

With

$$\underset{\sim}{N}^T = (N_{11}, N_{22}, N_{12}),$$

$$\underset{\sim}{M}^T = (M_{11}, M_{22}, M_{12}),$$

$$\underset{\sim}{Q}^T = (Q_{13}, Q_{23}),$$

$$\underset{\sim}{\varepsilon}^T = (\frac{\partial \overline{u}_1}{\partial x_1}, \frac{\partial \overline{u}_2}{\partial x_2}, \frac{\partial \overline{u}_1}{\partial x_2} + \frac{\partial \overline{u}_2}{\partial x_1}),$$

$$\underset{\sim}{\chi}^T = (\frac{\partial \psi}{\partial x_1}, \frac{\partial \varphi}{\partial x_2}, \frac{\partial \psi}{\partial x_2} + \frac{\partial \varphi}{\partial x_1}),$$

$$\underset{\sim}{\gamma}^T = (\frac{\partial \overline{u}_3}{\partial x_1} - \psi, \frac{\partial \overline{u}_3}{\partial x_2} - \varphi),$$

the isothermal constitutive equation of the laminate can be written in the following form:

$$\begin{pmatrix} \underset{\sim}{N} \\ \underset{\sim}{M} \end{pmatrix} = \begin{pmatrix} \underset{\approx}{A} & \underset{\approx}{B} \\ \underset{\approx}{B} & \underset{\approx}{D} \end{pmatrix} \begin{pmatrix} \underset{\sim}{\overline{\varepsilon}} \\ -\underset{\sim}{\chi} \end{pmatrix},$$

$$\underset{\sim}{Q} = \begin{pmatrix} \bar{A}_{55} & \bar{A}_{56} \\ \bar{A}_{56} & \bar{A}_{66} \end{pmatrix} \underset{\sim}{\gamma}.$$

The elements of the stiffness matrices can be expressed as:

$$[A_{ij}, B_{ij}, D_{ij}] = \frac{1}{k_{[A,B,D]}} \sum_{n=1}^{N} {}^{n}E_{ij}\left({}^{n}x_3^{\,k_{[A,B,D]}} - {}^{n-1}x_3^{\,k_{[A,B,D]}}\right),$$

with $k_{[A,B,D]} = [1,2,3]$ and $i,j = 1,2,3,$

$$\bar{A}_{kl} = \sum_{n=1}^{N} K_k K_l\, {}^{n}E_{kl}({}^{n}x_3 - {}^{n-1}x_3),$$

with $k,l = 5,6$ and K_k, K_l are the shear correction coefficients according to [12].

The characteristic isothermal behavior of a laminate with respect to different coupling effects depends on the matrices $\underset{\approx}{\mathbf{A}}$, $\underset{\approx}{\mathbf{B}}$ and $\underset{\approx}{\mathbf{D}}$, i.e. on their elements

$$A_{ij} = \sum_{n=1}^{N} {}^{n}E_{ij}({}^{n}x_3 - {}^{n-1}x_3)$$

$$B_{ij} = \frac{1}{2}\sum_{n=1}^{N} {}^{n}E_{ij}({}^{n}x_3^2 - {}^{n-1}x_3^2) \qquad (33),$$

$$D_{ij} = \frac{1}{3}\sum_{n=1}^{N} {}^{n}E_{ij}({}^{n}x_3^3 - {}^{n-1}x_3^3)$$

with ${}^{n}E_{ij}$ depending on the elements of the ${}^{n}\underset{\approx}{\mathbf{E}}_L$ matrices and the orientation of the principal material axes with respect to the x_1-axis, i.e. ${}^{n}\Theta$, corresponding to eqn. (7). Hence, by specific stacking sequences special laminates can be designed, like symmetric, quasi-orthotropic or quasi-isotropic laminates.

a) Symmetric Laminates

Consideration of eqns. (28) or (33) leads to the conclusion that the coupling matrix $\underset{\approx}{\mathbf{B}}$ can be made to be identically zero if every lamina above the midsurface has an identical lamina (same $\underset{\approx}{\mathbf{E}}_L$, Θ and thickness) an equal distance below the midsurface resulting in a so-called "symmetric laminate". Since the $\underset{\approx}{\mathbf{B}}$-matrix is zero, symmetric laminates do not show the bending-stretching coupling effects.

b) Quasi-Orthotropic Laminates

If the elements A_{14} and A_{24} (index notation corresponding to eqn (4)) become zero, the laminate behaves orthotropically with respect to pure in-plane loading, i.e. as a quasi-orthotropic layer. The ${}^{n}E_{ij}$ in eqn. (33) are obtained from the ${}^{n}(E_{ij})_L$ and the ply angles ${}^{n}\Theta$ via eqn (7). It is apparent from eqns. (7) and (8) that ${}^{n}E_{11}$, ${}^{n}E_{22}$, ${}^{n}E_{12}$ and ${}^{n}E_{44}$ are always greater than zero, and $({}^{n}x_3 - {}^{n-1}x_3)$ is positive, too. Accordingly, the terms A_{11}, A_{22}, A_{12} and A_{44} are all greater than zero. However, ${}^{n}E_{14}$ and ${}^{n}E_{24}$ are odd functions of ${}^{n}\Theta$ and, therefore, can be positive or negative. They are zero for orientations of 0° or 90°. Hence, the terms A_{14} and A_{24} are zero if for every lamina with $+\Theta$ there exists another identical lamina with $-\Theta$ which, for in-plane loading, leads to a laminate without coupling between normal stresses and shear strains, i.e. a "quasi-orthotropic lamina". Since the position of the laminae is immaterial with this respect symmetric, quasi-orthotropic laminates can be constructed with $\underset{\approx}{B}$ identical zero.

"Unidirectional laminates", "cross-ply laminates" with plies with 0° or 90° ply angles only and "angle-ply laminates" with an equal number of identical laminae oriented alternatingly at $\pm\Theta$ angle are typical quasi-orthotropic laminates.

Similar considerations as described above lead to the conclusion that the terms D_{14} and D_{24} vanish if every lamina above the midsurface with orientation Θ has an identical lamina (same $\underset{\approx}{E}_L$ and thickness) but orientation $-\Theta$ an equal distance below the midsurface. This, however, will not possess midplane symmetry, except for a cross-ply laminate. On the other hand, if an angle-ply laminate is constructed by many thin laminae, the terms D_{14} and D_{24} do become small. Those laminates behave like homogeneous orthotropic plates.

c) Quasi-Isotropic Laminates

If the matrix $\underset{\approx}{A}$ shows the structure of an isotropic material with respect to plane-stress conditions, i.e.

$$A_{11} = A_{22}\ , \quad A_{11} - A_{12} = 2A_{44}\ , \quad A_{14} = A_{24} = 0 \qquad (34),$$

then the laminate behaves isotropically for in-plane loading, i.e. it is a "quasi-isotropic laminate".

The requirements of eqns. (34) are fulfilled, if the total number of plies $N \geq 3$, the individual laminae have identical $\underset{\approx}{E}_L$-matrices and identical thicknesses, and, finally, the difference in ply-angle from one to the next ply must be constant, i.e. ${}^{n}\Theta - {}^{n-1}\Theta = \pi/N$. If the laminate is constructed from identical sets of sublaminates, the above conditions must be fulfilled for each sublaminate, or arbitrary but equal sublaminates can be bounded to form a quasi-isotropic laminate if the sublaminates, considered as layers, fulfill the above requirements.

2.3 Calculation of Stresses and Strains in the Laminae

If $\underset{\sim}{N}$, $\underset{\sim}{M}$ and $\underset{\sim}{Q}$ are determined and the temperature and moisture distributions are known eqns. (32) allow the determination of $\underset{\sim}{\bar{\varepsilon}}$, $\underset{\sim}{\chi}$ and $\underset{\sim}{\gamma}$ (the inversion of the $\underset{\approx}{A}$-, $\underset{\approx}{B}$- and $\underset{\approx}{D}$-supermatrices as well as of $\underset{\approx}{\bar{A}}$ is required, but can be performed analytically, see e.g. [13] for simple bending theory). The use of eqn (13) with the appropriate x_3 denoting the position of the lamina n with respect to the reference surface leads to the lamina strains, that is for laminates with many thin plies $x_3 = ({}^nx_3 - {}^{n-1}x_3)/2$. If more detailed information is required (for example to determine interlaminar stresses, see next section) x_3 can be used as ${}^nx_{3b} = {}^{n-1}x_3$, i.e. at the bottom of ply n, and ${}^nx_{3t} = {}^nx_3$, i.e. at the top of ply n.

These strains, which are related to the global reference system (x_1, x_2, x_3), are transformed to be related to the local, lamina specific reference system ${}^n(l, q, t)$ by the rotational transformation

$$ {}^n\underset{\sim}{\varepsilon}_L = \underset{\approx}{\overline{T}}({}^n\Theta)\,{}^n\underset{\sim}{\varepsilon}\,, \qquad {}^n\underset{\sim}{\gamma}_L = \underset{\approx}{\overline{T}}^t({}^n\Theta)\,{}^n\underset{\sim}{\gamma} \tag{35} $$

The transformation matrices $[\underset{\approx}{\overline{T}}, \underset{\approx}{\overline{T}}^t]({}^n\Theta)$ are described by eqn. (8). Now eqn. (4), applied to the ply n under consideration, renders the stresses ${}^n\underset{\sim}{\sigma}_L$ in the lamina related to the material axes ${}^n(l, q, t)$.

These stresses or strains, respectively, are determined under the assumption of Reissner kinematics, and, accordingly, they are not valid near free edges, where this assumption cannot be met. They can, however, be used as a starting point for the determination of the stresses at the edges. Except for the edge regions they can be compared to some ply failure criteria; other failure criteria are available for edge stresses. Edge effects and failure criteria are described in the following sections. Furthermore, it is assumed in the above stress analysis, that the laminate is free of residual stresses which, because of thermal mismatches at the ply level and at the micro-scale, is normally not the case. There are at least two sources for residual stresses: First, residual micro-stresses may be the result of the hygro-thermal mismatches between fiber and matrix, and, second, hygro-thermal mismatches between the individual laminae may lead to the development of residual stresses, often called "curing stresses" during cool-down from the fabrication temperature. (It should be noted that residual stresses after curing might be considered similar to thermal stresses due to thermal loading. As long as linear relations can be assumed these stresses can be superimposed.) The micro-stresses are discussed in Section 1; for more details and the more complex situation of an elastic-plastic matrix see e.g. [14].

2.4 Interlaminar Stresses and Free-Edge Effects

The laminate analysis described above assumes perfect bonding between neighbouring laminae and constant shear deformation over the thickness of the laminate.

However, different layers tend to slide over each other because of their different ply angles Θ and – in the case of hybrid laminates – of the differences in their $\underset{\approx}{\mathbf{E}}_L$-matrices. Furthermore, the stress components which at free edges should be zero are, when calculated by the above lamination theory, typically not zero, and, hence the boundary conditions are not met. This shows that at the boundaries the lamination theory is not satisfactory. These phenomena produce additional, so-called interlaminar stresses which are, however, negligible in regions away from the boundaries of the laminate. One of the first who studied this problem was Hayashi [15]. Here, we have to deal with a three-axial stress state instead of the generalized plane stress state used in lamination theory. Especially the interlaminar shear stresses τ_{31} for external loading in x_1-direction and τ_{32} for external loading in x_2-direction may be quite large near the edge and may lead to onset of failure by edge delamination. Figure 7 shows results derived by Pipes and Pagano in the classical paper [16] for a ±45° HM-graphite-epoxy laminate with E_l = 138 kN/mm², E_q = 14.5 kN/mm², ν_{lq} = 0.21, G_{lq} = 5.8 kN/mm².

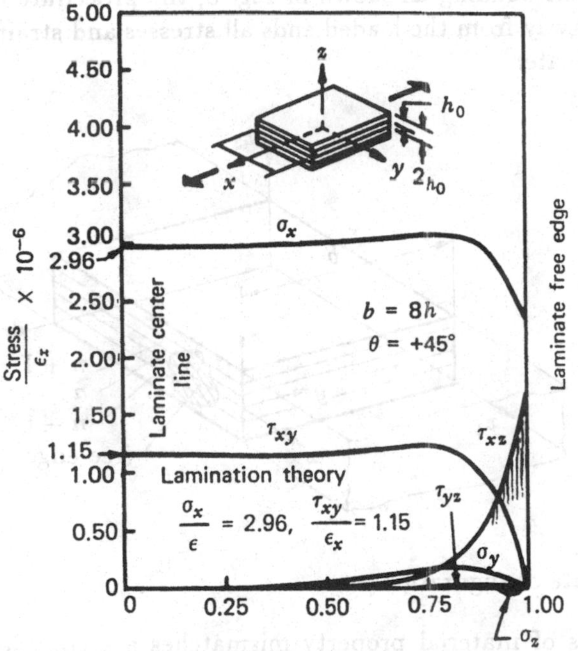

Fig. 7 Interlaminar stresses – classical results by [16]

It is noticable that at a distance from the edge inwards the edge effects deminish and the stresses approach values as obtained from the lamination theory. In general, the width of the region in which the stresses differ from those predicted by lamination theory is rather small, typically approximately equal to the thickness of the laminate. The magnitudes and distributions of the interlaminar stresses depend significantly on the stacking sequence, see e.g. [17].

A number of theories for approximating the three-axial stress state near the free edges are available which use the results obtained by the simple lamination theory and provide formulae for estimating the interlaminar stresses. One of these theories has been published by Whitney [18] who prescribed the distribution of some stress components ($\sigma_{22} = \sigma_{yy}$ and $\tau_{12} = \tau_{xy}$) by trial functions approximating the results of Pipes and Pagano and derived the other stress components by satisfying the local equilibrium condition $\sigma_{ij,j} = 0$. Improvements of this appoach have been published recently. For example, the work by Rose and Herakovich [19] which is an extension of the solution by Kassapoglou and Lagace [20] provides an efficient method for predicting interlaminar stresses for symmetric and unsymmetric (even hybrid) laminates under in-plane and bending loading. Their method is based on statically admissible stress fields which take into consideration local property mismatch effects and global equilibrium requirements. The unknown parameters in the assumed stress distributions are determined by minimizing the complementary energy of the laminate.

Assuming a long, finite width laminate strip subjected to uniform axial displacement or out-of-plane bending as shown in Fig. 8, the procedure is briefly described in the following. Away from the loaded ends all stresses and strains are independent of the axial coordinate.

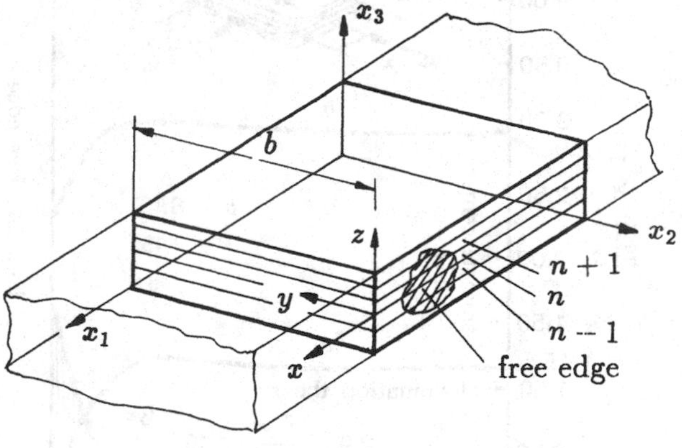

Fig. 8 Laminate configuration

In [19] two sets of material property mismatches are considered to be mainly responsible for edge effects; a mismatch in the coefficient of mutual influence,

$$\begin{aligned} \delta\eta_{xy,x}(n,1) &= {}^{n-1}\eta_{xy,x} - {}^{n}\eta_{xy,x} \\ \delta\eta_{xy,x}(n,2) &= {}^{n}\eta_{xy,x} - {}^{n+1}\eta_{xy,x} \\ &\text{with} \\ {}^{n}\eta_{xy,x} &= {}^{n}\left(\frac{\gamma_{xy}}{\varepsilon_{xx}}\right) \end{aligned} \qquad (36),$$

and a mismatch in Poisson's ratio (related to the laminate coordinate system and not to the individual lamina principal material axes),

$$\begin{aligned}\delta\nu_{xy}(n,1) &= {}^{n-1}\nu_{xy} - {}^{n}\nu_{xy}\\ \delta\nu_{xy}(n,2) &= {}^{n}\nu_{xy} - {}^{n+1}\nu_{xy}\\ &\text{with}\\ {}^{n}\nu_{xy} &= -{}^{n}\Big(\frac{\varepsilon_{yy}}{\varepsilon_{xx}}\Big)\end{aligned} \qquad (37).$$

The mismatch in the coefficient of mutual influence is primarily responsible for the development of the interlaminar shear stresses $\tau_{xy} = \sigma_{xy}$ and $\tau_{xz} = \sigma_{xz}$, and the interlaminar stress components σ_{yy}, $\tau_{yz} = \sigma_{yz}$ and σ_{zz} are mainly influenced by the Poisson's ratio mismatch.

Starting with results obtained by classical lamination theory, the edge effects are described in [19] by trial stress fields as expressed for individual laminae in the following form. For layer n:

$$\begin{aligned}{}^{n}\sigma_{ij} &= {}^{n}\sigma_{ij(E)} + {}^{n}\sigma_{ij(\delta\eta)} + {}^{n}\sigma_{ij(\delta\nu)}\\ &\text{with}\\ {}^{n}\sigma_{ij(E)} &= {}^{n}f_{ij}(y)\,{}^{n}g_{ij}(z)\\ {}^{n}\sigma_{ij(\delta\eta)} &= {}^{n}h_{ij}(y)\,{}^{n}l_{ij}(z)\\ {}^{n}\sigma_{ij(\delta\nu)} &= {}^{n}m_{ij}(y)\,{}^{n}n_{ij}(z) \qquad i = x,y,z, \quad j = y,z\end{aligned} \qquad (38).$$

The term ${}^{n}\sigma_{ij(E)}$ corresponds to the stress assumptions in [20] in which explicit sums of decaying exponential functions, with unknown decay rates, are assumed for ${}^{n}f_{ij}(y)$ and polynomial functions for ${}^{n}g_{ij}(z)$. These functions are chosen to satisfy local and global equilibrium, stress free boundary conditions, and traction continuity conditions at the ply interface. Hence, the additional terms, ${}^{n}\sigma_{ij(\delta\eta)}$ and ${}^{n}\sigma_{ij(\delta\nu)}$, which are associated with the mismatch in material properties between adjacent layers, must represent self-equilibrating stress fields. The functions ${}^{n}h_{ij}(y)$ and ${}^{n}m_{ij}(y)$ again consists of exponential functions, with unknown decay rates. ${}^{n}l_{ij}(z)$ as well as ${}^{n}n_{ij}(z)$ are polynomial functions; for details see [19]. These trial functions must recover classical lamination theory stress states away from the free edges. They contain – in addition to ply thicknesses ${}^{n}h$, the mismatches $\delta\eta_{xy,x}(n,1)$ and $\delta\eta_{xy,x}(n,2)$ as well as $\delta\nu_{xy}(n,1)$ and $\delta\nu_{xy}(n,2)$ – the results of the classical lamination theory and, of course, a number of unknown parameters, C_k. These unknown parameters are determined by minimizing the laminate's complementary energy, Π_c. The system of nonlinear simultaneous equations

$$\frac{\partial \Pi_c}{\partial C_k} = 0 \qquad (39)$$

will render non-unique solutions. That solution is relevant which contains only real C_k-values leading to the absolute minimum of Π_c.

The solution according to [20] represents a simplification to the above described procedure. It is presented in a form as summarized in [21]:

$$\begin{aligned}
{}^n\sigma_{yy} &= {}^n\bar{\sigma}_{yy}\left(1+\frac{\lambda}{1-\lambda}e^{-\phi y}+\frac{1}{\lambda-1}e^{-\lambda\phi y}\right)\\
{}^n\sigma_{zz} &= {}^n\bar{\sigma}_{yy}\left(\frac{\lambda}{1-\lambda}\phi^2 e^{-\phi y}+\frac{1}{\lambda-1}\lambda^2\phi^2 e^{-\lambda\phi y}\right)(z^2/2+z\,{}^nB_4+{}^nB_5)\\
{}^n\tau_{xy} &= {}^n\bar{\tau}_{xy}(1-e^{-\phi y})\\
{}^n\tau_{xz} &= {}^n\bar{\tau}_{xy}\phi e^{-\phi y}(z+{}^nB_2)\\
{}^n\tau_{yz} &= -{}^n\bar{\sigma}_{yy}\left(\frac{\lambda}{1-\lambda}\phi e^{-\phi y}+\frac{1}{\lambda-1}\lambda\phi e^{-\lambda\phi y}\right)(z+{}^nB_4)\\
{}^n\sigma_{xx} &= \frac{1}{{}^nS_{11}}(\bar{\varepsilon}_{xx}-{}^nS_{12}\,{}^n\sigma_{yy}-{}^nS_{13}\,{}^n\sigma_{zz}-{}^nS_{16}\,{}^n\tau_{xy})
\end{aligned} \qquad (40),$$

where ${}^nS_{ij}$ are the ply compliances, and

$$\begin{aligned}
{}^nB_2 &= -\frac{1}{{}^n\bar{\tau}_{xy}}\sum_{j=1}^{n-1}({}^j\bar{\tau}_{xy}\,{}^jh)-{}^nz'\\
{}^nB_4 &= -\frac{1}{{}^n\bar{\sigma}_{yy}}\sum_{j=1}^{n-1}({}^j\bar{\sigma}_{yy}\,{}^jh)-{}^nz'\\
{}^nB_5 &= -\frac{1}{{}^n\bar{\sigma}_{yy}}\sum_{j=1}^{n-1}({}^j\bar{\sigma}_{yy}\,{}^jh\,{}^jz)+{}^nz'^2
\end{aligned} \qquad (41).$$

The terms ${}^n\bar{\sigma}_{ii}$, ${}^n\bar{\tau}_{ij}$ or $\bar{\varepsilon}_{xx}$, respectively, denote lamination theory solutions. ${}^nz$ denotes the z-coordinate of the mid-plane of ply n, i.e. ${}^nz = {}^nx_3 - {}^nh/2$, and ${}^nz'$ denotes the position of the top surface of ply n with respect to mid-surface of the laminate, that is ${}^nz' = {}^nx_3$.

The interlaminar stresses calculated this way can be evaluated with respect to interlaminar failure, i.e. delamination, by using a proper failure criterion as will be discussed in the following section.

3 Failure Criteria

In this section the failure criteria are discussed only very briefly. Several papers are dealing with failure criteria, and reviews of failure theories are presented for example in [12,22,23].

Due to the nature of FRP composite material the failure behavior is very complex with numerous different failure modes possible. Therefore, an accurate prediction of failure requires micro-mechanical considerations in combination with damage mechanics. Here a simple way of dealing with this kind of behavior is chosen:

First, failure within the layers is considered. Delamination will be treated later.

3.1 Ply Failure

Assuming a linear elastic stress–strain relationship up to failure (which is valid for many FRPs) one can use proper strength criteria to determine onset of failure. Although the following procedures are based on "homogenized" material, the local damage of the composite, i.e. matrix or fiber cracking, can be estimated. In a simple approach [3] FRP ply failure is indicated by a combination of two failure criteria, and two distinct failure modes are assumed:

A quadratic strength criterion, the Tsai–Wu-criterion, see e.g. [24], serves for predicting failure for stress states with relatively large transverse stress components which affect the matrix material rather than the fibers. According to this criterion endurable stress states lie within a failure surface in the stress space, i.e.

$$F_{01}\sigma_{ll} + F_{11}\sigma_{ll}^2 + F_{12}\sigma_{ll}\sigma_{qq} + F_{02}\sigma_{qq} + F_{22}\sigma_{qq}^2 + F_{44}\tau_{lq}^2 < 1 \qquad (42),$$

with

$$\begin{aligned} F_{01} &= \frac{1}{\sigma_{lTu}} + \frac{1}{\sigma_{lCu}} & F_{11} &= \frac{-1}{\sigma_{lTu}\sigma_{lCu}} & F_{44} &= \frac{1}{\tau_{lqu}^2} \\ F_{02} &= \frac{1}{\sigma_{qTu}} + \frac{1}{\sigma_{qCu}} & F_{22} &= \frac{-1}{\sigma_{qTu}\sigma_{qCu}} & F_{12} &= -\sqrt{F_{11}F_{22}} \end{aligned} \qquad (43).$$

The following notations are used:

σ_{lCu} maximum endurable, i.e. ultimate uniaxial compression stress in fiber direction,
σ_{lTu} ultimate tensile stress in fiber direction,
σ_{qCu} ultimate compression stress normal to fiber direction,
σ_{qTu} ultimate tensile stress normal to fiber direction,
τ_{lqu} ultimate shear stress.

In some cases it might be more convenient to express this criterion in the strain space:

$$G_{01}\varepsilon_{ll} + G_{11}\varepsilon_{ll}^2 + G_{12}\varepsilon_{ll}\varepsilon_{qq} + G_{02}\varepsilon_{qq} + G_{22}\varepsilon_{qq}^2 + G_{44}\gamma_{lq}^2 < 1 \qquad (44),$$

with

$$\begin{aligned} G_{01} &= F_{01}\mathrm{E}_{11} + F_{02}\mathrm{E}_{12} \\ G_{02} &= F_{02}\mathrm{E}_{22} + F_{01}\mathrm{E}_{12} \\ G_{11} &= F_{11}\mathrm{E}_{11}^2 + F_{22}\mathrm{E}_{12}^2 + F_{12}\mathrm{E}_{11}\mathrm{E}_{12} \\ G_{22} &= F_{22}\mathrm{E}_{22}^2 + F_{11}\mathrm{E}_{12}^2 + F_{12}\mathrm{E}_{22}\mathrm{E}_{12} \\ G_{12} &= 2\mathrm{E}_{12}(F_{11}\mathrm{E}_{11} + F_{22}\mathrm{E}_{22}) + F_{12}(\mathrm{E}_{12}^2 + \mathrm{E}_{11}\mathrm{E}_{22}) \\ G_{44} &= F_{44}\mathrm{E}_{44}^2 \end{aligned} \qquad (45).$$

Since in many cases the Tsai–Wu-criterion overestimates the strength of the UD-layer in the case of stresses acting predominantly in fiber direction, a maximum stress limit in fiber direction is imposed:

$$\sigma_{lCu} < \sigma_{ll} < \sigma_{lTu} \tag{46}$$

leading finally to a failure surface as shown in Fig. 9.

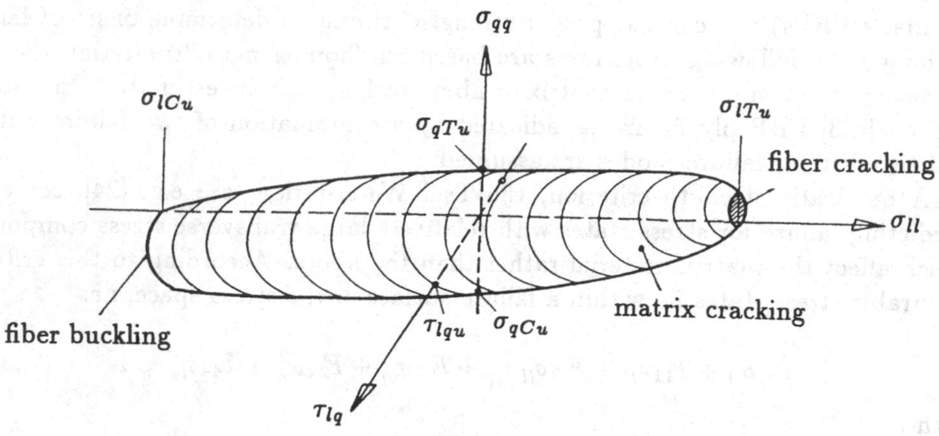

Fig. 9 Combined failure criterion

A rough distinction, whether fiber or matrix failure has appeared, can be made by considering which of the two criteria has been violated first. In the case of violation of eqn. (42) or (45), respectively, the failure mode is rather matrix-cracking, and if eqn. (46) is violated first, fiber-cracking or fiber-buckling has appeared. More precise distinctions are possible by using direct mode determining failure criteria, see e.g. [23].

Post-cracking stiffnesses are introduced according to the two assumed failure modes. In the case of matrix cracking they take the form:

$$\hat{\underset{\approx}{\mathbf{E}}}_L = \begin{pmatrix} (\mathrm{E}_{11})_L & (\mathrm{E}_{12})_L & 0 & 0 & 0 \\ & (\mathrm{E}_{22})_L & 0 & 0 & 0 \\ & sym. & (\mathrm{E}_{44})_L & 0 & 0 \\ & & & (\mathrm{E}_{55})_L & 0 \\ & & & & (\mathrm{E}_{66})_L \end{pmatrix} \xrightarrow{\text{matrix-cracking}}$$

$$\begin{pmatrix} \beta_E(\mathrm{E}_{11})_L & 0 & 0 & 0 & 0 \\ & 0 & 0 & 0 & 0 \\ & sym. & 0 & 0 & 0 \\ & & & (\mathrm{E}_{55})_L & 0 \\ & & & & (\mathrm{E}_{66})_L \end{pmatrix} \tag{47}$$

$(E_{12})_L$, $(E_{22})_L$ and $(E_{44})_L$ are set to zero, which is a simple representation of the matrix stiffness being removed. The reason for introducing a correction factor β_E is twofold: First, a damaged matrix also reduces the stiffness in fiber direction and, second, for nonstraight (e.g. wavy) fibers the reduced support by the damaged matrix can lead to a further loss of stiffness. β_E depends on the composition of the layer and must be given as an additional material parameter.

In the case of fiber failure the reduced stiffnesses are modeled by:

$$\hat{\underset{\approx}{E}}_L = \begin{pmatrix} (E_{11})_L & (E_{12})_L & 0 & 0 & 0 \\ & (E_{22})_L & 0 & 0 & 0 \\ & sym. & (E_{44})_L & 0 & 0 \\ & & & (E_{55})_L & 0 \\ & & & & (E_{66})_L \end{pmatrix} \xrightarrow{\text{fiber-failure}} \begin{pmatrix} (E_{22})_L & (E_{12})_L & 0 & 0 & 0 \\ & (E_{22})_L & 0 & 0 & 0 \\ & sym. & (E_{44})_L & 0 & 0 \\ & & & (E_{55})_L & 0 \\ & & & & (E_{66})_L \end{pmatrix} \quad (48).$$

As can be seen from eqn. (48), $(E_{11})_L$ is set equal to $(E_{22})_L$. This is a simple representation of the assumption that locally after fiber failure only the matrix stiffness remains. Alternative stiffness reductions are proposed for example in [23].

By assuming thin layers it is sufficient to investigate the stress state in the layers' midsurface only. In addition, if cracking occurs the corresponding failure mode is taken to be valid through the entire thickness of the layer. Thick layers with significant stress gradients in thickness direction can be treated by subdividing them into several "thin" sublayers.

Equations (47,48) represent "secant"-stiffnesses of the material (compare experimental results in Fig. 10) and, therefore, stresses can be computed directly from the total strains.

3.2 Interlaminar Failure

Using the stresses near the edges of the laminate as for example estimated by procedures like described in Subsection 2.4 the onset of interlaminar failure can be checked by proper failure criteria.

Like in ply failure analysis a quadratic failure criterion can be applied. If the stress point for the interface between layers n and $n+1$ in the stress space does not lie within an area bounded by the failure surface as proposed in [25], i.e.

$$\left(\frac{{}^{n}\tau'_{xz}}{{}^{\;\;n}_{n+1}\tau_{xz\,u}}\right)^2 + \left(\frac{{}^{n}\tau'_{yz}}{{}^{\;\;n}_{n+1}\tau_{yz\,u}}\right)^2 + \left(\frac{{}^{n}\sigma'_{zz}}{{}^{\;\;n}_{n+1}\sigma_{zz\,u}}\right)^2 < 1 \quad (49),$$

interlaminar failure will appear at that interface. The denominators of eqn. (49) represent empirically derived values for the interlaminar failure stresses. For ${}^{\;\;n}_{n+1}\sigma_{zz\,u}$

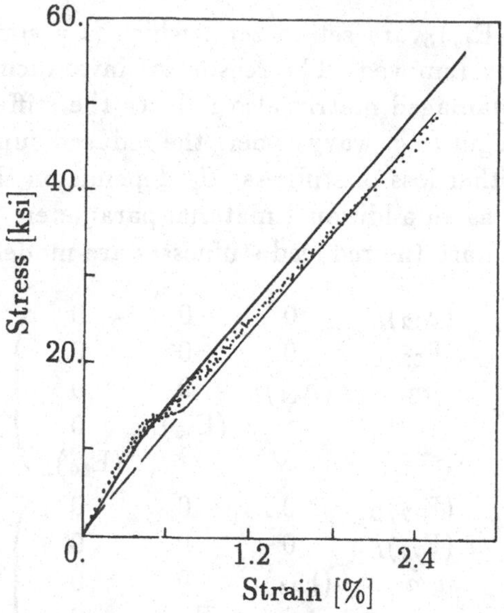

Fig. 10 Typical post-failure behavior of a glass-epoxy $[0/90_2]_s$ laminate; according to [13]

two values must be considered: ${}^{n}_{n+1}\sigma_{zz\,Tu}$ for ${}^{n}\sigma'_{zz} > 0$ and ${}^{n}_{n+1}\sigma_{zz\,Cu}$ for ${}^{n}\sigma'_{zz} < 0$. Since the analytical approach predicts very high (singular) interlaminar stresses which do not, in practice, occur, 'effective stress components' ${}^{n}\sigma'_{ij}$ are used in applying failure criteria. Different ways are proposed to derive ${}^{n}\sigma'_{ij}$ from ${}^{n}\sigma_{ij}$. In [26] σ'_{ij} are proposed to be the interlaminar stresses evaluated at a characteristic distance a^* (which is for thin plies equal to one ply thickness) from the free edge. Average stresses are proposed in [21], i.e.:

$$ {}^{n}\sigma'_{ij} = \frac{1}{d}\int_0^d {}^{n}\sigma_{ij}\, dy \tag{50}$$

d is a characteristic 'averaging distance'. The choice of the definition of the 'effective stress components' must be compatible with the determination of the failure stresses.

4 Sandwich Plates and Shells

Sandwich structures are used in extreme lightweight applications. They represent specific layered composites. Here we are dealing with symmetric sandwiches with very thin face layers and a relatively thick core which has negligible in-plane stiffnesses. The face layers are assumed to behave isotropically (for example they are built in form of a thin quasi-isotropic laminate, see Subsection 2.2), and the core material is orthotropic (which, for example, is a reasonable approximation for honeycomb cores).

4.1 Lamination Theory for Antiplane Core Conditions

Under the assumptions described above and using the notations given in Fig. 11 the local material matrices (which in the linear regime correspond to the local elasticity matrices) for the individual (four) layers (two face layers and two core half-layers) are, in accordance with antiplane core conditions, given as follows.

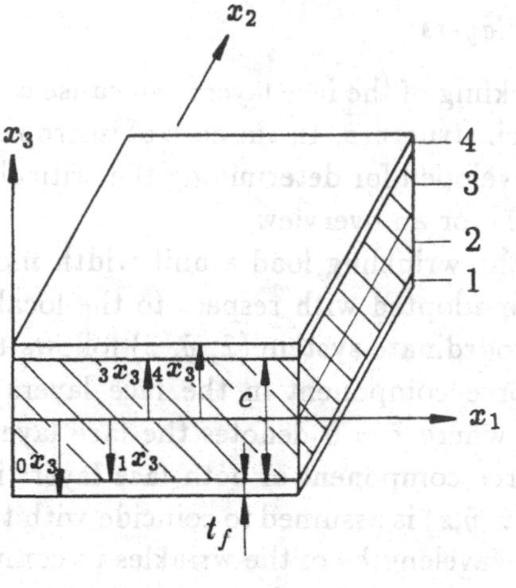

Fig. 11 Definition of the sandwich layer model

For the isotropic face layer, denoted by subscript $_f$:

$$^1\hat{\underset{\approx}{\mathbf{E}}}_L = {}^4\hat{\underset{\approx}{\mathbf{E}}}_L = \hat{\underset{\approx}{\mathbf{E}}}_f = \begin{pmatrix} \frac{E_f}{1-\nu_f^2} & \frac{\nu_f E_f}{1-\nu_f^2} & 0 & 0 & 0 \\ \frac{\nu_f E_f}{1-\nu_f^2} & \frac{E_f}{1-\nu_f^2} & 0 & 0 & 0 \\ 0 & 0 & G_{f,lq} & 0 & 0 \\ 0 & 0 & 0 & 0 & 0 \\ 0 & 0 & 0 & 0 & 0 \end{pmatrix} \qquad (51),$$

and the material matrix $\hat{\underset{\approx}{\mathbf{E}}}_c$ for the orthotropic core material are defined with respect to the orthotropy axes:

$$^2\hat{\underset{\approx}{\mathbf{E}}}_L = {}^3\hat{\underset{\approx}{\mathbf{E}}}_L = \hat{\underset{\approx}{\mathbf{E}}}_{c,L} = \begin{pmatrix} 0 & 0 & 0 & 0 & 0 \\ 0 & 0 & 0 & 0 & 0 \\ 0 & 0 & 0 & 0 & 0 \\ 0 & 0 & 0 & G_{c,lt} & 0 \\ 0 & 0 & 0 & 0 & G_{c,qt} \end{pmatrix} \qquad (52).$$

Since the out-of-plane shear stresses can be assumed to remain constant over the core thickness, no correction factors are required for the shear rigidities. The transverse shear moduli of the core material are either determined experimentally or estimated by micromechanical methods; for honeycomb cores see e.g. [27].

4.2 Additional Failure Modes in Sandwich Shells

Failure modes which are typical for sandwiches with extremely thin face layers are related to local instabilities, e.g. wrinkling of the face layers or buckling of the cell walls in honeycomb cores.

a) Wrinkling of the face layers

Short wavelength buckling of the face layers can cause considerable loss in stiffness for a low-density sandwich structure. In the case of isotropic core materials a number of methods have been developed for determining the critical loads that induce wrinkling, see [Starlinger, 1991] for an overview.

In the derivation of the wrinkling load a unit width model is assumed [28]. The following conventions are adopted with respect to the local coordinate system. The $\tilde{x}$-direction of the local coordinate system ($\tilde{x}$, $\tilde{y}$, $\tilde{z}$) follows the direction of the larger compressive principal force component in the face layers. $\tilde{z}$ is directed from the face layer into the core, where $\tilde{z} = 0$ denotes the face layer in which the maximum compressive principal force component of both face layers is found. Specifically, the local coordinate system ($\tilde{x}$,$\tilde{y}$,$\tilde{z}$) is assumed to coincide with the axes of orthotropy (i.e. l, q, t). Due to the short wavelengths of the wrinkles in comparison to the dimensions of a typical sandwich structure, the analysis of critical loads reduces to the analysis of an infinite plate. For shell structures, the influence of curvature on the local instability limit is neglected if the radii of shell curvatures are significantly larger than the wavelengths of the wrinkles.

Sinusoidal wrinkling deformations of the face layers are assumed, thus

$$u(\tilde{x},\tilde{y},\tilde{z}) = \tilde{u}(\tilde{z}) \cos \frac{\pi \tilde{x}}{a_{\tilde{x}}} \sin \frac{\pi \tilde{y}}{a_{\tilde{y}}} \qquad (53),$$

$$v(\tilde{x},\tilde{y},\tilde{z}) = \tilde{v}(\tilde{z}) \sin \frac{\pi \tilde{x}}{a_{\tilde{x}}} \cos \frac{\pi \tilde{y}}{a_{\tilde{y}}} \qquad (54),$$

$$w(\tilde{x},\tilde{y},\tilde{z}) = \tilde{w}(\tilde{z}) \sin \frac{\pi \tilde{x}}{a_{\tilde{x}}} \sin \frac{\pi \tilde{y}}{a_{\tilde{y}}} \qquad (55).$$

$a_{\tilde{x}}$ is the half wavelength of the wrinkle in the direction $\tilde{x}$, and $a_{\tilde{y}}$ is the half wavelength of the wrinkle in the direction $\tilde{y}$. In order to characterize the wrinkling deformation pattern the following parameters are introduced:

$$\alpha = \frac{\pi}{a_{\tilde{x}}}, \qquad \beta = \frac{\pi}{a_{\tilde{y}}} \qquad (56).$$

The complete elasticity matrix $\underset{\approx}{\tilde{\mathbf{E}}}_{c,L}$ for the orthotropic core materials (axes of orthotropy l, q, t) must be used

$$\tilde{\underset{\approx}{E}}_{c,L} = \begin{pmatrix} E_{ll} & E_{lq} & E_{lt} & 0 & 0 & 0 \\ E_{lq} & E_{qq} & E_{qt} & 0 & 0 & 0 \\ E_{lt} & E_{qt} & E_{tt} & 0 & 0 & 0 \\ 0 & 0 & 0 & G_{qt} & 0 & 0 \\ 0 & 0 & 0 & 0 & G_{lt} & 0 \\ 0 & 0 & 0 & 0 & 0 & G_{lq} \end{pmatrix} \quad (57),$$

$E_{ll}, E_{lq}, \ldots, E_{tt}$ being defined by the direction dependent moduli of elasticity and the direction dependent Poisson's ratios which can be the results of micro-mechanical considerations or of macroscopic measurements.

By using Hooke's Law the assumed wrinkling pattern in combination with the local equilibrium condition leads to the following set of coupled linear differential equations dependent on $\bar{z}$ ($'$ defines derivatives with respect to $\bar{z}$), see [29]:

$$\underset{\approx}{U}_1 \underset{\sim}{y}'' - \underset{\approx}{U}_2 \underset{\sim}{y}' - \underset{\approx}{U}_3 \underset{\sim}{y} = \underset{\sim}{0} \quad (58),$$

where $\underset{\sim}{y}$ represents the amplitudes $\underset{\sim}{y}^T = (\bar{u}, \bar{v}, \bar{w})$ depending on $\bar{z}$. The coefficient matrices $\underset{\approx}{U}_i$ are defined as:

$$\underset{\approx}{U}_1 = \begin{pmatrix} G_{\bar{x}\bar{z}} & 0 & 0 \\ 0 & G_{\bar{y}\bar{z}} & 0 \\ 0 & 0 & C_{\bar{z}\bar{z}} \end{pmatrix} \quad (59),$$

$$\underset{\approx}{U}_2 = \begin{pmatrix} 0 & 0 & -\alpha(C_{\bar{x}\bar{z}} + G_{\bar{x}\bar{z}}) \\ 0 & 0 & -\beta(C_{\bar{y}\bar{z}} + G_{\bar{y}\bar{z}}) \\ \alpha(C_{\bar{x}\bar{z}} + G_{\bar{x}\bar{z}}) & \beta(C_{\bar{y}\bar{z}} + G_{\bar{y}\bar{z}}) & 0 \end{pmatrix} \quad (60),$$

$$\underset{\approx}{U}_3 = \begin{pmatrix} \alpha^2 C_{\bar{x}\bar{x}} + \beta^2 G_{\bar{x}\bar{y}} & \alpha\beta(C_{\bar{x}\bar{y}} + G_{\bar{x}\bar{y}}) & 0 \\ \alpha\beta(C_{\bar{x}\bar{y}} + G_{\bar{x}\bar{y}}) & \alpha^2 G_{\bar{x}\bar{y}} + \beta^2 C_{\bar{y}\bar{y}} & 0 \\ 0 & 0 & \alpha^2 G_{\bar{x}\bar{z}} + \beta^2 G_{\bar{y}\bar{z}} \end{pmatrix} \quad (61).$$

The following substitution can be used for solving this system of differential equations:

$$\underset{\sim}{y} = \underset{\sim}{\mu}\, e^{\varrho \bar{z}} \quad (62).$$

After inserting eqn. (62) into eqn. (58) one obtains

$$\underset{\approx}{\mathrm{U}}\ \underset{\sim}{\mu}\ e^{\varrho\tilde{z}} = \underset{\sim}{0} \tag{63}$$

where $\underset{\approx}{\mathrm{U}}$ is defined by the following eigenvalue problem, which results from the requirement to find a nontrivial solution of eqn. (63),

$$\underset{\approx}{\mathrm{U}}\ \underset{\sim}{\mu} = \left(\underset{\approx}{\mathrm{U}}_1\ \varrho^2 - \underset{\approx}{\mathrm{U}}_2\ \varrho - \underset{\approx}{\mathrm{U}}_3\right)\ \underset{\sim}{\mu} = \underset{\sim}{0} \tag{64}$$

Here, $\underset{\sim}{\mu}$ is an eigenvector, and ϱ is the corresponding eigenvalue. With

$$\Lambda = \varrho^2 \tag{65}$$

the characteristic equation of the eigenvalue problem can be found to be

$$det\ \underset{\approx}{\mathrm{U}} = \Theta_6\Lambda^3 + \Theta_4\Lambda^2 + \Theta_2\Lambda + \Theta_0 = 0 \tag{66}$$

The parameters Θ_i depend on the material properties and on the buckling wavelength parameters α and β. The roots of this cubic polynomial can be found analytically. For the special case of three real positive roots (i.e. six real solutions $\pm F_i$ for ϱ), $\underset{\sim}{\mathrm{y}}$ can be expressed as

$$\begin{aligned}\underset{\sim}{\mathrm{y}} = {} & D_1\ e^{F_1(\tilde{z}-c)}\ \underset{\sim}{\mu_1} + D_2\ e^{-F_1\tilde{z}}\ \underset{\sim}{\mu_2} + D_3\ e^{F_2(\tilde{z}-c)}\ \underset{\sim}{\mu_3} \\ & + D_4\ e^{-F_2\tilde{z}}\ \underset{\sim}{\mu_4} + D_5\ e^{F_3(\tilde{z}-c)}\ \underset{\sim}{\mu_5} + D_6\ e^{-F_3\tilde{z}}\ \underset{\sim}{\mu_6}\end{aligned} \tag{67}$$

Other cases for roots of eqn. (66), i.e. complex or negative roots, are discussed in [29]; they are, however, not of physical relevance. The six independent coefficients D_i in eqn. (67) are determined by satisfying certain boundary conditions which depend on the local loading situation, i.e. bending or pure compression of the shell, and on the wrinkling mode, i.e. symmetrical or antisymmetrical wrinkling or wrinkling under bending.

In any case five of the six boundary conditions are:

$$u(\tilde{x},\tilde{y},\tilde{z}=0) = v(\tilde{x},\tilde{y},\tilde{z}=0) = u(\tilde{x},\tilde{y},\tilde{z}=c) = v(\tilde{x},\tilde{y},\tilde{z}=c) = 0 \tag{68}$$

$$\sigma_{zz}(\tilde{x},\tilde{y},\tilde{z}=0) = -\sigma_0 \sin\alpha\tilde{x}\sin\beta\tilde{y} \tag{69}$$

The sixth boundary condition depends on the wrinkling mode:

In the case of pure bending the $\tilde{z}$-displacements are supposed to vanish completely at the opposite side of the core, because the resulting tensile forces do not induce buckling at the opposite face layer:

$$w(\tilde{x},\tilde{y},\tilde{z}=c) = 0 \tag{70}$$

In the case of pure compressive forces the sixth boundary condition for antisymmetrical wrinkling is

$$\sigma_{zz}(\tilde{x}, \tilde{y}, \tilde{z} = c) = \sigma_0 \sin \alpha\tilde{x} \sin \beta\tilde{y} \qquad (71).$$

If a symmetrical wrinkling pattern is assumed, the remaining boundary condition is determined by the condition of symmetry:

$$w(\tilde{x}, \tilde{y}, \tilde{z} = c/2) = 0 \qquad (72).$$

These boundary conditions involve a system of six linear equations in terms of the coefficients D_i.

The vector $\underset{\sim}{y}$ is now determined and the deformation behavior of the core is known in dependence on the wavelengths $a_{\tilde{x}}$ and $a_{\tilde{y}}$. If the deformations of the face layers and of the core are supposed to be identical at the interface $\tilde{z} = 0$, i.e. $w_f(\tilde{x}, \tilde{y}, \tilde{z} = 0) = w(\tilde{x}, \tilde{y}, \tilde{z} = 0)$, corresponding to a perfect interface between the face layers and the core, the equivalent stiffness of the core C_{core} is given by

$$C_{core} \;=\; \frac{\sigma_0}{\tilde{w}(\tilde{z} = 0)} \qquad (73).$$

If the loading conditions differ from pure bending and pure compression the following approximation for C_{core} is proposed in analogy to [30]

$$C_{core} = \frac{1}{2}\left[(1+\eta)\, C_{core}^{compr} + (1-\eta)\, C_{core}^{bending}\right] \qquad (74),$$

where

$$\eta = \frac{P_{\tilde{x}}^{lf}}{P_{\tilde{x}}^{uf}} \qquad (75),$$

with $P_{\tilde{x}}^{uf}$ being the membrane force in the "upper" (i.e. $\tilde{z} = 0$) face layer relevant for wrinkling, and $P_{\tilde{x}}^{lf}$ is the membrane force in the "lower" (i.e. $\tilde{z} = c$) face layer.

Under general loading conditions the equivalent stiffness of the core depends on the parameter η and on the wavelengths $a_{\tilde{x}}$ and $a_{\tilde{y}}$.

Thus, the equation for the considered face layer modelled as a plate on an elastic foundation can be formulated

$$K_f^{local}\nabla\nabla w_f + P_{\tilde{x}}\frac{\partial^2 w_f}{\partial \tilde{x}^2} + P_{\tilde{y}}\frac{\partial^2 w_f}{\partial \tilde{y}^2} + C_{core}(a_{\tilde{x}}, a_{\tilde{y}})\, w_f = 0 \qquad (76).$$

Since the axes $\tilde{x}$ and $\tilde{y}$ are assumed to coincide with the principal stress axes at the considered point on the face layer, the shear force term is not present in eqn. (76). Introduction of the definition

$$\xi = \frac{P_{\tilde{y}}}{P_{\tilde{x}}} \qquad (77)$$

allows the reduction of the description of biaxial loading by one parameter, and one obtains:

$$P^c_{\tilde{x}} = \frac{(\alpha^2+\beta^2)^2 K_f^{local} + C_{core}(\alpha,\beta,\eta)}{\alpha^2+\xi\beta^2} \qquad (78).$$

This equation for $P^c_{\tilde{x}}$ still contains the wavelength parameters α and β. The smallest value of $P^c_{\tilde{x}}$ defined as P^{wrinkl}_{crit} has to be determined by minimization in the parameter space $(\alpha \times \beta)$. Since the equivalent stiffness of the core C_{core} is not an explicit function of these parameters, the extremum cannot be determined by a closed formula, but a numerical procedure must be used, see [29].

As an example, a wrinkling analysis is performed for a locally uniaxial load case (compression in l-direction) in a sandwich shell with an isotropic core (PU-foam) with $E_c = 10\ N/mm^2$, $\nu_c = 0.3$, and isotropic face layers (aluminum) with $E_f = 70000\ N/mm^2$, $\nu_f = 0.3$, $t_f = 0.1\ mm$. The core thickness c is varied. In Fig. 12(a) the critical load P^{wrinkl}_{crit} is plotted as a function of the core thickness c, assuming symmetrical and antisymmetrical buckling patterns. For a comparison, analytical approximations according to [1] are also shown in this figure:

$$P^{wrinkl}_{crit} = Min\left(P^{shear} \approx \frac{1}{2}G_{c,lt}\,c;\ P^{wrinkl} \approx 0.85 t_f (E_f\ E_{c,tt}\ G_{c,lt})^{\frac{1}{3}}\right) \qquad (79).$$

for uniaxial loading in l-direction.

The antisymmetrical buckling modes yield smaller values for the critical loads for this isotropic core. This agrees with the results obtained for isotropic cores in [30]. For thin or very thick cores, respectively, the asymptotic solutions, eqn. (79) are retained.

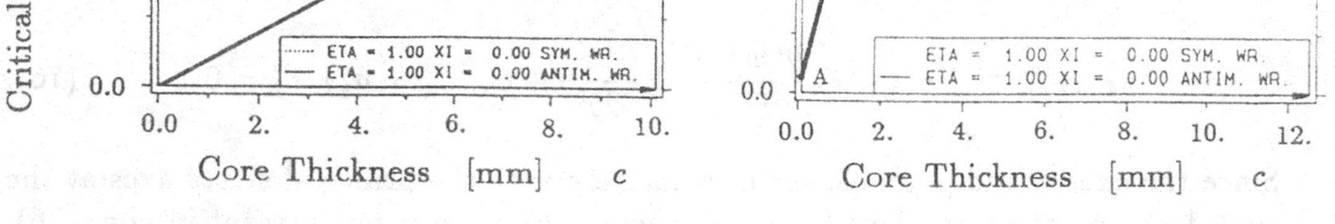

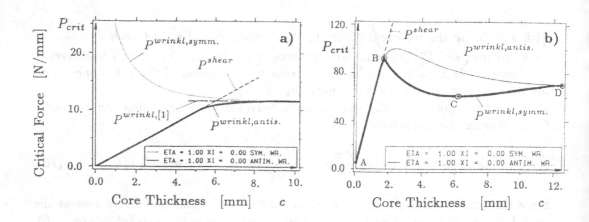

Fig. 12 P^{wrinkl}_{crit} over core thickness c for $\eta = 1.0$, $\xi = 0.0$; a) for isotropic, b) for orthotropic core material

In order to show the influence of orthotropy of the core material on the critical load, P_{crit}^{wrinkl}, the core material parameters are changed as follows: Aramid honeycomb core with $E_{c,ll} = E_{c,qq} = 1.0\ N/mm^2$, $E_{c,tt} = 310\ N/mm^2$, $\nu_{c,lq} = 0.3$, $\nu_{c,lt} = \nu_{c,qt} = 0.0$, $G_{c,lq} = 1.0\ N/mm^2$, $G_{c,qt} = 220\ N/mm^2$, $G_{c,lt} = 110\ N/mm^2$, and the isotropic face layers are the same as above. The results for uniaxial compression loading in l-direction are shown in Fig. 12(b). Beginning with a small core thickness c, the critical load, P_{crit}^{wrinkl}, for the antisymmetrical wrinkling pattern is smaller than the value for the symmetrical wrinkling pattern. In this regime the critical load varies linearly with the core thickness c and can be calculated by the linear relation for shear buckling (i.e. $P_{crit}^{shear} = \frac{1}{2} G_{c,lt}\, c$, see eqn. (79). Thus, the case of local shear buckling is accounted for by taking the minimum of the rhs. of eqn. (78) with respect to (α, β). When the core thickness c is increased, the critical load for symmetrical buckling patterns becomes smaller than the load for antisymmetrical buckling. The minimum value of the critical load, P_{crit}^{wrinkl}, is reached for $c \approx 6\ mm$. If thick cores (i.e. $c \geq 12\ mm$) are considered, the critical load for symmetrical buckling approximately coincides with the critical load for antisymmetrical buckling. Due to the decay of disturbances in the core, the face layers are sufficiently far apart and do not influence each other. Thus, they can be analyzed separately. The phenomenon of changing buckling patterns can only be observed in orthotropic core materials.

b) Intracell buckling

In honeycomb cores the local instability phenomenon of intracell buckling of the face layers can also occur. The face layers which are supported by cell walls can exhibit a dimpling pattern. Since the out-of-plane buckling deformations vanish at the cell walls, the analysis is confined to the hexagonal area described by the supporting cell walls. For example in [30] the phenomenon of intracell buckling assuming a nonregular hexagonal shape for the cells is considered and practicable formulae for estimating critical membrane loads in the face layers, P_{crit}^{icb}, are presented.

In order to include the various forms of hexagons used in practical design, the following cell size parameters are introduced (see Fig. 13) and characterized by γ_{icb}:

$$\gamma_{icb} = \frac{a_{hc}}{l_{hc}} \quad (80).$$

Assuming a local buckling pattern

$$w_f = \sin\frac{\pi\, x\, m}{l_{hc}}\ \sin\frac{\pi\, y\, n}{a_{hc}} \quad (81)$$

the following equation can be obtained for the critical force:

$$P_{crit}^{icb} = \Gamma_{icb}\ \frac{K_f^{local}}{a_{hc}^2} \quad (82),$$

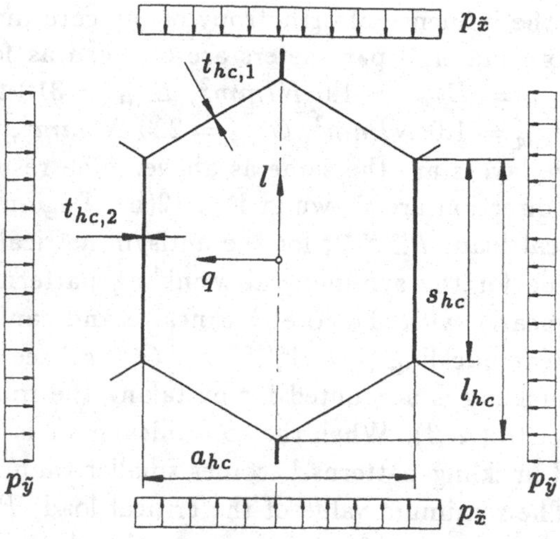

Fig. 13 Cell size parameters for nonregular hexagonal honeycomb cells

where

$$\Gamma_{icb} = \pi^2 \frac{\left(\gamma_{icb}^2\, m^2 + n^2\right)^2}{\gamma_{icb}^2\, m^2 + \xi^p\, n^2} \qquad m, n = 1, 2, \ldots$$

$$\xi^p = \frac{p_{\bar{y}}}{p_{\bar{x}}} \tag{83}$$

Since eqn. (83) still contains the buckling wavenumbers (m, n), the critical value has to be determined by minimization with respect to these parameters.

c) Failure due to transverse normal stresses

In sandwich shells, which are curved by their initial geometry or due to loading, bending moments lead to transverse normal stresses in the core perpendicular to the shell's mid-surface. Due to the limited resistance of the sandwich core to deformations perpendicular to the face layers, local thickness changes may appear. This local phenomenon is similar to the Brazier effect. Transverse tensile stresses in the core may lead to a thickness increase, to tensile cracking of the core and to debonding between core and face layers. Transverse compressive stresses may cause buckling of the cell walls in the case of honeycomb cores or to considerable local thickness reductions which may result in a loss in bending stiffness; see Fig. 14.

A detailed study of this phenomenon is presented in [31]. In order to investigate the influence of this effect in sandwich shells in a more simplified manner, the strains in the direction of the shell normal have to be evaluated. With M as the global bending moment of the shell (positive if R is decreased), and R as the radius of shell curvature in the deformed state the transverse normal stress in the core can be obtained by

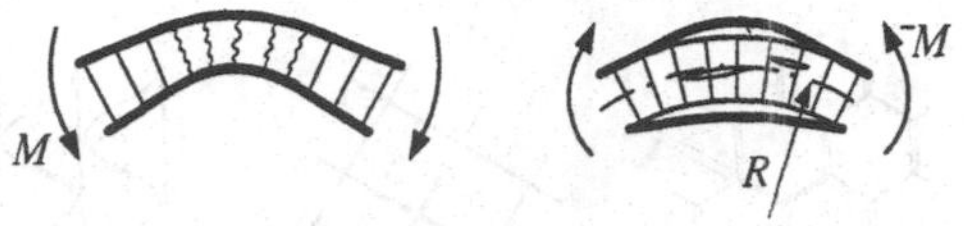

Fig. 14 Core failure due to bending

$$\sigma_n = -\frac{M}{R\,(c+t_f)} \tag{84}$$

Since there exist two principal curvatures in each shell point, the resulting stress in the shell's normal direction can be approximated by their sum

$$\sigma_n^{bend} = \sigma_{n,1} + \sigma_{n,2} \tag{85}$$

In the case of a sandwich structure loaded by external pressure, additional normal stresses appear in the core. Due to the constant shear stresses in the core these normal stresses decrease linearly from the loaded face layer ($\hat{z} = 0$) to the unloaded one ($\hat{z} = c$):

$$\sigma_n^{pr} = -p\left(\frac{\hat{z}}{c} - 1\right) \tag{86}$$

The core stress in the direction of the shell's normal is determined by superposition of the contributions due to bending and pressure loading. This core stress can be evaluated with respect to failure at the interface between core and face layer (especially in the case of positive σ_n-values) or with respect to buckling of the honeycomb cell walls (in the case of negative σ_n-values). It should be mentioned, that these σ_n-values represent 'effective', i.e. smeared-out, stresses which must be re-weighted by

$$\sigma_{zz}^{local} = \sigma_n \frac{A}{\sum s_i\, t_i} \tag{87}$$

i.e. according to unit-cell cross section, A, of the honeycomb vs. wall section area, $\sum s_i\, t_i$, within this unit-cell cross section, before comparing with critical cell wall stresses; see Fig. 15.

Similar to the discussion of the post-cracking behaviour of FRP laminates the behavior of sandwich shells after face layer wrinkling is considered. As a rough approximation it is assumed in [29] that in the wrinkled face layer area the post-critical membrane compression force perpendicular to the wrinkles remains constant.

As in the case of ply failure in FRP laminae the tangential membrane rigidity of the face layer is reduced, the elasticity parameters have to be adapted. By setting – as an approximation based on the stable post-buckling behavior of plates – the

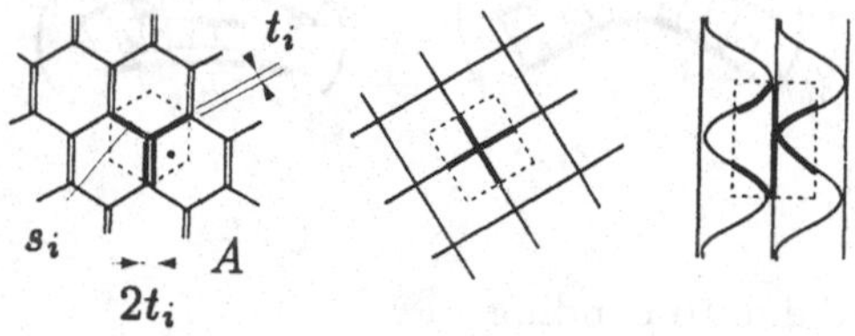

Fig. 15 Typical core unit-cells

tangential Young's modulus of elasticity of the facing material E_f^ℓ to zero in the direction ℓ, where local buckling occurred, the material matrix can be updated. The shear terms in the material matrix are assumed not to be influenced by local buckling.

In the case of local buckling in direction ℓ the updated tangential material matrix ${}^m\breve{\mathbf{E}}_f$ for the buckled face layer takes the form

$$\breve{\mathbf{E}}_f^\ell = \begin{pmatrix} 0 & 0 & 0 & 0 & 0 \\ 0 & \frac{E_f}{1-\nu_f^2} & 0 & 0 & 0 \\ 0 & 0 & G_{f,LW} & 0 & 0 \\ 0 & 0 & 0 & 0 & 0 \\ 0 & 0 & 0 & 0 & 0 \end{pmatrix} \tag{88}$$

Since this matrix is defined in a local coordinate system of the face layer with the main axis ℓ pointing into the direction of the local buckling load (i.e. the extension of wrinkles is perpendicular to this direction ℓ), a transformation into the laminate's coordinate system x_1, x_2, x_3 is necessary.

In the case of biaxial local buckling in one face layer (i.e. buckling in both principal directions, which is only possible for pure biaxial compression: $\xi = 1.0$) all tangential stiffness material parameters are set to zero, except the shear modulus $G_{f,LW}$.

By this approach stress redistribution during progressive wrinkling can be approximated. A more detailed consideration of this issue can be found in [31], where the stiffness loss due to core thickness reduction caused by wrinkling is taken into account, too.

Acknowledgement

Some protions of this Chapter are taken from the paper
"Combined micro- and macromechanical considerations of layered composite shells" by F.G.Rammerstorfer, A.Starlinger and K.Dorninger, to appear in Int.J. Numer. Meth. Engng., Vol.37 (1994). The permission by John Wiley & Sons, Ltd. is gratefully acknowledged.

References

1. Rammerstorfer F.G.: *Repetitorium Leichtbau.* Oldenbourg Verlag, Vienna, 1992.
2. Rammerstorfer F.G., Böhm H.J.: *Micromechanics for Macroscopic Material Description;* in this book, 1994.
3. Dorninger K.: *Entwicklung von nichtlinearen FE–Algorithmen zur Berechnung von Schalenkonstruktionen aus Faserverbundschalen.* VDI–Fortschrittsberichte **18/65**, VDI–Verlag, Düsseldorf, 1989.
4. Reissner E.: *The Effect of Transverse Shear Deformation on the Bending of Elastic Plates;* J.Appl.Mech. **12**, 69–77, 1945.
5. Noor A.K., Peters J.M.: *A Posteriori Estimates for Shear Correction Factors in Multi-Layered Composite Cylinders;* J.Engng.Mech. **115**, 1225–1244, 1989.
6. Reddy J.N.: *A Refined Nonlinear Theory of Plates With Transverse Shear Deformations;* Int.J.Sol.Struct. **20**, 881–896, 1984.
7. Başar Y.: *Finite-Rotation Theories for Composite Laminates;* Acta Mech. **98**, 159–176, 1993.
8. Başar Y., Yunhe Ding, Schultz R.: *Refined Shear Deformation Models for Composite Laminates with Finite Rotations;* Int.J.Sol.Struct. **30**, 2611–2638, 1993.
9. Reddy J.N.: *A Simple Higher-Order Theory for Laminated Composite Plates;* J.Appl. Mech. **51**, 745–752, 1984.
10. Heuer R.: *Static and Dynamic Analysis of Transversely Isotropic, Moderately Thick Sandwich Beams by Analogy;* Acta Mech. **91**, 1–9, 1992.
11. Lee K.H., Xavier P.B., and Chew C.H.: *Static Response of Unsymmetric Sandwich Beams Using an Improved Zig-Zag Model;* Compos.Engng. **3**, 235–248, 1993.
12. Reddy J.N., Pandey A.K.: *A First-Ply Failure Analysis of Composite Laminates;* Comput.Struct. **25**, 371–393, 1987.
13. Agarwal B.D., Broutman L.J.: *Analysis and Performance of Fiber Composites;* John Wiley & Sons, New York, NY, 1990.
14. Böhm H.J., Rammerstorfer F.G.: *Micromechanical Investigation of the Processing and Loading of Fibre-Reinforced Metal Matrix Composites;* Mater.Sci.Engng. **A135**, 185–188, 1991.
15. Hayashi T.: *Analytical Study of Interlaminar Shear Stresses in a Laminated Composite Plate;* Trans.Japan Soc.Aerosp.Sci. **10**, 43–48, 1967.
16. Pipes R.B., Pagano N.J.: *Interlaminar Stresses in Composite Laminates Under Uniform Axial Extension;* J.Compos.Mater. **4**, 538–548, 1970.

17. Pagano N.J., Pipes R.B.: *The Influence of Stacking Sequence of Laminate Strength;* J.Compos.Mater. **5**, 50–58, 1971.

18. Whitney J.M.: *Free-Edge Effects in the Characterization of Composite Materials;* in *"Analysis of the Test Methods for High Modulus Fibers and Composites"* ASTM STP 521, American Society for Testing and Materials, Philadelphia, PA, 1973.

19. Rose C.A., Herakovich C.T.: *An Approximate Solution for Interlaminar Stresses in Composite Laminates;* Compos.Engng. **3**, 271–285, 1993.

20. Kassapoglou C., Lagace P.A.: *Closed Form Solutions for the Interlaminar Stress Fields in Angle-Ply and Cross-Ply Laminates;* J.Compos.Mater. **21**, 292–308, 1987.

21. Morton S.K., Webber J.P.H.: *Interlaminar Failure due to Mechanical and Thermal Stresses at the Free Edges of Laminated Plates;* Compos.Sci.Technol. **47**, 1–13, 1993.

22. Hashin Z.: *Analysis of Composite Materials — A Survey;* J.Appl.Mech. **50**, 481–505, 1983.

23. Tolson S., Zabaras N: *Finite Element Analysis of Progressive Failure in Laminated Composite Plates;* Comput.Struct. **38**, 361–376, 1991.

24. Chawla K.K.: *Composite Materials.* Springer–Verlag, New York, NY, 1987.

25. Brewer J.C., Lagace P.A.: *Quadratic Stress Criterion for Initiation of Delamination;* J.Compos.Mater. **22**, 1141–1155, 1988.

26. Garg A.C.: *Delamination — A Damage Mode in Composite Structures;* Engng. Fract.Mech. **29**, 557–584, 1988.

27. Grediac M.: *A Finite Element Study of the Transverse Shear in Honeycomb Cores;* Int.J.Solids Structures **30**, 1777–1788, 1993.

28. Starlinger A., Rammerstorfer F.G.: *A Finite Element Formulation for Sandwich Shells Accounting for Local Failure Phenomena;* Proc. 2nd Int. Conf. on Sandwich Construction, EMAS, Warley, UK, 1992.

29. Starlinger A.: *Development of Efficient Finite Shell Elements for the Analysis of Sandwich Structures Under Large Deformations and Global as Well as Local Instabilities.* VDI–Fortschrittsberichte **18/93**, VDI–Verlag, Düsseldorf, 1991.

30. Stamm K., Witte H.: *Sandwichkonstruktionen — Berechnung, Fertigung, Ausführung.* Springer–Verlag, Vienna, 1974.

31. Kühhorn A.: *Geometrisch nichtlineare Theorie für Sandwichschalen unter Einbeziehung des Knitterphänomens.* VDI–Fortschrittsberichte **18/100**, VDI–Verlag, Düsseldorf, 1991.

CHAPTER 5

FRACTURE AND DAMAGE OF COMPOSITE LAMINATES

P. Gudmundson
Royal Institute of Technology, Stockholm, Sweden

ABSTRACT

Models for prediction of fracture and damage in composite laminates are presented. Present knowledge on the controlling mechanisms of fracture and damage are as well discussed. Failure criteria and failure mechanisms are first presented for unidirectional laminates. Models for failure prediction of unidirectional laminates under multiaxial stress states are explained and recommendations are made. The first ply failure and last ply failure criteria for composite laminates composed of plies with varying orientations are then discussed. Fracture mechanics of composite laminates, in particular delaminations, are analysed in detail. Elastic crack tip stress and strain fields are presented for cracks in anisotropic materials as well as for delaminations between dissimilar materials. The last section covers damage mechanics concepts in application to composite laminates. Advantages and disadvantages of the phenomenological and the micromechanical approach are discussed.

1. INTRODUCTION

In order to design a load bearing structural component in an efficient way, it is of importance to have detailed knowledge of the material behaviour. The behaviour prior to fracture is generally described by a constitutive law giving relations between stresses and strains in the material. The failure of the material is modelled by a failure criterion in terms of stresses or strains. In recent years the distinction between constitutive laws and failure criteria has become less pronounced. Many materials have successfully been described by use of damage models. In these cases the stress analysis is carried out utilizing a damage dependent constitutive law in combination with a damage evolution law. Failure is predicted when the damage reaches a critical level. For composite laminates, models for certain types of damages have been developed in recent years. The research on damage mechanics is today very active and one can expect that in the future generally applicable models will be developed. Until these models have been fully established it is however recommended to rely on more classical methods for design.

The most important difference between composite laminates and metals regarding structural design is the anisotropic material behaviour of the composite laminate. The failure loads of a composite laminate will depend on the direction of loading contrary to the case of isotropic metallic materials. For a metal the onset of plastic deformation is quite well described by the well known von Mises or Tresca criterion. The critical von Mises or Tresca stress can be determined from a single uniaxial tensile test. For a uniaxial fibre composite lamina the failure stress will differ by large amounts if it is loaded parallell or transverse to the fibres. In order to describe the failure characteristics of a unidirectional fibre composite laminae, at least tensile and compressive ultimate stresses parallell and transverse to fibres as well as the shear strength parallell to the fibrers must be measured. Thus, more experimental data are generally required in design with composite laminates. Another important difference between composites and metals is that the mechanical behaviour of the composite easily can be altered by changes in the fibre volume content, by use of different matrix and fibre materials. This is in fact one of the main advantages with composite materials. It is possible to optimize the structural behaviour by clever design not only of the structural geometry but also of the composite material itself. By allowing for changes in the material composition the required amount of material data increases correspondingly. For each material combination at least five strength parameters are needed. It would be prohibitive to experimentally determine strength data for every possible combination of fibres and matrices. In these cases micromechanical models for strength and stiffness which are based on constituent material properties are of great value. If strength data for one unidirectional composite is experimentally determined then by use of micromechanical models the strength for other unidirectional laminates with similar compositions can be estimated without additional experiments.

A composite laminate is composed of stacked plies. The strength of the laminate depends of course on the choice of plies and their orientation. The designer has a large freedom in the selection of laminate layups. It would be impossible to experimentally determine the strength characteristics for every layup in consideration. Therefore, strength models which are based on strengths for the individual plies are very useful in design.

From the discussion above one realizes that there are several mechanisms which influence the failure strengths of a composite laminate. First of all the strength and stiffness characteristics of the fibres and the matrix themselves play a crucial role. These data together with the fibre volume fraction and the interface strength between fibres and matrix control the strength of a unidirectional ply. Micromechanical models are very useful in estimating the ply strengths from given constituent material properties. Experimental data are however usually required when it comes to building prototypes of a new structure. If the strength characteristics of a unidirectional ply are known either from experimental tests or estimated from micromechanical models then the next issue is the strength of a composite laminate composed of unidirectional plies with varying orientations. As will be shown later the laminate strength can be quite well estimated from the ply properties. However, also in this case experimental verifications are needed when it comes to implementation in a new product. For strength design of composite structures, the combined use of micromechanical models and experimental characterizations is probably the only way to achieve optimal use of materials.

The lectures on failure models for composite laminates will cover a review on micromechanical models and design philosophies which are used today in the composites industry. Recommendations for further reading and research will also be given.

The final failure of a composite laminate is almost always caused by a crack which has grown and become critical. Fracture mechanics therefore plays an important role in the understanding of failure mechanisms in composite laminates. In the lectures on fracture mechanics the stress and strain fields which result close to a crack tip in a linear elastic anisotropic material or between two dissimilar materials will be discussed. Especially the oscillating stress fields which occur in front of a bimaterial crack will be discussed in detail. The case of delamination buckling and delamination crack growth under compressive loading will be investigated as an example of bimaterial crack problems.

From failure mechanisms of metallic materials we know that notches through their stress concentration factors may strongly influence the strength of a structure. The same is true for composite laminates. The most severe type of notch is a sharp crack. For reliable design of composite laminates it is hence of importance to be able to judge the effect of notches and cracks. A method for prediction of notch strengths will briefly be described.

As was mentioned in the beginning of the Introduction the distinction between constitutive models and failure criteria becomes less distinct with the introduction of damage mechanics. Damage mechanics of composite laminates is today a field of intense research. In the lectures on damage mechanics of composite laminates the principal equations for damage modelling will be presented. As a special case transverse matrix cracking will be discussed in detail. A micromechanically based damage model for transverse matrix cracking in composite laminates will be presented.

2. FAILURE MODELS FOR COMPOSITE LAMINATES

2.1 Strength of unidirectional laminates

The strength of unidirectional laminates form the basis for strength of composite laminates. An absolute minimum of five strength parameters are required in order to characterize the strength characteristics of a unidirectional ply: σ_{Lt} = Longitudinal tensile strength, σ_{Lc} = Longitudinal compressive strength, σ_{Tt} = Transverse tensile strength, σ_{Tc} = Transverse compressive strength and τ_{LT} = Longitudinal shear strength. The longitudinal direction is parallell to the fibres (1-direction), the transverse direction is perpendicular to the fibres (2-direction) and the shear is measured in the 12-direction. Alternatively the strengths can be characterized by their corresponding ultimate strains: ε_{Lt} = Longitudinal ultimate tensile strain, ε_{Lc} = Longitudinal ultimate compressive strain, ε_{Tt} = Transverse ultimate tensile strain, ε_{Tc} = Transverse ultimate compressive strain and γ_{LT} = Longitudinal ultimate shear strain. Some typical data for unidirectional composites are given in Table 1.

	CFRP	GFRP	KFRP
σ_{Lt} (MPa)	1500	1100	1400
σ_{Lc} (MPa)	1200	610	235
σ_{Tt} (MPa)	40	31	12
σ_{Tc} (MPa)	246	120	53
τ_{LT} (MPa)	68	72	34
ε_{Lt} (%)	0.83	2.75	1.84
ε_{Lc} (%)	0.66	1.58	0.31
ε_{Tt} (%)	0.39	0.38	0.22
ε_{Tc} (%)	2.39	1.43	0.96
γ_{LT} (%)	0.95	1.74	1.48

Table 1. *Strength data for typical carbon fibre reinforced epoxy (CFRP), E-glass fibre reinforced epoxy (GFRP) and Kevlar fibre reinforced epoxy (KFRP).*

One observes from Table 1 that the transverse and shear strengths are much lower than the longitudinal properties. The reason is that the transverse and shear strengths are to a large extent controlled by the matrix properties whereas the longitudinal strength is controlled by the fibres.

In a general loading situation a unidirectional composite ply experiences a combination of longitudinal, transverse and shear stresses. An important question then is if the failure strength in a general loading situation can be predicted from the five strength parameters discussed above. Before this subject will be discussed, micromechanical models for the five strengths will shortly be reviewed.

2.1.1 Longitudinal tensile strength

The longitudinal stress in a unidirectional ply reinforced by long fibres can quite accurately be estimated from the rule of mixtures:

$$\sigma_{11} = \sigma_f V_f + \sigma_m (1 - V_f) \tag{1}$$

where V_f denotes fibre volume fraction, σ_f fibre stress and σ_m matrix stress. The longitudinal strain in the fibres and in the matrix are approximately the same. Since the failure strain of the matrix for most polymer based fibre composites is larger than the ultimate strain for the fibres, the longitudinal tensile strength for the composite can be estimated by eq.(1) with $\sigma_f = \sigma_{fu}$ and $\sigma_m = \sigma_m(\varepsilon_{fu})$, where σ_{fu} is the ultimate tensile strength of the fibres and $\sigma_m(\varepsilon_{fu})$ is the matrix stress evaluated at the ultimate tensile strain of the fibres ε_{fu}. At equal strain in fibres and matrix the stresses in the fibres are much higher because of their higher stiffness. Hence, the longitudinal tensile strength is to a large extent controlled by the fibres.

2.1.2 Longitudinal compressive strength

Longitudinal compressive failure of unidirectional composites may depend on different micromechanical mechanisms. It is therefore more difficult to predict in comparison to the longitudinal tensile strength. Basically there are four different failure mechanisms which can play a role: 1. fibre microbuckling, 2. fibre yielding or crushing, 3. transverse tensile failure, 4. shear failure.

Rosen (1965) derived an equation for fibre microbuckling in an elastic matrix. He predicted the compressive longitudinal strength to be:

$$\sigma_{Lc} = \frac{G_m}{(1 - V_f)} \tag{2}$$

where G_m denotes the matrix shear modulus. The compressive strength is however often overestimated by eq.(2). Argon (1972) also investigated the microbuckling problem but instead for a rigid-perfectly plastic matrix. In order to predict a compressive strength he had to introduce an initial fibre misalignment angle. Budiansky (1983) performed the microbuckling analysis for an elastic-perfectly plastic matrix and Budiansky and Fleck (1993) the same for an elastic-plastic strain hardening matrix. In their analyses, the initial misalignment angle of the fibres also play a crucial role. For vanishing misalignment however,

Rosen's elastic result is recovered. The predictions of compressive strength decrease with increasing misalignment.

For fibres which are weak in compression, the composite may fail simply by fibre compressive failure. This is generally the case for Kevlar reinforced polymers. Kevlar yields at a relatively low compressive stress.

In case of a brittle matrix a unidirectional composite under compressive load may split due to transverse failure of the matrix. The transverse strain in the composite resulting from the compressive load is:

$$\varepsilon_2 = -\nu_{12}\frac{\sigma_1}{E_1} \tag{3}$$

Since σ_1 is negative the transverse strain ε_2 will be tensile. The compressive strength may then be estimated from the ultimate transverse tensile strain and eq.(3), thus

$$\sigma_{Lc} = \frac{E_1 \varepsilon_{Tt}}{\nu_{12}} \tag{4}$$

The compressive stress results in a shear stress of $\sigma_1/2$ at 45 degrees to the fibre direction . If this shear stress reaches the shear strength of the composite in this direction, the compressive failure will be controlled by this shear mechanism (Ewins and Ham (1973)).

From the discussion above it is seen that different failure mechanisms are of importance for compressive strength. It is therefore often difficult to predict the compressive strength in an accurate way.

2.1.3 Transverse tensile strength

The transverse tensile strength is mainly controlled by the matrix properties and the fibre/matrix interface strength. The usually much stiffer fibres act like rigid inclusions in the matrix under transverse tensile loading. Thereby they introduce stress and strain concentrations in the matrix which may lower the strength. Experimentally it has been observed that the transverse tensile strength of a composite is lower than the strength of the matrix itself. The large strength reduction in comparison to the strength for a pure matrix material can not entirely be attributed to the strain concentration in the composite. Properties which are of importance are the fibre/matrix interface strength as well as matrix properties under a triaxial stress state. The local stress state in a transversely loaded composite is triaxial. Recent experimental results by Asp, Berglund and Gudmundson (1993) indicate that the triaxial ultimate tensile strain in typical epoxy systems are much smaller than their corresponding values for uniaxial stress states.

There are today really no reliable models available for prediction of transverse tensile strengths. Most polymer based composite materials have ultimate transverse tensile strains of the order of 0.2-0.5%.

2.1.4 Transverse compressive strength

The main mechanism for transverse compressive failure is shear on planes directed 45 degrees in relation to the loading direction and parallell to the fibre direction. The shear stress on this plane is 50% of the applied compressive stress. This failure mode is controlled by the matrix properties.

For quantitative predictions of transverse compressive strength there are no reliable micromechanical models available.

2.1.5 Longitudinal shear strength

The longitudinal shear strength is as transverse tensile and compressive strength controlled by the matrix behaviour. The fibres introduce stress concentrations which lower the shear strength in comparison to the shear strength of the pure matrix material. Nonlinear plastic and creep deformations in the matrix significantly influence both the transverse and shear behaviour of the composite.

For accurate predictions of longitudinal shear strength there are no reliable micromechanical models available.

2.2 Failure prediction in unidirectional composites

In design it is of interest to predict the strength of a unidirectional composite under multiaxial loading conditions. Generally only the five strength parameters discussed above are known. The question is then to predict the failure load under combined loadings. There are several methods which are used today. All of these methods are more or less empirical. It would be preferred if a micromechanically based criterion was available and applicable to general loading conditions. A comprehensive review on failure criteria for composites has been given by Nahas (1986). Below the most commonly applied failure criteria are presented.

2.2.1 Maximum stress criterion

The stresses are divided into normal stresses in the 1- and 2-directions and a shear stress in the 12-direction. Each of these stresses are compared to their corresponding ultimate strengths. The criterion can be stated as:

$$-\sigma_{Lc} \leq \sigma_1 \leq \sigma_{Lt} \tag{5}$$

$$-\sigma_{Tc} \leq \sigma_2 \leq \sigma_{Tt} \tag{6}$$

$$|\tau_{12}| \leq \tau_{LT} \tag{7}$$

An advantage with the maximum stress criterion is that the mode of failure can be identified. The criterion does however not take coupling between different failure modes into account.

2.2.2 *Maximum strain criterion*

The strains are divided into normal strains in the 1- and 2-directions and shear strain in the 12-direction. Each of the strains are compared to their corresponding ultimate strains. The criterion can be stated as:

$$-\varepsilon_{Lc} \leq \varepsilon_1 \leq \varepsilon_{Lt} \tag{8}$$

$$-\varepsilon_{Tc} \leq \varepsilon_2 \leq \varepsilon_{Tt} \tag{9}$$

$$|\gamma_{12}| \leq \gamma_{LT} \tag{10}$$

Also for the maximum strain criterion the failure mode can be identified. Due to the Poisson effect, the maximum stress and maximum strain criteria will generally not result in the same prediction for strength.

2.2.3 *Azzi-Tsai-Hill criterion*

There are a number suggestions in the literature for more or less empirical failure criteria which take interaction between different failure modes into account. Two of the most commonly applied criteria of these will be presented below.

Hill (1950) generalized the von Mises criterion for plastic deformation to a corresponding criterion for orthotropic materials. Azzi and Tsai (1965) applied this criterion to transversely isotropic unidirectional composites under plane stress. In Hill's formulation it was assumed that there was no difference in magnitude between tensile and compressive critical stresses. Azzi and Tsai state that the criterion is still valid for a material with different tensile and compressive strengths.

The stresses are divided into normal stresses in the 1- and 2-directions and a shear stress in the 12-direction. The criterion can be stated as:

$$\left(\frac{\sigma_1}{\sigma_{Lt}}\right)^2 + \left(\frac{\sigma_2}{\sigma_{Tt}}\right)^2 - \frac{|\sigma_1\sigma_2|}{\sigma_{Lt}^2} + \left(\frac{\tau_{12}}{\tau_{LT}}\right)^2 \leq 1 \tag{11}$$

The criterion stated in eq.(11) is valid for tensile longitudinal and transverse tensile stresses. In case of compressive stresses the corresponding ultimate tensile strength in eq.(11) is changed to its compressive strength value.

2.2.4 Tsai-Wu criterion

By assuming that the failure of a unidirectional laminate is governed by a polynomial of second order in the stresses, Tsai and Wu (1971) derived a failure criterion which can be expressed by six parameters, F_{11}, $F_{12}=F_{21}$, F_{22}, F_{66}, F_1, F_2 according to

$$F_{ij}\sigma_i\sigma_j + F_i\sigma_i \leq 1 \tag{12}$$

where

$$F_{11} = \frac{1}{\sigma_{Lt}\sigma_{Lc}} \tag{13}$$

$$F_{22} = \frac{1}{\sigma_{Tt}\sigma_{Tc}} \tag{14}$$

$$F_{66} = \frac{1}{\tau_{LT}^2} \tag{15}$$

$$F_1 = \frac{1}{\sigma_{Lt}} - \frac{1}{\sigma_{Lc}} \tag{16}$$

$$F_2 = \frac{1}{\sigma_{Tt}} - \frac{1}{\sigma_{Tc}} \tag{17}$$

and

$$-(F_{11}F_{12})^{1/2}/2 \leq F_{12} \leq 0 \tag{18}$$

In eq.(12), σ_6 denotes τ_{12}. It is observed that F_{12} is not determined from uniaxial test data. Biaxial tests must be conducted for the determination of F_{12}. If biaxial data are not available the lower limit in eq.(18) is recommended.

2.2.5 Comparison between different failure criteria

Off-axis tensile loading of a unidirectional laminate creates longitudinal, transverse and shear stresses. The criteria above will result in different predictions for strength which can be compared to experimental data. A prediction according to the maximum strength criterion is presented in Figure 1. It is seen that longitudinal fracture is predicted for loading directions between 0 and 4 degrees from the fibre direction, shear failure is predicted for loading directions between 4 and 24 degrees and transverse tensile fracture is predicted for load angles larger than 24 degrees. In Figure 2 comparisons between the maximum stress, maximum strain, the Azzi-Tsai-Hill criterion and experimental data are shown. The Azzi-Tsai-Hill criterion gives a slightly better fit to experimental data for load angles around 25 degrees, whichs correspond to a change between shear and transverse tensile failure according to the maximum stress theory. On the other hand, the Azzi-Tsai-Hill criterion can not give any prediction for the failure mode.

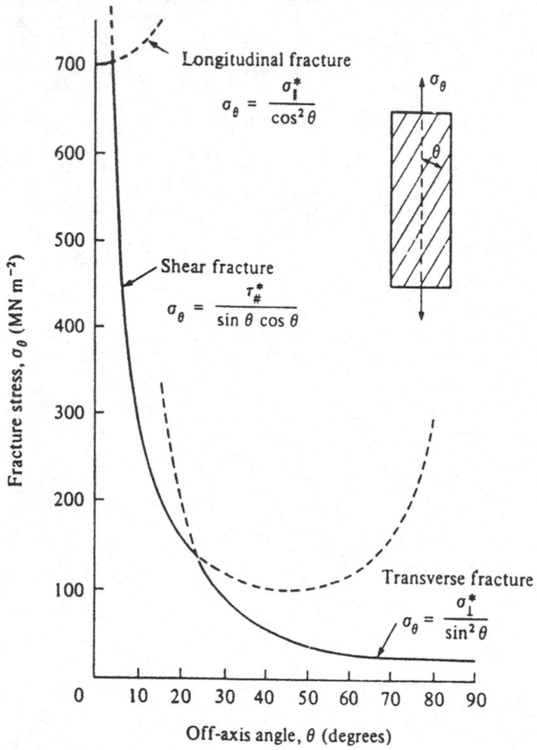

Figure 1. The maximum strength criterion for off-axis loading on a unidirectional glass/polyester laminate. (From Hull (1981)).

Biaxial test data can be compared to the failure predictions resulting from the different models described above. One way to compare these data is to plot the failure envelope in the stress space. In Figure 3 the failure envelope at vanishing shear stress is plotted for the

Tsai-Wu , the maximum strain and the Azzi-Tsai-Hill criteria. These predictions are compared to experimental data. There are no significant differences between the methods. Perhaps the Tsai-Wu criterion gives the best fit in this particular case.

2.2.6 Recommendations

The differences between the presented failure models are not extreme. The interactive models (Azzi-Tsai-Hill and Tsai-Wu) are convinient to use but they are purely empirical. They are just different forms of interpolations between experimental data points. The maximum stress and maximum strain theories are also empirical in a certain sense. There is no micromechanical justification for these theories either. The advantage of the maximum stress and strain theories is that they give a hint on the critical failure mode. According to Nahas (1986) the most frequently applied criteria in practice are in order of popularity: the maximum strain criterion, the maximum stress criterion, the Azzi-Tsai-Hill criterion, the Tsai-Wu criterion.

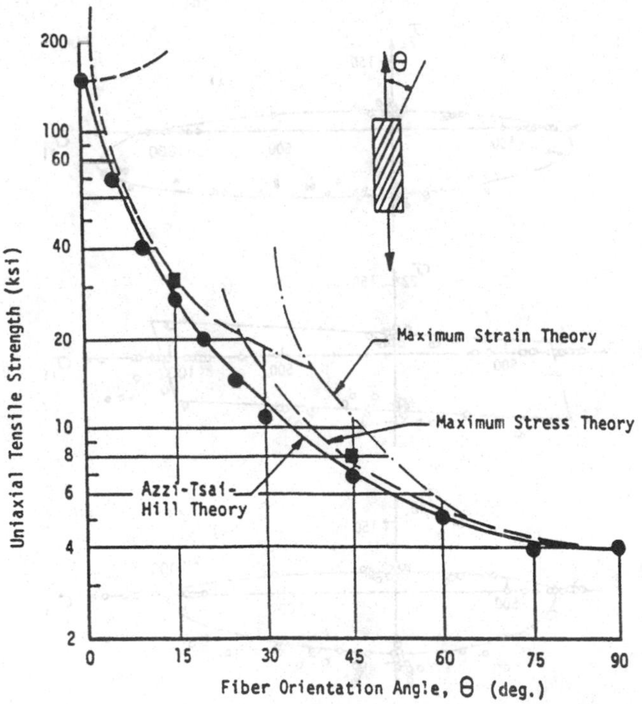

Figure 2. Comparison between the maximum stress, maximum strain, Azzi-Tsai-Hill criteria and experimental data for off-axis loading of a unidirectional glass/epoxy laminate. (From Mallik (1988)).

2.3 Strength of composite laminates

By use of one of the above presented failure criteria for a uniaxial ply, for example the maximum strain criterion, the failure load of an individual ply may be estimated. By use of laminate theory the local stresses and strains in all plies of a mechanically loaded laminate may be predicted. It is then possible to predict which of the plies that will fail first. The corresponding load is called the First Ply Failure (FPF) load. In many cases the laminate can carry much higher loads than the FPF load. It is often the case that a sequence of ply failures occur under increasing stresses before final failure of the whole laminate.

There are different design philosophies which are applied in industry today. The most conservative one is to design against first ply failure. This can however be too conservative. Consider for example a cross ply laminate composed of plies with low transverse tensile ultimate strain in comparison the longitudinal ultimate tensile strain. The laminate will experience a first ply tensile failure in the transverse plies at a loading which is much lower than the ultimate strength of the laminate. The first ply failure for glass reinforced cross plies can be as low as 15-20% of the ultimate strength. An example of a first ply failure

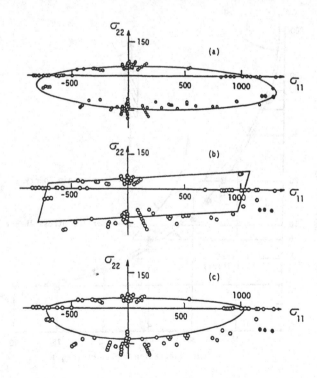

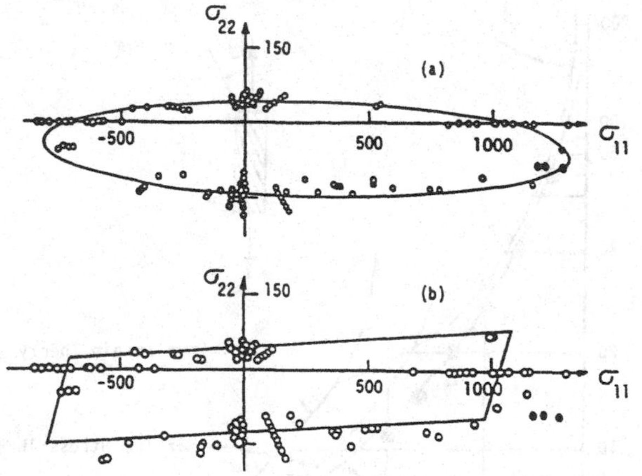

Figure 3. Failure envelopes in stress space at vanishing shear stress for the Tsai-Wu (a), maximum strain (b) and Azzi-Tsai-Hill (c) criteria. Comparisons to experimental data for a uniaxial carbon reinforced epoxy laminate. (From Mallik (1988)).

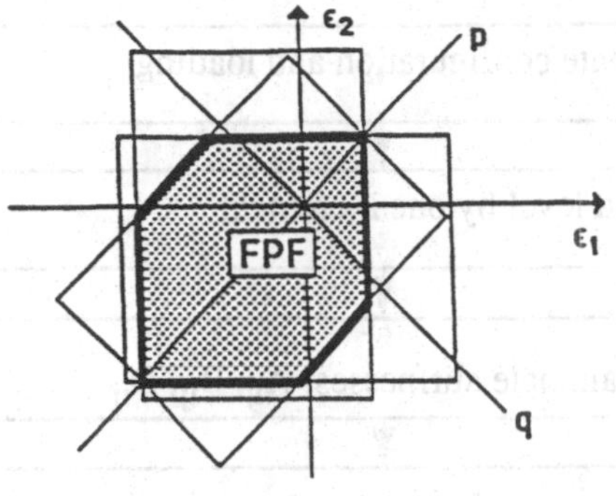

Figure 4. First ply failure envelope of a quasi-isotropic $(0/90/45/-45)_s$ laminate using the maximum strain criterion. (From Tsai (1987)).

analysis for a quasi-isotropic $(0/90/45/-45)_s$ laminate is presented in Figure 4. In Figure 4 the first ply failure envelope is shown in the strain space (ε_1, ε_2) at vanishing shear strain.

A better choice for design is to base the analysis on Last Ply Failure (LPF). In this type of analysis one takes ply failures into consideration in the calculations. There are basically two ways to do this analysis. In the total ply discount approach, it is assumed that the failed ply looses all its stiffness. The partial ply discount method however takes the failure mode into consideration. If a matrix controlled failure is predicted, the transverse modulus and the shear modulus are set to zero. The longitudinal modulus however maintains its value. If a fibre failure is predicted, all moduli of the failed ply are set to zero. After a ply has failed according to the applied ply failure criterion, the laminate stiffness has to be recalculated with the reduced ply stiffnesses according to either the total ply discount method or the partial ply discount method. A check is then performed whether the laminate can still carry some further load. If this is not the case, then the last ply failure load has been reached, otherwise the analysis proceeds until the next ply failure occurs. A flow chart of the last ply failure method is presented in Figure 5.

As an example of a last ply failure calculation, a carbon fibre reinforced epoxy laminate $(90/-45/45)_s$ is considered. The laminate is uniaxially loaded in the 0 degree direction. The stiffnesses of the plies are given by: $E_1 = 140$ GPa, $E_2 = 10$ GPa, $G_{12} = 5$ GPa, $\nu_{12} = 0.3$. The ply strengths are: $\sigma_{Lt} = 1500$ MPa, $\sigma_{Lc} = 1200$ MPa, $\sigma_{Tt} = 50$ MPa, $\sigma_{Tc} = 250$ MPa, $\tau_{LT} = 70$ MPa. Details of the calculations are given by Datoo (1991), pages 316-332. The analysis is based on the maximum stress criterion and the partial ply discount method. The stiffness of the undamaged laminate can be predicted by laminate theory to be 25.5 GPa. At a tensile load of 139 MPa the first ply failure load is reached. The failure is associated with tensile transverse failure in the 90 degree plies. The transverse and shear moduli of the 90 degree plies are then set to zero and the laminate stiffness is recalculated. The new stiffness is 22.7 GPa, a decrease of 11%. The ply stresses are recalculated to check if further plies fail at the

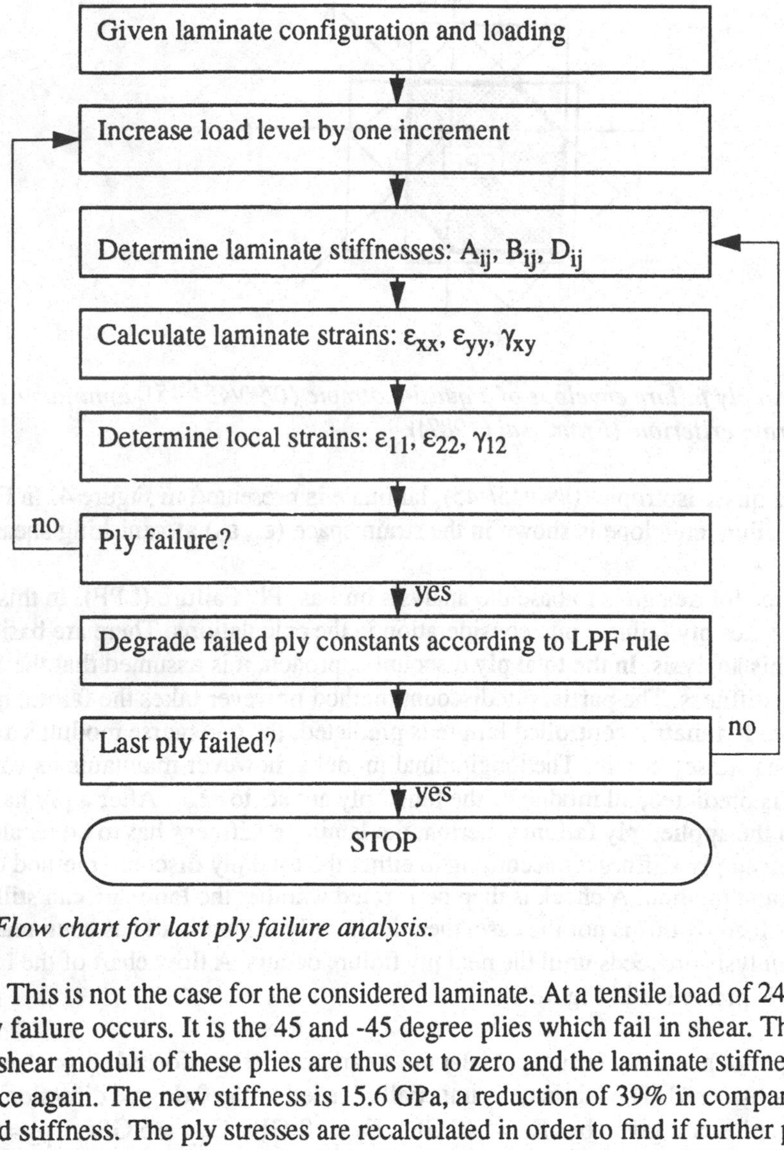

Figure 5. Flow chart for last ply failure analysis.

same load. This is not the case for the considered laminate. At a tensile load of 244 MPa, the second ply failure occurs. It is the 45 and -45 degree plies which fail in shear. The transverse and shear moduli of these plies are thus set to zero and the laminate stiffness is recalculated once again. The new stiffness is 15.6 GPa, a reduction of 39% in comparison to the undamaged stiffness. The ply stresses are recalculated in order to find if further plies fail at the same load. This is not the case in this example. At a tensile stress of 400 MPa, the compressive strength of the 90 degree ply is reached. The longitudinal modulus of the 90 degree ply is thus set to zero. Further calculation shows that the laminate can not take any further loads. Thus, the last ply failure stress is 400 MPa and the first ply failure stress is 139 MPa. In this case there is a substantial difference between FPF and LPF. An analysis based on FPF would be too conservative in this case. In Figure 6, a stress-strain diagram for the considered example is presented.

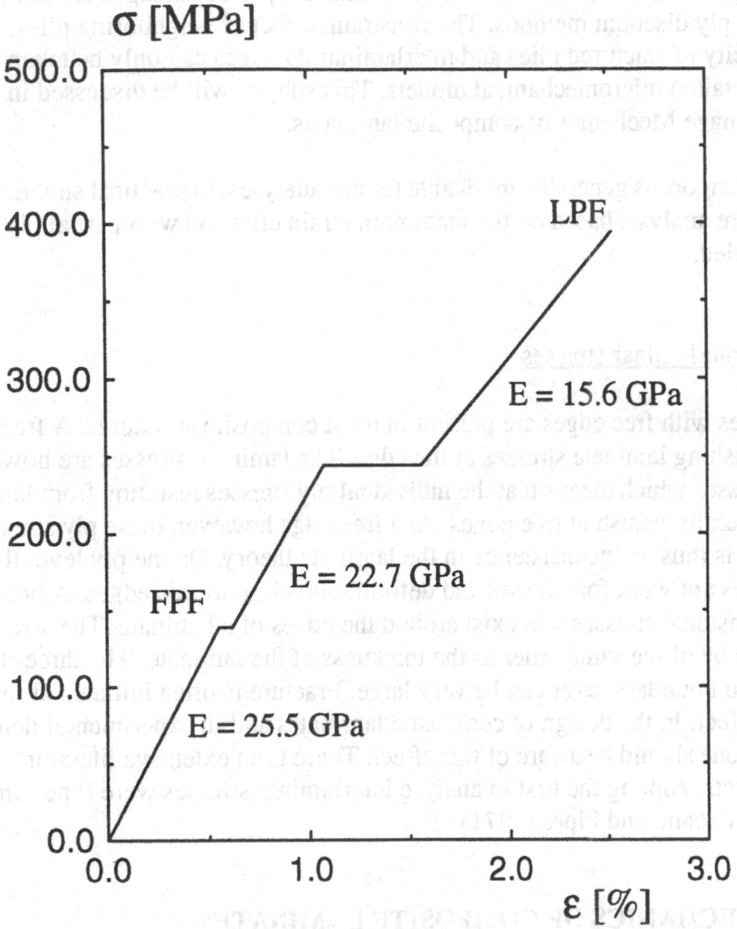

Figure 6. Predicted stress-strain behaviour according to the partial ply discount method for a $(90/\text{-}45/45)_s$ carbon fibre reinforced epoxy laminate loaded in the 0 degree direction.

2.3.1 Comments and recommendations

The first ply failure and last ply failure criterion are convinient to use and they are easily implemented in a computer program. One must remember however that they are based on approximative ply failure criteria which in addition do not take interactions between neighboring plies in a laminate into consideration. It has for example been observed that the transverse ply failure stress depends on the thickness of the ply at least below a certain thickness. This is due to the so called constraint effect of neighboring plies. Furthermore, the ply discount method assumes that a failed ply cannot take any further load. In reality a fractured ply will still take a not negligible load. Laminates also often show interlaminar damages, for example in the form of delaminations. These types of damages are not taken into account by the ply discount methods. The constraint effect of neighboring plies, the load carrying capacity of fractured plies and interlaminar damages can only be taken into account by more detailed micromechanical models. This subject will be discussed in detail in Section 4 on Damage Mechanics of composite laminates.

Due to lack of more rigorous generally applicable failure analyses, in practical situations the use of last ply failure analysis based on the maximum strain criterion with partial ply discount is recommended.

2.4 Edge effects - Interlaminar stresses

Composite laminates with free edges are present in most composite structures. A free edge corresponds to vanishing laminate stresses at the edge. The laminate stresses are however averages of ply stresses which means that the individual ply stresses resulting from laminate analysis do not generally vanish at free edges. At a free edge however, these ply stresses must vanish. There is thus an inconsistency in the laminate theory. On the ply level the laminate theory does not work for stresses and deformation close to free edges. A boundary layer of three dimensional stresses will exist around the edges of a laminate. The size of the boundary layer will be of the same order as the thickness of the laminate. The three-dimensional stresses in the boundary layer can be very large. Fracture is often initiated at the edges due to this effect. In the design of composite laminates and in experimental determinations of strength one should be aware of this effect. There is an extensive literature on the subject of edge effects. Among the first to analyse interlaminar stresses were Pipes and Pagano (1970) and Pagano and Pipes (1971).

3. FRACTURE MECHANICS OF COMPOSITE LAMINATES

Fracture of composite laminates is generally a result of cracks which have initiated, grown and become unstable. The are many types of cracks associated with composites failure; transverse matrix cracks, fiber/matrix debonding, delaminations, fibre cracks,... The analysis of cracks is therefore of importance for a deeper understanding of failure mechanisms in composites. The application of fracture mechanics to composite failure is however not simple. A composite laminate is an inhomogeneous material on different scales. Through the thickness of the laminate the material properties vary from ply to ply. Also within each ply the fibres and matrix makes the ply inhomogeneous on scales of order of fibre diameters. When fracture mechanics is applied one must be aware of these inhomogenities. Even though a ply or a laminate may be treated as a homogeneous material, it is generally anisotropic. The anisotropic behaviour complicates the analysis to some degree in comparison to isotropic materials for which well established knowledge exists today. Another feature of fracture mechanics in application to composites problems is that cracks at bimaterial interfaces often are of interest. As will be seen later on, bimaterial interface cracks introduce oscillating stress fields which do not exist in homogeneous materials. Similarly to fracture mechanics for metallic materials, significant nonlinearities can appear close to crack tips. These nonlinearities may limit the applicability of linear fracture mechanics concepts.

In spite of the difficulties mentioned above, fracture mechanics concepts have proved to be very useful in understanding composite failure mechanisms. Qualitative models can be derived which explains the dependence of different material parameters on composite properties. In certain type of problems also quantitative models have successfully been derived.

The lectures on fracture mechanics of composite laminates will first discuss elastic crack tip solutions and energy release rates for anisotropic and bimaterial problems. As an application example delamination growth will be discussed.

3.1 Cracks in anisotropic materials

The stress fields around a crack in a homogeneous anisotropic material have been derived by Sih, Paris and Irwin (1965). Similarly to the isotropic case the stresses have a square root singularity and close to the crack tip they are completely determined by three stress intensity factors K_I, K_{II}, K_{III} which are defined by:

$$K_I = \lim_{r \to 0} \sigma_{yy} \sqrt{2\pi r} \tag{19}$$

$$K_{II} = \lim_{r \to 0} \sigma_{xy} \sqrt{2\pi r} \tag{20}$$

$$K_{III} = \lim_{r \to 0} \sigma_{yz} \sqrt{2\pi r} \tag{21}$$

where the coordinate system is defined in Figure 7 and the equations (19-21) are evaluated for the angle $\theta = 0$. The stresses and displacements close to a crack tip are given by the stress intensity factors and corresponding universal functions of the coordinates in a similar way as for the isotropic case. The universal functions depend however in a quite complicated way on the elastic constants of the anisotropic material. For example, the stresses σ_{yy} and the displacements u_y can for a mode I loading be written as

$$\sigma_{yy} = \frac{K_I}{\sqrt{2\pi r}} \mathrm{Re} \left(\frac{1}{\mu_1 - \mu_2} (\mu_1 F_2 - \mu_2 F_1) \right) \tag{22}$$

$$u_y = \frac{K_I \sqrt{2\pi r}}{\pi} \mathrm{Re} \left(\frac{1}{\mu_1 - \mu_2} \left(\frac{\mu_1 p_2}{F_2} - \frac{\mu_2 p_1}{F_1} \right) \right) \tag{23}$$

where

$$F_1 = \frac{1}{\sqrt{\cos(\theta) + \mu_1 \sin(\theta)}} \tag{24}$$

$$F_2 = \frac{1}{\sqrt{\cos(\theta) + \mu_2 \sin(\theta)}} \tag{25}$$

and μ_1, μ_2, p_1, p_2 are complex constants depending on the elastic properties of the material, see Sih, Paris and Irwin (1965). It is observed that the displacements vary like the square root of distance from the crack tip in the same way as for the isotropic case.

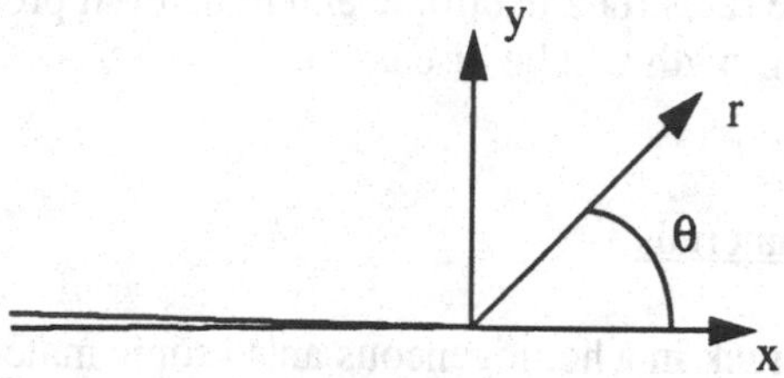

Figure 7. Coordinate system for the crack problem.

Knowing the stress and displacement fields in terms of the stress intensity factors, it is possible to determine the energy release rate at crack growth. The energy release rate is usually denoted as G and it defines the energy released at the crack tip for a unit crack advance and a unit thickness of the material. In the general anisotropic case it can be shown that the energy release rate will be a quadratic form in the stress intensity factors involving coupling terms between the different modes. In the special case of an orthotropic material with the crack on a plane of symmetry however, the three modes are uncoupled and the total energy release rate can be expressed as a sum over the three modes, i.e. $G = G_I + G_{II} + G_{III}$, where

$$G_I = K_I^2 \sqrt{\frac{a_{11} a_{22}}{2}} \left[\sqrt{\frac{a_{22}}{a_{11}}} + \frac{2a_{12} + a_{66}}{2a_{11}} \right]^{1/2} \tag{26}$$

$$G_{II} = K_{II}^2 \frac{a_{11}}{\sqrt{2}} \left[\sqrt{\frac{a_{22}}{a_{11}}} + \frac{2a_{12} + a_{66}}{2a_{11}} \right]^{1/2} \tag{27}$$

$$G_{III} = \frac{K_{III}^2}{2\sqrt{G_{13} G_{23}}} \tag{28}$$

The constants a_{ij} can be expressed in terms of elastic constants. For a plane stress state they are given by

$$a_{11} = 1/E_{11} \tag{29}$$

$$a_{22} = 1/E_{22} \tag{30}$$

$$a_{12} = -\nu_{12}/E_{11} \tag{31}$$
$$a_{66} = 1/G_{12} \tag{32}$$

and for a plane strain state by

$$a_{11} = (1-\nu_{13}\nu_{31})/E_{11} \tag{33}$$
$$a_{22} = (1-\nu_{23}\nu_{32})/E_{22} \tag{34}$$
$$a_{12} = -(\nu_{12}+\nu_{32}\nu_{13})/E_{11} \tag{35}$$
$$a_{66} = 1/G_{12} \tag{36}$$

If the nonlinear material behaviour is limited to a small region close to the crack tip one can expect that the stress intensity factors control the stresses and deformations close to the crack tip. Hence, in these cases the stress intensity factors should be the parameters that control crack growth. The stress intensity factors are determined from full solutions of the boundary value problem at hand. For a single crack of finite length in an infinite plate loaded by prescribed homogeneous stresses at infinity, the stress intensity factors have been derived by Sih, Paris and Irwin (1965). They showed that the stress intensity factors for this problem are the same as in the case of an isotropic medium, thus for mode I loading

$$K_I = \sigma_0\sqrt{\pi a} \tag{37}$$

where σ_0 is the normal stress at infinity and 2a is the crack length. In a general case the stress intensity factors will be of order applied stress times square root of a typical crack length times a nondimensional constant of the order of unity.

The theory for cracks in a uniform anisotropic medium is useful for example in application to cracks running parallell to the fibres in unidirectional composites. The case of delaminations can however not be treated by the theory above, since delaminations generally appear between two layers of different elastic properties.

3.2 Delaminations - Bimaterial interfaces

A frequently encountered damage mode in composite laminates is delaminations. They can be caused by transverse matrix cracks which are kinked into the interface between two plies. In several experimental investigations these kinds of so called local delaminations or matrix crack induced delaminations have been observed, see for example Jones and Hull (1979). Another source for delaminations is low energy impact damage. The resulting damage from low energy impact of composite laminates is generally a combination of matrix cracks and delaminations, see for example Abrate (1991). The residual strength of a laminate is quite sensitive to the size of the impact induced delaminations, in particular for compressive residual strength. One reason for this behaviour is delamination buckling, see Nilsson (1992).

Concerning damage growth and failure of composite laminates, the problem of delamination growth is therefore of large importance. A delamination may be treated as a crack between two different layers. This is a quite different problem in comparison to a crack in a homogeneous material as will be seen below.

Similarly to a crack in a homogeneous medium, a bimaterial crack will give rise to square root singular stresses close to the crack tip. The determination of the near tip stress and deformation fields for a certain combination of materials may be performed by methods similar to those applied by Williams (1952) in his original work on corner singularities.

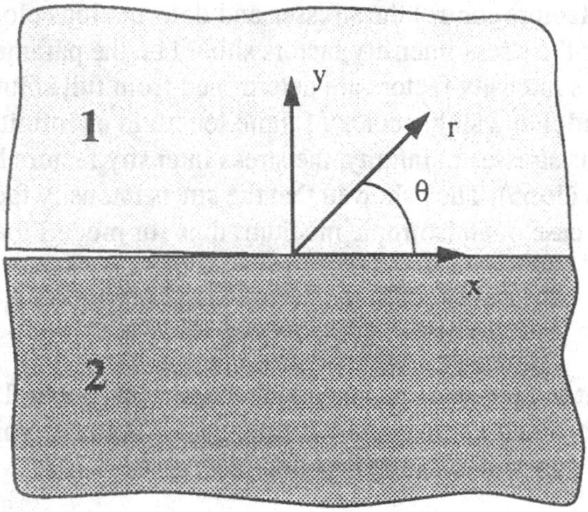

Figure 8. Coordinate system for the bimaterial interface crack problem.

The stresses close to a bimaterial crack tip can be written as

$$\sigma_{ij} = \mathrm{Re}\,(k r^{-1/2+i\varepsilon} g_{ij}(\theta)) \tag{38}$$

where k is a constant which is proportional to the external loading, r is the distance from the crack tip, θ is the angle according to Figure 8, g_{ij} are functions of θ which in the general case are complex valued. The main difference in comparison to the crack tip fields for an isotropic crack is the parameter ε which introduces severe oscillations in the solution close to the crack tip as can been seen if the real part of eq.(38) is evaluated.

$$\sigma_{ij} = |k g_{ij}| r^{-1/2} \cos(\varepsilon \ln(r) + \arg(k g_{ij})) \tag{39}$$

The term $\varepsilon\ln(r)$ causes rapid oscillations when r decreases. A similar equation results for the displacements which implies that the crack surfaces will penetrate each other. This is of course not realistic. In most practical cases however, this region of predicted interpenetration is extremely small in comparison to the crack length. This issue will be further discussed below.

The parameters in eqs.(38, 39) depend on the material properties below and above the delamination. When the stresses and displacements are known, the energy release rate can quite easily be computed. Instead of discussing the general bimaterial crack problem of two anisotropic materials, the proceeding discussion will be focused on the problem of a delamination between two isotropic materials. A thorough treatment of this problem has been given by Rice (1988). The in-plane stresses for this problem may be written as

$$(\sigma_{22} + i\sigma_{12})_{\theta=0} = Kr^{i\varepsilon}/(\sqrt{2\pi r}) \tag{40}$$

where K is the so called complex stress intensity factor. The parameter ε is given by

$$\varepsilon = (1/2\pi)\ln[(\kappa_1/\mu_1 + 1/\mu_2)/(\kappa_2/\mu_2 + 1/\mu_1)] \tag{41}$$

where μ_1 is the shear modulus of the upper half plane and μ_2 is the shear modulus of the lower half plane, $\kappa = 3 - 4\nu$ for plane strain and $(3-\nu)/(1+\nu)$ for plane stress, ν is the Poisson ratio. The crack opening displacements will be proportional to K and the square root of r. The oscillations in displacements will be similar to those for the stresses. As is discussed by Rice (1988) the awkward complex dimension of K makes it inconvenient for applications. Also the complex behaviour makes it impossible to decouple the state of stress close to the crack tip into a mode I and a mode II case respectively. If an arbitrary length scale is introduced Rice shows that real valued stress intensity factors may be defined.

The complex stress intensity factor controls the crack tip fields if the area of nonlinear material behaviour is small in comparison to the region which is dominated by the square root singularity. In this case one can expect that the complex stress intensity factor will control delamination crack growth. It is however impossible to define a pure mode I loading for example. For the interface crack of length L subject to remotely uniform stress σ_{yy}

$$K = \sigma_{yy}(1 + 2i\varepsilon)L^{-i\varepsilon}\sqrt{\pi L/2} \tag{42}$$

at the right-hand crack tip. It is observed that the parameter ε generally introduces an imaginary part of K. Not only the magnitude of K will vary with the crack length but also the phase. An unambiguous definition of mode I and mode II is therefore impossible to do.

The energy release rate for interface crack growth can as well be determined, see Rice (1988). It is proportional to the square of the magnitude of the complex stress intensity factor K.

Rice (1988) makes an estimate of the contact zone size in front of the crack tip. It is found that for realistic material combinations the contact zone can be neglected except for almost pure shear loadings. Comninou (1977) took contact into account in her analysis of a bimaterial interface crack. It was shown that other types of singular crack tip solutions appear within the contact region, but since the size of the contact region generally is extremely small one can expect that these crack tip fields are completely controlled by the complex stress intensity factor.

3.3 Determination of energy release rates for delamination problems

If the delamination size is large in comparison to the thickness of the composite laminate then plate and shell theory may be applied for modelling of stresses and deformations. Plate or shell theory can however not resolve the singular stresses close to the delamination front but the global deformations can generally be accurately predicted. Since the total energy release rate for delamination growth can be determined from the total potential energy of the whole structure containing the delamination under consideration, plate and shell theories can be applied to determine G. The method is similar to the well known compliance method for determination of stress intensity factors in split beams, for example the DCB specimen. Storåkers and Andersson (1988) showed how the energy release rate for delamination growth can be determined also for plates subjected to large rotations according to the von Karman plate theory. They derived a convinient equation for calculation of G from a von Karman solution for plates containing delaminations. The energy release rate at a delamination growth $\delta a(\Gamma)$ along the delamination front Γ was expressed as

$$-\delta U = \int_{\Gamma} \| P_{\alpha\beta} \| n_{\alpha} n_{\beta} \delta a \, d\Gamma \qquad (43)$$

where $P_{\alpha\beta}$ is an analogue of Eshelby's energy momentum tensor and n_{α} is the normal vector to the crack front pointing out from the delamination. The double bracket in eq.(43) denotes the difference in $P_{\alpha\beta}$ between the uncracked side and the cracked side of the delamination front. The tensor $P_{\alpha\beta}$ is given by

$$P_{\alpha\beta} = W\delta_{\alpha\beta} - S_{\alpha\gamma} u_{\gamma,\beta} + M_{\alpha\gamma} u_{3,\gamma\beta} - Q_{3,\beta} \qquad (44)$$

where W denotes the strain energy density, $S_{\alpha\beta}$ the membrane forces, u_{α} the transverse displacement, $M_{\alpha\beta}$ the bending moments and Q_{α} the transverse shear forces. The equation (43) is very easily implemented in a plate finite element program. Thus the total energy release for delaminations which are much larger than plate thicknesses can easily be determined from the plate solution to the delamination problem. In the thesis by Nilsson (1992) several problems have been solved using this approach.

Referring to the previous Section the total energy release rate is proportional to the square of the magnitude of the complex stress intensity factor. Thus, the magnitude of the complex stress intensity factor can be determined from knowledge of the total energy release rate but not the phase. If the failure criterion depends on the phase of the complex stress intensity factor, then this phase has to be determined. Suo and Hutchinson (1990) have by numerical solution of an integral equation been able also to determine the phase of K for a bimaterial crack between two isotropic materials.

3.4 Prediction of delamination growth

In the previous two sections the stress and strain fields close to a delamination front as well as the corresponding energy release rate have been discussed. We have also seen that the total energy release can be determined from plate theory provided that the delamination is much larger than the plate thickness. When predictions of delamination growth are of interest, there is a question whether linear fracture mechanics can be applied. One can assume that it is possible if the nonlinear material behaviour close to the delamination tip is restricted to a small area in comparison to the region in which the fields are dominated by the elastic stress singularity. This means that the nonlinear zone must be much smaller than plate thickness and delamination thickness. If fibre bridging occur at crack growth, linear elastic fracture mechanics can only be applied for limited delamination growth. Assuming that linear elastic fracture mechanics can be applied then there is the question if the phase of the complex stress intensity factor is of importance or not. If the fracture criterion would be independent of the phase and thus only dependent on the total energy release rate then the problem would be much simplified. Unfortunately, experimental observations indicate a phase dependence on delamination toughness. The mode II fracture toughness is generally larger than the mode I toughness. Even though the total energy release rate would be a relevant parameter for delamination growth one can expect that the critical energy release rate may vary with the angle of the delamination front in relation to the orientations of the neighboring generally anisotropic plies. If this is the case, extensive experimental tests for determinations of critical energy release rates at varying delamination front orientations are necessary. In many practical situations however the delamination growth criterion can to a first approximation be considered to be governed by a constant critical energy release rate. Nilsson et al (1993) performed compressive tests on a laminates with prefabricated circular delaminations. The experiments were set up so that the delaminations buckled out of the laminate and subsequently growed in a stable manner. Nilsson et al (1993) were able to get very good agreements between the experimental observations and numerical simulations of delamination growth based on a critical total energy release rate.

3.5 Notches and stress concentrations

The case of a crack can be looked upon as the most severe type of stress concentration. In between the case of a crack and a homogeneously stressed material different kinds of stress concentrations are present in real structures. For steel structures it known that stress concentrations must be carefully evaluated. The knowledge on design with stress concentrations in composite materials is limited. Whitney and Nuismer (1974) have suggested two methods which can be used to assess stress concentrations. In their first method they state that fracture can be predicted if the stress at a critical distance d_0 ahead of the notch reaches the unnotched tensile strength. Their second criterion states that fracture can be predicted if the average stress over a distance a_0 ahead of the notch reaches the unnotched tensile strength. Typically the material dependent distances are of the order $a_0 = 4$mm and $d_0 = 1$mm for composite laminates. The Whitney-Nuismer models are empirical in character and should be applied with caution.

4. DAMAGE MECHANICS OF COMPOSITE LAMINATES

It is a well established experimental fact that composite laminates develop different kinds of damage with increasing loads. Typical damage mechanisms are transverse matrix cracking, fibre/matrix debonding, local interlaminar delaminations, local fibre fractures. These damages not only influence the strength of the composite but also the stiffness characteristics. This is different from plastic deformations in metallic materials. Typically the damages are interrelated in the way that one damage mode may initiate the growth of a another damage mode. For design of composite laminates, the damage evolution plays an important role. Since the damage not only influences the strength but also the stiffness, damage dependent constitutive laws must be developed in order to simulate damage growth in a structure. Thus, the stiffness of a laminate will depend on parameters describing the actual state of damage in the laminate. Also the strength of the laminate will depend on the damage state. The theory of damage evolution and damage dependent material properties is called Damage Mechanics. One of the pioneers in this field was Kachanov who first treated creep problems in metallic materials with damage mechanics concepts. His original work is published in Russian, but he has later written an English book on the subject, Kachanov (1986). Comprehensive treatments of damage mechanics can be found in Lemaitre and Chaboche (1990) as well as in the review by Krajcinovic (1989).

In application of damage mechanics to composite laminates, the main problem is to find the dependence of damage parameters on the stiffness characteristics and the failure criterion. There are basically two different approaches which have been applied, either a phenomenological or a micromechanical approach.

In the phenomenological approach it is assumed that the damage can be represented by some damage parameters (λ_k, k = 1, N), where each λ_k can be a tensor valued quantity. The damage parameters are identified with certain observed damage mechanisms, for example matrix cracking or fibre/matrix debonding. The damage dependent constitutive law is then formally written as

$$\dot{\varepsilon}_{ij} = S_{ijkl}\dot{\sigma}_{kl} + \dot{\varepsilon}_{ij}^{D} \tag{45}$$
$$\dot{\varepsilon}_{ij}^{D} = \sum_{k} Q_{ij}^{k}\dot{\lambda}_{k} \tag{46}$$
$$\dot{\lambda}_{k} = R_{ij}^{k}\dot{\sigma}_{ij} \tag{47}$$

where the constitutive parameters Q, R, S may depend on the current stress, strain and damage state. By consideration of eventual material symmetries, continuum mechanics and thermodynamic principles, the structure of eqs.(45-47) may be derived. The explicit form of the damage dependence must however be assumed in some way. Experimental measurements are performed in order to determine unknown parameters in the constitutive law. Talreja (1985) used this approach in application to transverse matrix cracking in composite laminates. He assumed a linear dependence of a scalar damage parameter. An advantage of the phenomenological approach is that it is quite general. Disadvantages are the arbitrariness of the assumed damage dependent constitutive law as well as the experimental difficulties in determining constitutive parameters. Especially in applications to composite laminates where different layups may be considered, the determination of constitutive parameters at varying layup configurations will be very costly.

In the micromechanical approach, the different damage mechanisms are modelled on a micromechanical level and then the macromechanical behaviour such as eqs.(45-47) is derived by homogenization techniques. The main difficulty lies in the micromechanical modelling. Often these models depend on some micromechanical properties which are difficult to obtain. However, if it is possible to model the damage on a micromechanical level, the micromechanical approach has distinct advantages in comparison to the phenomenological approach. Usually fewer experimental data are required, the model is more flexible regarding varying laminate layups for example and the model is based on actual micromechanical processes instead of curve fits. Unfortunately, only a limited number of damage mechanisms can, with the present knowledge, be handled by micromechanical models in a convenient way. Even though the micromechanics could be modelled in a reliable way, the model itself should not be too complex and calculation intensive in order for the macromechanical analysis to be practically feasable. In conclusion, micromechanically based models should be used if it is possible otherwise phenomenological models should be applied.

The micromechanical damage mechanics approach for composite laminates will be demonstrated in Chapter 5 for the case of transverse matrix cracking.

5. REFERENCES

5.1 Textbooks

Agarwal B.D. and Broutman L.J., 1990, "Analysis and performance of fiber composites", John Wiley, ISBN 0-471-51152-8

Chou T.-W., 1992, "Microstructural design of fiber composites", Cambridge University Press, ISBN 0-521-35482-X

Christensen R.M., 1979, "Mechanics of composite materials", John Wiley, ISBN 0-471-05167-5

Composite Materials: Fatigue and Fracture (Third Volume), ed. T.K. O'Brien, ASTM STP 1110, ISBN 0-8031-1419-2

Composite Materials Series, 1989, "Application of fracture mechanics to composite materials", Vol.6, ed. K. Friedrich, Series editor R.B. Pipes, Elsevier, ISBN 0-444-87286-8 (Vol. 6), ISBN 0-444-42525-X (Series)

Composite Materials Series, 1991, "Fatigue of composite materials", Vol.4, ed. K.L. Reifsnider, Series editor R.B. Pipes, Elsevier, ISBN 0-444-70507-4 (Vol. 4), ISBN 0-444-42525-X (Series)

Datoo M.H., 1991, "Mechanics of fibrous composites", Elsevier, ISBN 1-85166-600-1

Harris B., 1986, "Engineering composite materials", The Institute of Metals, 1 Carlton House Terrace, London SW1Y 5DB, ISBN 0-901462-28-4

Hill R., 1950, "The mathematical theory of plasticity", Oxford University Press

Hull D., 1981, "An introduction to composite materials", Cambridge University Press, ISBN 0-521-28392-2

Kachanov L.M., 1986, "Introduction to continuum damage mechanics", Martinus Nijhoff, Dordrecht, The Netherlands

Lemaitre J. and Chaboche J.-L., 1990, "Mechanics of solid materials", Cambridge University Press, ISBN 0-521-32853-5

Mallik P.K., 1988, "Fiber-reinforced composites; materials, manufacturing, and design", Marcel Dekker, ISBN 0-8247-7796-4

Morley J.G., 1987, "High-performance fibre composites", Academic Press, ISBN 0-12-506445-4

Nilsson K.-F., 1992, "On combined buckling and interface crack growth", Doctoral thesis, Department of Solid Mechanics, Royal Institute of Technology, S-100 44 Stockholm, Sweden

Tsai S.W., 1987, "Composites design", Third edition, Think Composites, ISBN 0-9618090-0-0

Vinson J.R. and Sierakowski R.L., 1986, The behaviour of structures composed of composite materials", Martinus Nijhoff Publishers, ISBN 90-247-3125-9

5.2 Articles

Abrate S., 1991, "Impact on laminated composite materials", Appl. Mech. Rev., Vol. 44, 155-190

Argon A.S., 1972, Treatise of Materials Science and Technology, Vol. 1 p.79, Academic Press, New York

Asp L., Berglund L. and Gudmundson P., 1993, in preparation

Azzi V.D. and Tsai S.W., 1965, "Anisotropic strength of composites", Experimental Mechanics, Vol. 5, 283-288

Budiansky B., 1983, "Micromechanics", Computers & Structures, Vol. 16, 3-12

Budiansky B. and Fleck N.A., 1993, "Compressive failure of fibre composites", J. Mech. Phys. Solids, Vol. 41, 183-211

Comninou M, 1977, "The interface crack", J. Appl. Mech., Vol. 44, 631-636

Ewins P.D. and Ham A.C., 1973, "The nature of compressive failure in unidirectional carbon fibre reinforced plastics", Royal Aircraft Technical Report 73057

Gudmundson P. and Östlund S., 1992a, "Firts order analysis of stiffness reduction due to matrix cracking", J. Composite Materials, Vol. 26, 1009-1030

Gudmundson P. and Östlund S., 1992b, "Thermoelastic properties of composite laminates with matrix cracks", Composite Science and Technology, Vol. 44, 95-105

Gudmundson P. and Zang W., 1993, "An analytic model for thermoelastic properties of composite laminates containing transverse matrix cracks", To appear in Int. J. Solids Structures

Jones M.L.C and Hull D., 1979, Microscopy of failure mechanisms in filament wound pipe", J. Mater. Sci., Vol. 14, 165-174

Krajcinovic D., 1989, "Damage mechanics", Mechanics of Materials, Vol. 8, 117-197

Nahas M.N., 1986, "Survey of failure and post-failure theories of laminated fiber-reinforced composites", J. Composites Technology & Research, Vol. 8, 138-153

Nilsson K.-F., Thesken J.C., Sindelar P., Giannakopoulos A.E., Storåkers B., 1993, "A theoretical and experimental investigation of buckling induced delamination growth", J. Mech. Phys. Solids, Vol. 41, 749-782

Pagano N.J. and Pipes R.B., 1971, "The influence of stacking sequence of laminate strength", J. Composite Materials, Vol. 5, 50-57

Pipes R.B. and Pagano N.J., 1970, Interlaminar stresses in composite laminates under uniform axial extension", J. Composite Materials, Vol. 4, 538-548

Rice J.R., 1988, "Elastic fracture mechanics concepts for interfacial cracks", J. Appl. Mech., Vol. 55, 98-103

Rosen B.W., 1965, "Mechanics of composite strengthening", In Fibre Composite Materials, ch.3, ASM, Metals Park, Ohio

Sih G.C., Paris P.C. and Irwin G.R., 1965, "On cracks in rectilinearly anisotropic bodies", Int. J. Fracture Mechanics, Vol. 1, 189-203

Storåkers B. and Andersson B, 1988, "Nonlinear plate theory applied to delamination in composites", J. Mech. Phys. Solids, Vol. 36, 689-718

Suo Z. and Hutchinson J.W., 1990, "Interface crack between two elastic layers", Int. J. Fracture Mechanics, Vol. 43, 1-18

Talreja R., 1985, "A continuum mechanics characterization of damage in composite materials", Proc. Roy. Soc. London, Vol. A399, 195

Tsai S.W. and Wu E.M., 1971, "A general theory of strength for anisotropic materials", J. Composite Materials, Vol. 5, 58-80

Whitney J.M. and Nuismer R.J., 1974, "Stress fracture criteria for laminated composites containing stress concentrations", J. Composite Materials, Vol. 8, 253-265

Williams M.L., 1952, "Stress singularities resulting from various boundary conditions in angular corners of plates in extension", J. Appl. Mech., Vol. 19, 526-528

Zang W. and Gudmundson P., 1993b, "Damage evolution and thermoelastic properties of composite laminates", Int. J. Damage Mechanics, Vol. 2, 290-308

Östlund S. and Gudmundson P., 1992, "Numerical analysis of matrix crack induced delaminations in (+/-55) GFRP laminates", Composites Engineering, Vol. 2, 161-175

CHAPTER 6

DAMAGE EVOLUTION AND DETERMINATION OF THERMOELASTIC PROPERTIES OF LAMINATES CONTAINING CRACKS

P. Gudmundson and W. Zang
Royal Institute of Technology, Stockholm, Sweden

ABSTRACT

This Chapter consists of two papers reprinted

from Int.J.Solids Structures, Vol.30, No.23,
P.Gudmundson and W.Zang: **An Analytical Model for Thermoelastic Properties of Composite Laminates Containing Transverse Matrix Cracks**, pp.3211–3231, 1993, with kind permission from Elsevier Science Ltd, The Boulevard, Langford Lane, Kidlington 0X5 1GB, UK,

and

from Int.J.Damage Mechanics, Vol.2, No.3,
W.Zang and P.Gudmundson: **Damage Evolution and Thermoelastic Properties of Composite Laminates**,
pp.290–308, 1993, with kind permission from Technomic Publishing Co.Inc., 851 New Holland Ave., Lancaster, PA 17604, USA.

CHAPTER 6
PART I

AN ANALYTIC MODEL FOR THERMOELASTIC PROPERTIES OF COMPOSITE LAMINATES CONTAINING TRANSVERSE MATRIX CRACKS

P. Gudmundson and W. Zang
Swedish Institute of Composites, Piteå, Sweden

Abstract—An analytical model for the prediction of the thermoelastic properties of composite laminates containing matrix cracks is presented. In particular, transverse matrix cracks with their crack surfaces parallel to the fibre direction and perpendicular to the laminate plane are treated. Two- and three-dimensional laminates of arbitrary layup configurations are covered by the model. The presented expressions for stiffnesses, thermal expansion coefficients, strain contributions from release of residual stresses and local average ply stresses and strains do solely contain known ply property data and matrix crack densities. The key to the model is the judicious use of a known analytical solution for a row of cracks in an infinite isotropic medium. The model has been verified against numerically determined stiffnesses, thermal expansion coefficients and local average ply stresses for matrix cracked angle-ply and cross-ply laminates. Comparisons to experimental data for cross-ply laminates are also presented. It is shown that the present model to a very good accuracy can predict thermoelastic properties of matrix cracked composite laminates at varying matrix crack densities and layup configurations.

1. INTRODUCTION

When composite laminates are mechanically loaded, different kinds of damage modes such as transverse matrix cracking, delaminations and fibre fractures develop before final failure of the laminate. Transverse matrix cracking is often the first observed damage mode. This mode is generally not critical from a final fracture point of view. The matrix cracks can however initiate more severe damage such as delaminations and fibre fractures. A consequence of matrix cracking is that both local and global stress and strain redistributions occur in a laminate. For example, local stress concentrations close to the tips of the matrix cracks may cause the initiation of local delaminations and/or fibre fractures. Since matrix cracks generally are initiated long before final fracture of a structure, they should be taken into account in the design in order to fully utilize the load bearing capacity of a composite structure.

In order to simulate the mechanical behaviour of a matrix cracking composite laminate, the constitutive law which defines the stress–strain relationship for the laminate must include the effects of transverse matrix cracks. Compared to linear, elastic laminate theory the constitutive law should basically be extended to include two main aspects. First of all, criteria for transverse matrix crack initiation and growth must be implemented. Secondly, at given matrix crack densities the model must enable reliable estimations of the relationship between global stresses and strains as well as means for the estimation of local ply stresses and strains. In this paper this second aspect is addressed.

The simplest way to model transverse matrix cracks in composite laminates is to completely neglect the transverse stiffnesses of cracked plies. This method is generally called the ply discount method. The ply discount method will underestimate the stiffnesses of cracked laminates, since cracked plies can take some loading. Therefore the gradual changes of laminate properties with increasing matrix crack densities can never be covered by the ply discount method.

A relatively simple way to include the effects of load transfer between micro cracked plies and their neighbours is to apply a so-called shear lag analysis. In this theory, the load transfer between plies is assumed to take place in shear layers between neighbouring plies. The thicknesses and stiffnesses of these shear layers are generally unknown. The variations in the thickness direction of local ply stresses and strains are also neglected in the shear lag theory. Another aspect of the shear lag theory is that it is not obvious how it should be applied for layup configurations other than cross plies. The shear lag theory has however successfully been applied to cross-ply laminates (Highsmith and Reifsnider, 1982; Lim and Hong, 1989; Han and Hahn, 1989; Tan and Nuismer, 1989).

By application of the principle of minimum complementary potential energy Hashin (1985, 1987, 1988) derived estimates for stiffnesses, thermal expansion coefficients as well as local ply stresses of matrix cracked cross-ply laminates. He showed that the estimates were in good agreement with experimental data. An advantage with Hashin's method is that strict lower bounds for stiffnesses are obtained. Varna and Berglund (1991) have later by use of more extensive trial stress functions made some improvements to the Hashin model. A disadvantage of the Hashin method is that it is extremely difficult to use for laminate layups other than cross plies. To the authors' knowledge the method has only been applied to cross plies.

Laws *et al.* (1983) and Dvorak *et al.* (1985) have estimated stiffnesses and thermal expansion coefficients of matrix cracked composite plies by use of self consistent approximations. The self consistent stiffnesses were derived for an infinite, homogeneous, matrix cracked material. Laminate stiffnesses can then be derived by use of laminate theory and appropriate self consistent ply stiffnesses.

An alternative way to describe the mechanical behaviour of matrix cracked laminates is to apply concepts of damage mechanics. Talreja (1985, 1986) and Allen *et al.* (1987a,b) have derived models for laminate stiffnesses in terms of internal damage state parameters. In order to apply the models, it is necessary to fit certain parameters to experimental or numerical data. For a matrix cracked cross-ply laminate Lee and Allen (1989) and Allen and Lee (1991) have derived approximate relations between the internal damage state parameter and laminate stiffnesses. They determined approximate solutions for local stresses and strains by use of the principle of minimum potential energy. In this way upper bounds for laminate stiffnesses could be derived.

Gudmundson and Östlund (1992a, b, c) and Gudmundson *et al.* (1992) have shown that for dilute and infinite matrix crack densities respectively asymptotic expressions of high accuracy for the laminate stiffnesses can be derived in closed form for laminates of arbitrary layups. Asymptotic expressions for thermal expansion coefficients, strain contributions from release of residual stresses as well as average local stresses and strains were also determined. The dilute formulation is principally based on knowledge of the average crack opening displacement of a single matrix crack in the laminate as a function of the applied loading. Gudmundson and Östlund (1992a) showed that this average crack opening displacement was to a very good approximation given by an equivalent crack in an infinite, transversely isotropic medium. By use of this approximate expression for average crack opening displacements, closed form expressions could be derived for laminate stiffnesses, thermal expansion coefficients, strain contributions from release of residual stresses as well as average local ply stresses and strains (Gudmundson and Östlund, 1992a, b, c; Gudmundson *et al.*, 1992). There was no restriction concerning laminate layup or whether internal or edge micro cracks were considered. The theory was formulated for a general three-dimensional laminate. Comparisons to numerically and experimentally determined laminate stiffnesses, thermal expansion coefficients and local stresses and strains for laminates of different layups proved that the dilute theory worked extremely well up to certain matrix crack densities and that estimates based on infinite crack densities were good for matrix crack densities above certain limits (Gudmundson and Östlund, 1992a, b, c). For intermediate crack densities the differences between theory and numerically or experimentally determined data were most significant. Intermediate crack densities are here considered to be around one crack per unit thickness of a cracked ply. The reason for the discrepancies at intermediate crack densities is that interactions between neighbouring cracks become of importance. This effect is not taken into account in the dilute theory. In addition, intermediately cracked plies do still carry some load transverse to the cracks and this effect is neglected in the theoretical estimate for infinite crack densities.

Experimental observations (Highsmith and Reifsnider, 1982) have shown that the matrix crack density generally reaches a saturation state which can be characterized as an intermediate crack density. It would therefore be of advantage if the dilute and infinite theory developed by Gudmundson and Östlund (1992a, b, c) and Gudmundson *et al.* (1992) could be improved in the range of intermediate crack densities, but still keeping the nice features such as closed form expressions and applicability to laminates of arbitrary layups. In the present paper a significant improvement to the previous theory will be presented. It will be shown that the modified theory coincides with the dilute theory at small crack densities and with the infinite theory at large crack densities. At intermediate crack densities, it will be proved that the present theory is in very good agreement with numerically obtained data for angle-ply and cross-ply laminates. The key to the present theory is the judicious use of an existing analytical solution for a row of cracks in an infinite, homogeneous, isotropic medium (Benthem and Koiter, 1972; Tada *et al.*, 1973). Laminate stiffnesses, thermal expansion coefficients, strain contributions from release of residual stresses, average local stresses and strains will be expressed in closed forms only involving algebraic manipulations of known quantities such as ply stiffnesses, ply thermal expansion coefficients and micro crack densities. Both internal and edge cracks can be considered. The theory is developed for a three-dimensional laminate. As a special case the expressions for a two-dimensional laminate are derived.

2. THEORETICAL BASIS

2.1. *Three-dimensional laminate theory without transverse matrix cracks*

The stiffness and compliance tensors of a three-dimensional laminate without transverse matrix cracks have previously been derived by other researchers [see for example Pagano (1974) and Sun and Li (1988)]. However, in order to make the subsequent theoretical developments in Sections 2.2–2.3 easier to follow, three-dimensional laminate theory in a compact notation will here briefly be summarized. Laminate theory is actually a homogenization process. Instead of using the properties of each ply, a set of effective properties are employed and the laminate structure is treated as if it were made of an equivalent homogeneous material. It should be stressed that the homogenized equations (laminate theory) do have certain limitations. The existence of boundary layer effects (for composite laminates often called edge effects) cannot be modelled. The theory also breaks down when characteristic length scales of homogenized deformation variations are of the same order as microstructural dimensions (ply thicknesses for composite laminates). There are two basic tasks of a laminate theory, (1) to establish the relations between ply material properties (such as compliances and thermal expansion coefficients) and the effective properties, (2) to recover the ply stresses and strains from known effective stresses and strains.

A general three-dimensional thick laminate without matrix cracks is considered here. The laminate consists of N different types of plies. A type of ply is defined by ply material properties, layup angle and thickness. For laminates without matrix cracks, the global average stresses $\bar{\boldsymbol{\sigma}}$ and strains $\bar{\boldsymbol{\varepsilon}}$ are defined as:

$$\left.\begin{aligned} \bar{\boldsymbol{\sigma}} &= \sum_{k=1}^{N} v^k \boldsymbol{\sigma}^k \\ \bar{\boldsymbol{\varepsilon}} &= \sum_{k=1}^{N} v^k \boldsymbol{\varepsilon}^k \\ \sum_{k=1}^{N} v^k &= 1 \end{aligned}\right\}, \tag{1}$$

where $\boldsymbol{\sigma}^k$ denotes the ply average stresses, $\boldsymbol{\varepsilon}^k$ the ply average strains and v^k the volume fraction of ply k. Throughout this paper, variables with superscript bars denote the global properties and variables with superscript letters denote ply properties. For uncracked laminates under homogeneous deformation states, the ply stresses and strains are constant within each ply. In this case, there is no difference between ply averages and local values of stresses and strains.

Two sets of coordinate systems will be employed in the present study. One is the global coordinate system with its axes X_1 and X_2 lying in the same plane as the plies and the axis X_3 perpendicular to the plane of plies. The other coordinate system is the local coordinate system for each ply with its axis Y_1 along the fibre direction, axis Y_2 perpendicular to the fibre direction but in the ply plane and axis Y_3 parallel to the axis X_3. In the following, the stresses, strains and thermal expansion coefficients will be partitioned into in-plane parts and out-of plane parts:

$$\boldsymbol{\sigma} = \begin{pmatrix} \boldsymbol{\sigma}_{\mathrm{I}} \\ \boldsymbol{\sigma}_{\mathrm{O}} \end{pmatrix}, \quad \boldsymbol{\varepsilon} = \begin{pmatrix} \boldsymbol{\varepsilon}_{\mathrm{I}} \\ \boldsymbol{\varepsilon}_{\mathrm{O}} \end{pmatrix}, \quad \boldsymbol{\alpha} = \begin{pmatrix} \boldsymbol{\alpha}_{\mathrm{I}} \\ \boldsymbol{\alpha}_{\mathrm{O}} \end{pmatrix}, \tag{2}$$

where

$$\boldsymbol{\sigma}_{\mathrm{I}} = \begin{pmatrix} \sigma_{11} \\ \sigma_{22} \\ \sigma_{12} \end{pmatrix}, \quad \boldsymbol{\varepsilon}_{\mathrm{I}} = \begin{pmatrix} \varepsilon_{11} \\ \varepsilon_{22} \\ 2\varepsilon_{12} \end{pmatrix}, \quad \boldsymbol{\alpha}_{\mathrm{I}} = \begin{pmatrix} \alpha_{11} \\ \alpha_{22} \\ 2\alpha_{12} \end{pmatrix} \tag{3}$$

are in-plane stresses, strains and thermal expansion coefficients and

$$\boldsymbol{\sigma}_{\mathrm{O}} = \begin{pmatrix} \sigma_{33} \\ \sigma_{13} \\ \sigma_{23} \end{pmatrix}, \quad \boldsymbol{\varepsilon}_{\mathrm{O}} = \begin{pmatrix} \varepsilon_{33} \\ 2\varepsilon_{13} \\ 2\varepsilon_{23} \end{pmatrix}, \quad \boldsymbol{\alpha}_{\mathrm{O}} = \begin{pmatrix} \alpha_{33} \\ 2\alpha_{13} \\ 2\alpha_{23} \end{pmatrix}, \tag{4}$$

are out-of-plane stresses, strains and termal expansion coefficients. Using these notations, the relationship between global effective stresses and strains reads:

$$\begin{aligned} \bar{\boldsymbol{\varepsilon}} = \begin{pmatrix} \bar{\boldsymbol{\varepsilon}}_{\mathrm{I}} \\ \bar{\boldsymbol{\varepsilon}}_{\mathrm{O}} \end{pmatrix} &= \bar{\mathbf{S}}\bar{\boldsymbol{\sigma}} + \bar{\boldsymbol{\alpha}}\Delta T \\ &= \begin{pmatrix} \bar{\mathbf{S}}_{\mathrm{II}} & \bar{\mathbf{S}}_{\mathrm{IO}} \\ (\bar{\mathbf{S}}_{\mathrm{IO}})^{\mathrm{T}} & \bar{\mathbf{S}}_{\mathrm{OO}} \end{pmatrix} \begin{pmatrix} \bar{\boldsymbol{\sigma}}_{\mathrm{I}} \\ \bar{\boldsymbol{\sigma}}_{\mathrm{O}} \end{pmatrix} + \begin{pmatrix} \bar{\boldsymbol{\alpha}}_{\mathrm{I}} \\ \bar{\boldsymbol{\alpha}}_{\mathrm{O}} \end{pmatrix} \Delta T. \end{aligned} \tag{5}$$

Similarly, the relation between the ply stresses and strains can be written as

$$\begin{aligned} \boldsymbol{\varepsilon}^k = \begin{pmatrix} \boldsymbol{\varepsilon}_{\mathrm{I}}^k \\ \boldsymbol{\varepsilon}_{\mathrm{O}}^k \end{pmatrix} &= \mathbf{S}^k(\boldsymbol{\sigma}^k - \boldsymbol{\sigma}^{k(r)}) + \boldsymbol{\alpha}^k\Delta T \\ &= \begin{pmatrix} \mathbf{S}_{\mathrm{II}}^k & \mathbf{S}_{\mathrm{IO}}^k \\ (\mathbf{S}_{\mathrm{IO}}^k)^{\mathrm{T}} & \mathbf{S}_{\mathrm{OO}}^k \end{pmatrix} \begin{pmatrix} \boldsymbol{\sigma}_{\mathrm{I}}^k - \boldsymbol{\sigma}_{\mathrm{I}}^{k(r)} \\ \boldsymbol{\sigma}_{\mathrm{O}}^k \end{pmatrix} + \begin{pmatrix} \boldsymbol{\alpha}_{\mathrm{I}}^k \\ \boldsymbol{\alpha}_{\mathrm{O}}^k \end{pmatrix} \Delta T, \end{aligned} \tag{6}$$

where $\boldsymbol{\sigma}^{k(r)}$ denote residual stresses which may be present due to other reasons than thermal mismatch, for example chemical shrinkage during the manufacturing process. In eqns (5), (6), $\bar{\mathbf{S}}$ and $\mathbf{S}^k$ are the 6×6 effective and ply compliances respectively and the superscript T indicates the transpose of a matrix. The compliances in eqns (5), (6) have been divided into 3×3 sub-matrices, $\bar{\mathbf{S}}_t$ and $\mathbf{S}_t^k$ (t = II, IO, OO). The effective compliance tensor $\bar{\mathbf{S}}$ and the effective thermal expansion tensor $\bar{\boldsymbol{\alpha}}$ are still to be defined. It should be pointed out that due to equilibrium, the out-of-plane residual stresses and the volume average of in-plane residual stresses do vanish.

From the compatibility and equilibrium conditions in the laminate, the following relations result:

$$\boldsymbol{\varepsilon}_{\mathrm{I}}^k = \bar{\boldsymbol{\varepsilon}}_{\mathrm{I}}, \quad \boldsymbol{\sigma}_{\mathrm{O}}^k = \bar{\boldsymbol{\sigma}}_{\mathrm{O}}. \tag{7}$$

Thus from eqns (6), (7)

$$\boldsymbol{\sigma}_{\mathrm{I}}^{k} = (\mathbf{S}_{\mathrm{II}}^{k})^{-1}(\bar{\boldsymbol{\varepsilon}}_{\mathrm{I}} - \mathbf{S}_{\mathrm{IO}}^{k}\bar{\boldsymbol{\sigma}}_{\mathrm{O}} - \boldsymbol{\alpha}_{\mathrm{I}}^{k}\Delta T) + \boldsymbol{\sigma}_{\mathrm{I}}^{k(r)}. \tag{8}$$

An application of eqn (1) and a rearrangement of the resulting equation yield

$$\bar{\boldsymbol{\varepsilon}}_{\mathrm{I}} = \bar{\mathbf{S}}_{\mathrm{II}}\bar{\boldsymbol{\sigma}}_{\mathrm{I}} + \bar{\mathbf{S}}_{\mathrm{IO}}\bar{\boldsymbol{\sigma}}_{\mathrm{O}} + \bar{\boldsymbol{\alpha}}_{\mathrm{I}}\Delta T, \tag{9}$$

where

$$\left.\begin{aligned} \bar{\mathbf{S}}_{\mathrm{II}} &= \left[\sum_{k=1}^{N} v^{k}(\mathbf{S}_{\mathrm{II}}^{k})^{-1}\right]^{-1} \\ \bar{\mathbf{S}}_{\mathrm{IO}} &= \bar{\mathbf{S}}_{\mathrm{II}}\left[\sum_{k=1}^{N} v^{k}(\mathbf{S}_{\mathrm{II}}^{k})^{-1}\mathbf{S}_{\mathrm{IO}}^{k}\right] \\ \bar{\boldsymbol{\alpha}}_{\mathrm{I}} &= \bar{\mathbf{S}}_{\mathrm{II}}\left[\sum_{k=1}^{N} v^{k}(\mathbf{S}_{\mathrm{II}}^{k})^{-1}\boldsymbol{\alpha}_{\mathrm{I}}^{k}\right] \end{aligned}\right\}. \tag{10}$$

In eqn (9), the fact that the volume average of residual stresses vanishes has been utilized. Similarly, the following equation results from eqns (6), (7):

$$\boldsymbol{\varepsilon}_{\mathrm{O}}^{k} = (\mathbf{S}_{\mathrm{IO}}^{k})^{\mathrm{T}}(\boldsymbol{\sigma}_{\mathrm{I}}^{k} - \boldsymbol{\sigma}_{\mathrm{I}}^{k(r)}) + \mathbf{S}_{\mathrm{OO}}^{k}\bar{\boldsymbol{\sigma}}_{\mathrm{O}} + \boldsymbol{\alpha}_{\mathrm{O}}^{k}\Delta T. \tag{11}$$

An application of eqn (1) yields

$$\bar{\boldsymbol{\varepsilon}}_{\mathrm{O}} = (\bar{\mathbf{S}}_{\mathrm{IO}})^{\mathrm{T}}\bar{\boldsymbol{\sigma}}_{\mathrm{I}} + \bar{\mathbf{S}}_{\mathrm{OO}}\bar{\boldsymbol{\sigma}}_{\mathrm{O}} + \bar{\boldsymbol{\alpha}}_{\mathrm{O}}\Delta T, \tag{12}$$

where

$$\left.\begin{aligned} \bar{\mathbf{S}}_{\mathrm{OO}} &= (\bar{\mathbf{S}}_{\mathrm{IO}})^{\mathrm{T}}(\bar{\mathbf{S}}_{\mathrm{II}})^{-1}\bar{\mathbf{S}}_{\mathrm{IO}} + \sum_{k=1}^{N} v^{k}[\mathbf{S}_{\mathrm{OO}}^{k} - (\mathbf{S}_{\mathrm{IO}}^{k})^{\mathrm{T}}(\mathbf{S}_{\mathrm{II}}^{k})^{-1}\mathbf{S}_{\mathrm{IO}}^{k}] \\ \bar{\boldsymbol{\alpha}}_{\mathrm{O}} &= (\bar{\mathbf{S}}_{\mathrm{IO}})^{\mathrm{T}}(\bar{\mathbf{S}}_{\mathrm{II}})^{-1}\bar{\boldsymbol{\alpha}}_{\mathrm{I}} + \sum_{k=1}^{N} v^{k}[\boldsymbol{\alpha}_{\mathrm{O}}^{k} - (\mathbf{S}_{\mathrm{IO}}^{k})^{\mathrm{T}}(\mathbf{S}_{\mathrm{II}}^{k})^{-1}\boldsymbol{\alpha}_{\mathrm{I}}^{k}] \end{aligned}\right\}. \tag{13}$$

In summary, eqns (10), (13) establish the relations between the local material properties and the effective laminate properties. The ply stresses and strains can be recovered from eqns (7), (8), (11).

It should be pointed out that the equations derived above yield as a special case the effective thermoelastic parameters given by the standard two-dimensional laminate theory. In this case, $\bar{\boldsymbol{\sigma}}_{\mathrm{O}} = \mathbf{0}$ and usually only the in-plane properties are considered.

2.2. *Thermoelastic properties for composite laminates containing transverse matrix cracks*

A general three-dimensional thick composite laminate containing transverse matrix cracks is considered (see Fig. 1). The ply material properties and the number of transverse

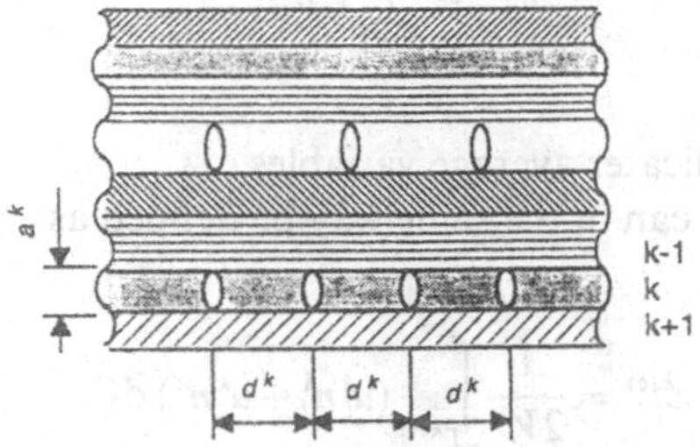

Fig. 1. A general three-dimensional laminate with micro cracks in ply k.

matrix cracks in each ply are assumed to be known. The matrix crack density in a typical ply k is denoted as ρ^k and is in this paper defined as the ratio between ply thickness and average distance between micro cracks

$$\rho^k = a^k/d_k. \tag{14}$$

The parameters in eqn (14) are defined in Fig. 1 in which a case of equidistant cracks is presented.

In the homogenization procedure for cracked laminates discussed below, a representative volume V [cf. Hill (1963)] which is large in comparison with ply thicknesses and distances between matrix cracks is implicitly considered. On the outer boundary Γ^{out} of V, displacements or tractions which are consistent with a homogeneous deformation field are prescribed. The displacement or traction boundary conditions will induce a surface layer effect, but within V a macroscopically homogeneous state will develop. Concerning volume averages, the effects of the surface layer can be made negligible by taking V large enough. In the analysis below, the various considered averages and effective properties should be interpreted in this sense.

The terms effective and average strains are often interchangeably referring to the same properties. When matrix cracks occur, however, these terms do have different meanings. In the present paper, effective strains are the strains which would be measured on a global scale and the average strains are the averages of strains experienced by the material in different plies. The difference between effective and average strains is due to the contribution from crack opening displacements. The global effective strains are defined as [cf. Hill (1963)]

$$\bar{\varepsilon}_{ij}^{(e)} = \frac{1}{2V}\int_{\Gamma^{out}} (u_i n_j + u_j n_i)\,d\Gamma, \tag{15}$$

where u_i denotes the displacements, n_i the unit normal vector on Γ^{out} (the outer boundary surface of a volume V which is large in comparison to distances between cracks and ply thicknesses) and the superscript (e) the effective variables. For stresses, there is no distinction between global effective and average stresses. This follows immediately from the relation between global effective and average stresses given by Hill (1963) [see also the review by Kachanov (1992)]. The global average stresses are defined as

$$\bar{\sigma}_{ij}^{(a)} = \sum_{k=1}^{N} v^k \sigma_{ij}^{k(a)}, \tag{16}$$

where the superscript (a) indicates average variables.

The ply effective strains can in the same way be defined as

$$\varepsilon_{ij}^{k(e)} = \frac{1}{2V^k} \int_{\Gamma^{kout}} (u_i^k n_j^k + u_j^k n_i^k)\, d\Gamma, \tag{17}$$

where V^k is the volume of ply k within V and the integral is only performed on the outer boundary surfaces of ply k. It is obvious that

$$\frac{1}{2V} \int_{\Gamma^{out}} (u_i n_j + u_j n_i)\, d\Gamma = \sum_{k=1}^{N} v^k \left[\frac{1}{2V^k} \int_{\Gamma^{kout}} (u_i^k n_j^k + u_j^k n_i^k)\, d\Gamma \right]. \tag{18}$$

Equations (15), (17), (18) thus imply that

$$\bar{\varepsilon}_{ij}^{(e)} = \sum_{k=1}^{N} v^k \varepsilon_{ij}^{k(e)}. \tag{19}$$

Equations (16), (19) establish two fundamental relations for a laminate theory taking matrix cracks into account.

By an application of the divergence theorem, the integral for the effective ply strains can be divided into two parts as

$$\varepsilon_{ij}^{k(e)} = \frac{1}{2V^k} \int_{V^k} (u_{i,j}^k + u_{j,i}^k)\, dV_k - \frac{1}{2V^k} \int_{\Gamma^{kc}} (u_i^k n_j^k + u_j^k n_i^k)\, d\Gamma, \tag{20}$$

where Γ^{kc} denotes the matrix crack surfaces in ply k and the positive normal directions are defined in Fig. 2. The first integral in eqn (20) is the ply average strain:

$$\varepsilon_{ij}^{k(a)} = \frac{1}{2V^k} \int_{V^k} (u_{j,i}^k + u_{i,j}^k)\, d\Gamma. \tag{21}$$

Since the normal vector on crack surfaces is constant, the second integral in eqn (20) can be evaluated as

$$\begin{aligned} \Delta\varepsilon_{ij}^k &= \frac{-1}{2V^k} \int_{\Gamma^{kc}} (u_i^k n_j^k + u_j^k n_i^k)\, d\Gamma \\ &= \frac{\rho^k}{2a^k} (\Delta\bar{u}_i^k n_j^{k(-)} + \Delta\bar{u}_j^k n_i^{k(-)}), \end{aligned} \tag{22}$$

where

$$\Delta\bar{u}_i^k = \frac{1}{a^k}\int_0^{a^k}(u_i^{k(+)} - u_i^{k(-)})\,da^k = \frac{1}{a^k}\int_0^{a^k}\Delta u_i^k\,da^k \tag{23}$$

is the average crack opening displacement for ply k and $\Delta\varepsilon_{ij}^k$ the average incremental strains due to crack opening displacements. In eqns (22), (23), the superscripts $(+)$ and $(-)$ denote the upper and lower crack surfaces respectively. The ply effective strains can thus be expressed as

$$\varepsilon_{ij}^{k(e)} = \varepsilon_{ij}^{k(a)} + \Delta\varepsilon_{ij}^k, \tag{24}$$

where the average ply strain $\varepsilon_{ij}^{k(a)}$ is given by eqn (21) and the strain increment due to matrix cracks $\Delta\varepsilon_{ij}^k$ by eqn (22). Expressions like eqn (24) have previously been derived and applied

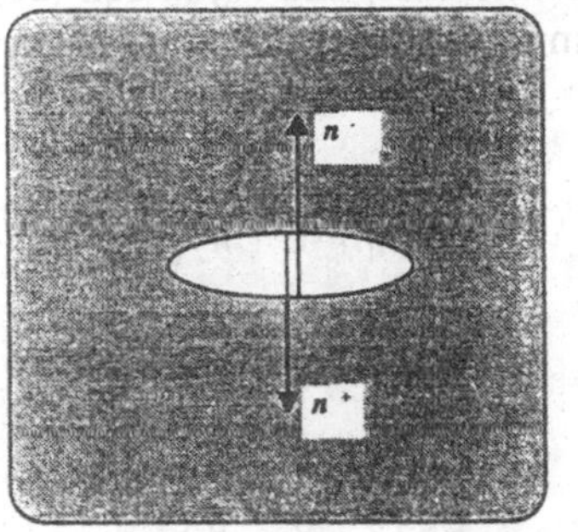

Fig. 2. Definition of the normal vectors on crack surfaces.

by other researchers, see for example the review on the effective elastic properties of cracked solids by Kachanov (1992).

The task is now to derive expressions for effective global strains, effective and average ply strains as well as average global and ply stresses for the laminate under the action of certain loading systems. From this information, effective global properties, e.g. effective compliances, effective thermal expansion coefficients and the contribution to the effective global strains from release of residual stresses, can be obtained. For this purpose, the laminate structure with matrix cracks is subjected to prescribed effective in-plane strains $\bar{\varepsilon}_{\mathrm{I}}^*$ and out-of-plane stresses $\bar{\sigma}_{\mathrm{O}}^*$ as well as a temperature change ΔT^*. In addition, a residual stress state $\sigma_{\mathrm{I}}^{k(r)}$ $(k = 1, 2, \ldots, N)$ is assumed to exist prior to micro cracking. Readers may ask why such a particular loading system ($\bar{\varepsilon}_{\mathrm{I}}^*$ and $\bar{\sigma}_{\mathrm{O}}^*$) has been chosen. There are basically two reasons. First, from compatibility and equilibrium the ply effective in-plane strains and ply average out-of-plane stresses are immediately defined. This simplifies the required algebraic manipulations in determination of thermoelastic properties. Secondly and most importantly, accurate estimations of average crack opening displacements can be derived for this particular loading system ($\bar{\varepsilon}_{\mathrm{I}}^*$, $\bar{\sigma}_{\mathrm{O}}^*$, ΔT^* and $\sigma_{\mathrm{I}}^{k(r)}$). This will be further discussed in Section 2.3.

Since only linear elasticity is considered, the problem can be solved by a superposition of two problems. In the first problem, the laminate without matrix cracks loaded by prescribed $\bar{\varepsilon}_{\mathrm{I}}^{*}$, $\bar{\sigma}_{\mathrm{O}}^{*}$, ΔT^{*} and $\sigma_{\mathrm{I}}^{k(r)}$ ($k = 1, 2, \ldots, N$) is considered. This problem can be solved by application of the ordinary laminate theory (see Section 2.1). In particular, the ply stresses can be expressed in terms of the prescribed loading. The tractions on prospective crack surfaces can be written as

$$\boldsymbol{\tau}^{k} = \mathbf{A}^{k}\bar{\boldsymbol{\varepsilon}}_{\mathrm{I}}^{*} + \mathbf{B}^{k}\bar{\boldsymbol{\sigma}}_{\mathrm{O}}^{*} + (\mathbf{C}^{k} - \mathbf{A}^{k}\bar{\boldsymbol{\alpha}}_{\mathrm{I}})\Delta\mathbf{T}^{*} + \boldsymbol{\tau}^{k(r)}, \tag{25}$$

where

$$\left.\begin{aligned} \mathbf{A}^{k} &= \mathbf{N}_{\mathrm{I}}^{k}(\mathbf{S}_{\mathrm{II}}^{k})^{-1} \\ \mathbf{B}^{k} &= -\mathbf{N}_{\mathrm{I}}^{k}(\mathbf{S}_{\mathrm{II}}^{k})^{-1}\mathbf{S}_{\mathrm{IO}}^{k} + \mathbf{N}_{\mathrm{O}}^{k} \\ \mathbf{C}^{k} &= \mathbf{A}^{k}(\bar{\boldsymbol{\alpha}}_{\mathrm{I}} - \boldsymbol{\alpha}_{\mathrm{I}}^{k}) \\ \boldsymbol{\tau}^{k(r)} &= \mathbf{N}_{\mathrm{I}}^{k}\boldsymbol{\sigma}_{\mathrm{I}}^{k(r)} \end{aligned}\right\}, \tag{26}$$

and $\bar{\boldsymbol{\alpha}}_{\mathrm{I}}$ is given in eqn (10). The matrices $\mathbf{N}_{\mathrm{I}}^{k}$ and $\mathbf{N}_{\mathrm{O}}^{k}$ in eqn (26) represent the unit normal vector n_j^k on the crack surfaces in ply k [the superscript $(-)$ according to eqn (22) is here omitted]:

$$\mathbf{N}_{\mathrm{I}}^{k} = \begin{pmatrix} n_1^k & 0 & n_2^k \\ 0 & n_2^k & n_1^k \\ 0 & 0 & 0 \end{pmatrix}, \tag{27}$$

$$\mathbf{N}_{\mathrm{O}}^{k} = \begin{pmatrix} 0 & 0 & 0 \\ 0 & 0 & 0 \\ 0 & n_1^k & n_2^k \end{pmatrix}. \tag{28}$$

According to the definition of the coordinate systems, the normal vector on crack surfaces always lies in the local 1, 2-plane, i.e. $n_3^k = 0$.

In the second problem, the tractions on crack surfaces resulting from the first problem [eqn (25)] are released under vanishing effective global in-plane strains ($\bar{\boldsymbol{\varepsilon}}_{\mathrm{I}}^{(e)} = \mathbf{0}$) and average global out-of-plane stresses ($\bar{\boldsymbol{\sigma}}_{\mathrm{O}}^{(a)} = \mathbf{0}$). The solution to this problem will enable the determination of average crack opening displacements. The average crack opening displacements in a typical ply k will in a general case linearly depend on all crack surface tractions. Thus

$$\Delta\bar{\mathbf{u}}^{k} = a^{k}\sum_{i=1}^{N}\boldsymbol{\beta}^{ki}\boldsymbol{\tau}^{i}, \tag{29}$$

where $\boldsymbol{\beta}^{ki}$ ($k, i = 1, 2, \ldots, N$) are 3×3 matrices which depend on laminate layup, ply properties and matrix crack densities. The determination of these matrices will be discussed in Section 2.3. The average increment strains due to crack opening displacements for ply k can, according to eqns (22), (25)–(29), be written as

$$\Delta\varepsilon_{\mathrm{I}}^{k} = \rho^k/a^k(\mathbf{N}_{\mathrm{I}}^k)^{\mathrm{T}}\Delta\bar{\mathbf{u}}^k$$
$$= \rho^k(\mathbf{N}_{\mathrm{I}}^k)^{\mathrm{T}} \sum_{i=1}^{N} \beta^{ki}[\mathbf{A}^i\bar{\varepsilon}_{\mathrm{I}}^* + \mathbf{B}^i\bar{\sigma}_{\mathrm{O}}^* + (\mathbf{C}^i - \mathbf{A}^i\bar{\alpha}_{\mathrm{I}})\Delta\mathbf{T}^* + \tau^{i(r)}], \tag{30a}$$

$$\Delta\varepsilon_{\mathrm{O}}^{k} = \rho^k/a^k(\mathbf{N}_{\mathrm{O}}^k)^{\mathrm{T}}\Delta\bar{\mathbf{u}}^k$$
$$= \rho^k(\mathbf{N}_{\mathrm{O}}^k)^{\mathrm{T}} \sum_{k=1}^{N} \beta^{ki}[\mathbf{A}^i\bar{\varepsilon}_{\mathrm{I}}^* + \mathbf{B}^i\bar{\sigma}_{\mathrm{O}}^* + (\mathbf{C}^i - \mathbf{A}^i\bar{\alpha}_{\mathrm{I}})\Delta\mathbf{T}^* + \tau^{i(r)}]. \tag{30b}$$

The effective ply strains in ply k (problem 1 + problem 2) are given by

$$\varepsilon_{(c)}^{k(e)} = \varepsilon_{(c)}^{k(a)} + \Delta\varepsilon^k, \tag{31}$$

where the subscript (c) denotes the variables for the cracked laminate (problem 1 + problem 2) and $\Delta\varepsilon^k$ is given in eqn (30). In addition, the relations between stresses and strains for a laminate with matrix cracks can be expressed as

$$\left.\begin{aligned} \bar{\varepsilon}_{(c)}^{(e)} &= \bar{\mathbf{S}}_{(c)}\bar{\sigma}_{(c)}^{(a)} + \bar{\alpha}_{(c)}\Delta T + \bar{\varepsilon}_{(c)}^{(r)} \\ \varepsilon_{(c)}^{k(a)} &= \mathbf{S}^k(\sigma_{(c)}^{k(a)} - \sigma^{k(r)}) + \alpha^k\Delta T \end{aligned}\right\}, \tag{32}$$

where the effective compliance ($\bar{\mathbf{S}}_{(c)}$), thermal expansion vectors ($\bar{\alpha}_{(c)}$) and the global effective strains due to release of residual stresses ($\bar{\varepsilon}_{(c)}^{(r)}$) remain to be defined. In eqn (32), the out-of-plane residual stresses are zero ($\sigma_{\mathrm{O}}^{k(r)} = \mathbf{0}$) due to equilibrium and the in-plane residual stresses $\sigma_{\mathrm{I}}^{k(r)}$ are assumed to be known. Furthermore, compatibility and equilibrium conditions for the laminate with micro cracks under the present loading system read

$$\left.\begin{aligned} \varepsilon_{\mathrm{I(c)}}^{k(e)} &= \bar{\varepsilon}_{\mathrm{I(c)}}^{(e)} = \bar{\varepsilon}_{\mathrm{I}}^* \\ \sigma_{\mathrm{O(c)}}^{k(a)} &= \bar{\sigma}_{\mathrm{O(c)}}^{(a)} = \bar{\sigma}_{\mathrm{O}}^* \end{aligned}\right\}. \tag{33}$$

By a substitution of eqns (31), (33) into eqn (32b) and a rearrangement of the resulting equations, the following expressions can be derived:

$$\sigma_{\mathrm{I(c)}}^{k(a)} = (\mathbf{S}_{\mathrm{II}}^k)^{-1}[(\bar{\varepsilon}_{\mathrm{I}}^* - \Delta\varepsilon_{\mathrm{I}}^k) - \mathbf{S}_{\mathrm{IO}}^k\bar{\sigma}_{\mathrm{O}}^* - \alpha_{\mathrm{I}}^k\Delta T] + \sigma_{\mathrm{I}}^{k(r)}, \tag{34a}$$

$$\begin{aligned} \varepsilon_{\mathrm{O(c)}}^{k(e)} &= \varepsilon_{\mathrm{O(c)}}^{k(a)} + \Delta\varepsilon_{\mathrm{O}}^k \\ &= \{(\mathbf{S}_{\mathrm{IO}}^k)^{\mathrm{T}}(\mathbf{S}_{\mathrm{II}}^k)^{-1}(\bar{\varepsilon}_{\mathrm{I}}^* - \Delta\varepsilon_{\mathrm{I}}^k) + [\mathbf{S}_{\mathrm{OO}}^k - (\mathbf{S}_{\mathrm{IO}}^k)^{\mathrm{T}}(\mathbf{S}_{\mathrm{II}}^k)^{-1}\mathbf{S}_{\mathrm{IO}}^k]\bar{\sigma}_{\mathrm{O}}^* \\ &\quad + [\alpha_{\mathrm{O}}^k - (\mathbf{S}_{\mathrm{IO}}^k)^{\mathrm{T}}(\mathbf{S}_{\mathrm{II}}^k)^{-1}\alpha_{\mathrm{I}}^k]\Delta T\} + \Delta\varepsilon_{\mathrm{O}}^k. \end{aligned} \tag{34b}$$

With the help of eqns (16), (19), (26), (30)–(34), the effective compliance tensor ($\bar{\mathbf{S}}_{(c)}$), the thermal expansion coefficients ($\bar{\alpha}_{(c)}$) and the global effective strains due to release of residual stresses ($\bar{\varepsilon}_{(c)}^{(r)}$) can be derived in the same way as was summarized in Section 2.1 for the ordinary laminate theory. The resulting expressions are given below:

$$\bar{\mathbf{S}}_{\mathrm{II(c)}} = \left((\bar{\mathbf{S}}_{\mathrm{II}})^{-1} - \sum_{k=1}^{N} v^k \rho^k (\mathbf{A}^k)^{\mathrm{T}} \sum_{i=1}^{N} \boldsymbol{\beta}^{ki} \mathbf{A}^i \right)^{-1}, \tag{35a}$$

$$\bar{\mathbf{S}}_{\mathrm{IO(c)}} = \bar{\mathbf{S}}_{\mathrm{II(c)}} \left[(\bar{\mathbf{S}}_{\mathrm{II}})^{-1} \bar{\mathbf{S}}_{\mathrm{IO}} + \sum_{k=1}^{N} v^k \rho^k (\mathbf{A}^k)^{\mathrm{T}} \sum_{i=1}^{N} \boldsymbol{\beta}^{ki} \mathbf{B}^i \right], \tag{35b}$$

$$\bar{\mathbf{S}}_{\mathrm{OO(c)}} = (\bar{\mathbf{S}}_{\mathrm{IO(c)}})^{\mathrm{T}} (\bar{\mathbf{S}}_{\mathrm{II(c)}})^{-1} \bar{\mathbf{S}}_{\mathrm{IO(c)}} - (\bar{\mathbf{S}}_{\mathrm{IO}})^{\mathrm{T}} (\bar{\mathbf{S}}_{\mathrm{II}})^{-1} \bar{\mathbf{S}}_{\mathrm{IO}} + \bar{\mathbf{S}}_{\mathrm{OO}} + \sum_{k=1}^{N} v^k \rho^k (\mathbf{B}^k)^{\mathrm{T}} \sum_{i=1}^{N} \boldsymbol{\beta}^{ki} \mathbf{B}^i, \tag{35c}$$

$$\bar{\boldsymbol{\alpha}}_{\mathrm{I(c)}} = \bar{\boldsymbol{\alpha}}_{\mathrm{I}} + \bar{\mathbf{S}}_{\mathrm{II(c)}} \sum_{k=1}^{N} v^k \rho^k (\mathbf{A}^k)^{\mathrm{T}} \sum_{i=1}^{N} \boldsymbol{\beta}^{ki} \mathbf{C}^i, \tag{35d}$$

$$\bar{\boldsymbol{\alpha}}_{\mathrm{O(c)}} = \bar{\boldsymbol{\alpha}}_{\mathrm{O}} + (\bar{\mathbf{S}}_{\mathrm{IO(c)}})^{\mathrm{T}} (\bar{\mathbf{S}}_{\mathrm{II(c)}})^{-1} (\bar{\boldsymbol{\alpha}}_{\mathrm{I(c)}} - \bar{\boldsymbol{\alpha}}_{\mathrm{I}}) + \sum_{k=1}^{N} v^k \rho^k (\mathbf{B}^k)^{\mathrm{T}} \sum_{i=1}^{N} \boldsymbol{\beta}^{ki} \mathbf{C}^i, \tag{35e}$$

$$\bar{\boldsymbol{\varepsilon}}^{(r)}_{\mathrm{I(c)}} = \bar{\mathbf{S}}_{\mathrm{II(c)}} \sum_{k=1}^{N} v^k \rho^k (\mathbf{A}^k)^{\mathrm{T}} \sum_{i=1}^{N} \boldsymbol{\beta}^{ki} \boldsymbol{\tau}^{i(r)}, \tag{35f}$$

$$\bar{\boldsymbol{\varepsilon}}^{(r)}_{\mathrm{O(c)}} = (\bar{\mathbf{S}}_{\mathrm{IO(c)}})^{\mathrm{T}} (\bar{\mathbf{S}}_{\mathrm{II(c)}})^{-1} \bar{\boldsymbol{\varepsilon}}^{(r)}_{\mathrm{I(c)}} + \sum_{k=1}^{N} v^k \rho^k (\mathbf{B}^k)^{\mathrm{T}} \sum_{i=1}^{N} \boldsymbol{\beta}^{ki} \boldsymbol{\tau}^{i(r)}. \tag{35g}$$

In conclusion, eqn (35) defines exact expressions for the thermoelastic properties of composite laminates containing matrix cracks, provided that the $\boldsymbol{\beta}^{ki}$ matrices are known. Under a given loading system, the global effective strains ($\bar{\boldsymbol{\varepsilon}}^{(e)}_{(c)}$) and the global average stresses ($\bar{\boldsymbol{\sigma}}^{(a)}_{(c)}$) for the cracked laminate can be obtained from eqn (32). These results provide the global effective in-plane strains ($\bar{\boldsymbol{\varepsilon}}^*_{\mathrm{I}} = \bar{\boldsymbol{\varepsilon}}^{(e)}_{\mathrm{I(c)}}$) and the global average out-of-plane stresses ($\bar{\boldsymbol{\sigma}}^*_{\mathrm{O}} = \bar{\boldsymbol{\sigma}}^{(a)}_{\mathrm{O(c)}}$). Finally, eqns (31)–(34) can be applied to recover the ply average stresses and strains. Thus, the laminate theory taking matrix cracks into account is complete. It is observed that the thermoelastic properties of a composite laminate containing micro cracks can be expressed in terms of thermoelastic properties for an uncracked laminate, the micro crack densities and the $\boldsymbol{\beta}^{ki}$ matrices which relate the average crack opening displacements to the crack surface tractions. The determination of these $\boldsymbol{\beta}^{ki}$ matrices will be discussed in the next section.

It should be pointed out that the above equations are valid also for a two-dimensional thin laminate containing matrix cracks. In this case, $\bar{\boldsymbol{\sigma}}_{\mathrm{O}} = \mathbf{0}$ and usually only the in-plane properties are considered.

2.3. *Determination of average crack opening displacements*

The theoretical development above has shown that the thermoelastic properties as well as average local stresses and strains for a micro cracked laminate can be exactly determined provided that the exact solution for the average crack opening displacements or equivalently the $\boldsymbol{\beta}^{ki}$ matrices are known. Exact analytical solutions for the $\boldsymbol{\beta}^{ki}$ matrices are however impossible to derive except for extremely simplified cases. In order to predict thermoelastic properties of micro cracked laminates, approximate solutions for average crack opening displacements must be derived. The quality of the resulting theory therefore strongly depends on the accuracy of the approximate solutions for $\boldsymbol{\beta}^{ki}$.

In the present paper transverse matrix cracks in composite laminates are considered. Thus, the crack surfaces are parallel to the fibre direction in each ply and perpendicular to the laminate plane (the Y_1–Y_3-plane according to the definition of the local coordinate system in Section 2.1). In the case of dilute crack density ($\rho \ll 1$), Gudmundson and Östlund (1992a) showed that the average crack opening displacements in angle plies and cross plies to a surprisingly good accuracy could be determined from the well-known analytical solution of a single crack in an infinite, homogeneous transversely isotropic medium. Thus the average crack opening displacements were independent on the orientation of the neighbouring plies. The average crack opening displacements were generally found to be slightly overestimated by use of the approximate analytical solution. The effect of using approximate average crack opening displacements is however much reduced in application to the determination of thermoelastic properties [see Gudmundson and Östlund (1992a, b, c)].

The fact that the average crack opening displacements for the dilute case showed a very limited dependence on the orientation of the neighbouring plies gives hope for finding closed form but still accurate approximate solutions for nondilute matrix crack densities. If this assumption holds, then the average crack opening displacements for a row of matrix cracks in a ply should be well approximated by the average crack opening displacements for the same row of cracks in an infinite homogeneous transversely isotropic medium which has the same properties as the ply in consideration (see Fig. 3). The stress intensity factors for an infinite row of equidistant cracks in an infinite homogeneous isotropic medium under the action of uniform tractions on crack surfaces are given by Benthem and Koiter (1972) and Tada *et al.* (1973). These stress intensity factor solutions are also valid for the same crack problem in a transversely isotropic medium. The well-known relation between strain energy release rate and stress intensity factors make it possible to determine expressions for average crack opening displacements. Thus, the $\boldsymbol{\beta}^{ki}$ $(k, i = 1, 2, \ldots, N)$ matrices defined in eqn (29) can be approximately determined in this manner. It should be pointed out that the use of these approximate $\boldsymbol{\beta}^{ki}$ $(k, i = 1, 2, \ldots, N)$ matrices in the present model is the single approximation in the determination of thermoelastic properties of composite laminates containing matrix cracks. The use of the analytical solution for an infinite row of equidistant cracks implies of course that the effects of non-equidistant crack spacings cannot be covered by the present model. Experimental observations have however shown that the assumption of equidistant matrix cracks generally is a good representation of the reality [see for example Highsmith and Reifsnider (1982)]. The assumption above implies that there will be no coupling between the crack opening displacements of different plies and that the $\boldsymbol{\beta}^{ki}$ matrices must be diagonal, thus

$$\boldsymbol{\beta}^{ki} = \mathbf{0}, \quad \text{for all } k \neq i, \tag{36a}$$

$$\boldsymbol{\beta}^{kk} = \begin{pmatrix} \beta_1^k & 0 & 0 \\ 0 & \beta_2^k & 0 \\ 0 & 0 & \beta_3^k \end{pmatrix}. \tag{36b}$$

Equation (29) for the average crack opening displacements in ply k can thus be rewritten as

$$\Delta\bar{\mathbf{u}}^k = a^k \boldsymbol{\beta}^{kk} \boldsymbol{\tau}^k \quad \text{(no sum over } k\text{)}. \tag{37}$$

The diagonal components of the $\boldsymbol{\beta}^{kk}$ matrix can be expressed in closed form as will be shown below. Therefore, the thermoelastic parameters for laminates containing micro

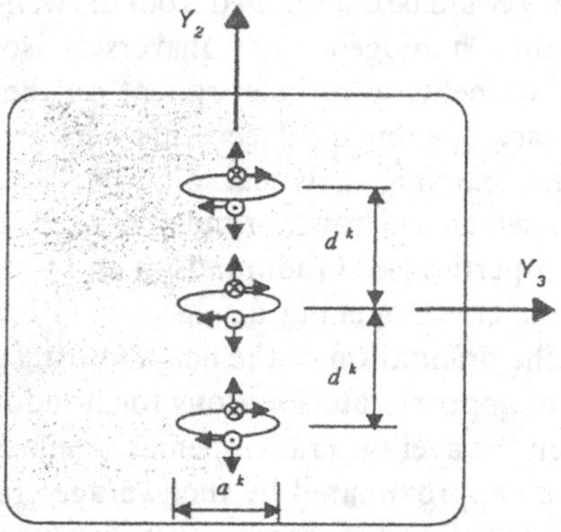

Fig. 3. An infinite row of cracks subjected to crack surface tractions in an infinite transversely isotropic plane.

cracks presented in Section 2.2. can also be written in closed form. The accuracy of the proposed model can only be checked by comparisons to numerical solutions or experimental results. In Section 3, extensive comparisons to numerical finite element results as well as experimental results will be presented.

The stress intensity factors for the crack problem shown in Fig. 3 can be expressed as

$$\left.\begin{aligned} K_{\rm I} &= f_2\tau_2^k \\ K_{\rm II} &= f_3\tau_3^k \\ K_{\rm III} &= f_1\tau_1^k \end{aligned}\right\}, \tag{38}$$

where f_i $(i = 1,2,3)$ can be found in the papers by Benthem and Koiter (1972) and Tada *et al.* (1973). Due to symmetries of the crack problem in Fig. 3, it can be shown that the effective strains in the 1–2-plane for the cracked layer are zero. Similarly, equilibrium enforces the average stress σ_{23} in the cracked layer to vanish. The effective strain and average stress state are thus in exact agreement with the loading conditions for the crack problem in Section 2.2. This is of course the reason for the particular loading system chosen for the crack problem in Section 2.2.

The well-known relation between strain energy release rate and stress intensity factors can now be applied to derive an equation for the average crack opening displacements

$$\tfrac{1}{2}a^k(\Delta\bar{\mathbf{u}}^k)^{\rm T}\boldsymbol{\tau}^k = \int_0^{a^k} [\gamma_1(f_1\tau_1^k)^2 + \gamma_2(f_2\tau_2^k)^2 + \gamma_3(f_3\tau_3^k)^2]\,{\rm d}a^k, \tag{39}$$

where

$$\left.\begin{aligned} \gamma_1 &= 1/(2G_{\rm TL}^k) \\ \gamma_2 = \gamma_3 &= (1-\nu_{\rm LT}^k\nu_{\rm TL}^k)/E_{\rm T}^k \end{aligned}\right\}. \tag{40}$$

In eqn (40), E_T^k denotes the transverse E-modulus, G_{TL}^k the out-of-plane shear modulus and ν_{TL}^k, ν_{LT}^k the Poisson ratios. Equations (37)–(39) yield relations between the coefficients β_j^k and f_j,

$$\left.\begin{aligned} \beta_1^k &= \frac{2}{(a^k)^2}\gamma_1 \int_0^{a^k} (f_1)^2 \, da^k \\ \beta_2^k &= \frac{2}{(a^k)^2}\gamma_2 \int_0^{a^k} (f_2)^2 \, da^k \\ \beta_3^k &= \frac{2}{(a^k)^2}\gamma_3 \int_0^{a^k} (f_3)^2 \, da^k \end{aligned}\right\}. \tag{41}$$

The first integral in eqn (41) was analytically evaluated. Numerical integration was performed on the other two integrals, and a curve fitting technique with an error less than 0.5% was then employed to generate the resulting expression for β_2^k and β_3^k. Finally, the diagonal components of the $\boldsymbol{\beta}^{kk}$ matrix can be expressed as

$$\left.\begin{aligned} \beta_1^k &= \frac{4}{\pi}\gamma_1 \ln[\cosh(\rho^k\pi/2)]/(\rho^k)^2 \\ \beta_2^k &= \frac{\pi}{2}\gamma_2 \sum_{j=1}^{10} a_j/(1+\rho^k)^j \\ \beta_3^k &= \frac{\pi}{2}\gamma_3 \sum_{j=1}^{9} b_j/(1+\rho^k)^{j-2} \end{aligned}\right\}, \tag{42}$$

where a_j and b_j are given in Table 1.

Table 1. Numerical parameters used in eqns (42), (43)

J	a	b	c
1	0.63666	0.63662	0.25256
2	0.51806	−0.08945	0.27079
3	0.51695	0.15653	−0.49814
4	−1.04897	0.13964	8.62962
5	8.95572	0.16463	−51.24655
6	−33.09444	0.06661	180.96305
7	74.32002	0.54819	−374.29813
8	−103.06411	−1.07983	449.59474
9	73.60337	0.45704	−286.51016
10	−20.34326	—	73.84223

In some cases, surface cracks occur in a laminate. For thin laminate structures, the effects due to surface cracks will become more important and have to be taken into account. In this case, the two-dimensional theory will be employed and only two components of the $\boldsymbol{\beta}^{kk}$ matrix (β_1^k and β_2^k) are required. These two components can be derived in the same manner as above and they can be written as

$$\left.\begin{aligned} \beta_1^{k(s)} &= \frac{8}{\pi}\gamma_1 \ln[\cosh(\rho^k\pi)]/(2\rho^k)^2 \\ \beta_2^{k(s)} &= 2(1.12)^2\left[\frac{\pi}{2}\gamma_2 \sum_{j=1}^{10} c_j/(1+\rho^k)^j\right] \end{aligned}\right\}, \tag{43}$$

where the superscript (s) indicates a surface crack and the parameters c_j are given in Table 1. It is easily shown that

$$\beta_1^{k(s)}(\rho^k) = 2\beta_1^k(2\rho^k), \tag{44a}$$

and that

$$\left.\begin{aligned} \beta_2^{k(s)} &= 2(1.12)^2\beta_2^k \quad \text{as} \quad \rho \to 0 \\ \beta_2^{k(s)} &= \beta_2^k \qquad\qquad\quad \text{as} \quad \rho \to \infty \end{aligned}\right\}. \tag{44b}$$

It is noticed that when ρ^k tends to zero, the coefficients β_j^k approach the dilute results given by Gudmundson and Östlund (1992a). For ρ^k tending to infinity, it can be shown that the results obtained by the present method are in agreement with the results from the infinite limit given by Gudmundson and Östlund (1992b). The present theory is thus in agreement with the asymptotic results for small and infinite matrix crack densities respectively. For intermediate values of ρ^k, the accuracy of the solutions obtained by the present theory has to be checked against numerical or experimental results.

3. RESULTS

In order to verify the efficiency and reliability of the present theory, a number of two- and three-dimensional problems have been studied by the present theory and compared either to finite element calculations or to experimental results presented in the literature. Two kinds of laminate systems have been considered, thin cross-ply laminates with micro cracks in one type of ply and angle-ply laminates with micro cracks in both types of plies. In the finite element calculations, periodic cells with appropriate periodic boundary conditions were employed and the FE program ABAQUS was used. A detailed description of the finite element modelling was given by Gudmundson and Östlund (1992c). The ply material properties for the laminates which were used in the verifications are presented in Table 2. Some typical results for cross-ply and angle-ply composite laminates will be presented in the following sections.

Table 2. Material properties for GFRP and CFRP unidirectional plies

Type	E_L (GPa)	E_T (GPa)	ν_{LT}	ν_{TT}	G_L (GPa)	Ply thickness	α_L $10^{-6}\,{}^\circ C^{-1}$	α_T $10^{-6}\,{}^\circ C^{-1}$
GFRP	41.7	13	0.3	0.42	3.4	0.203	6.72	29.3
CFRP	142	9.85	0.3	—	4.48	0.127	—	—

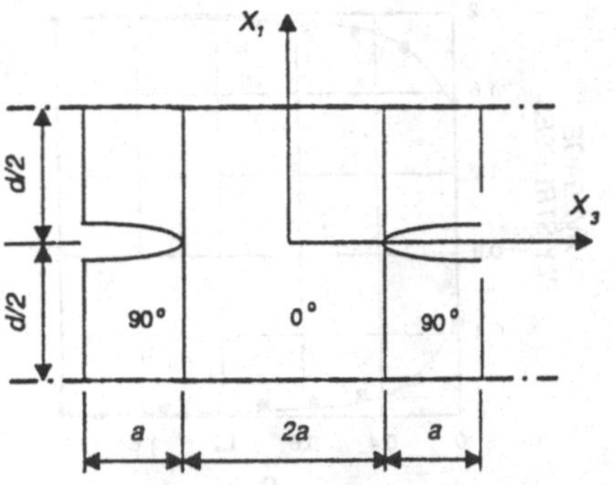

Fig. 4. A representative periodic cell for a cross-ply laminate with surface cracks.

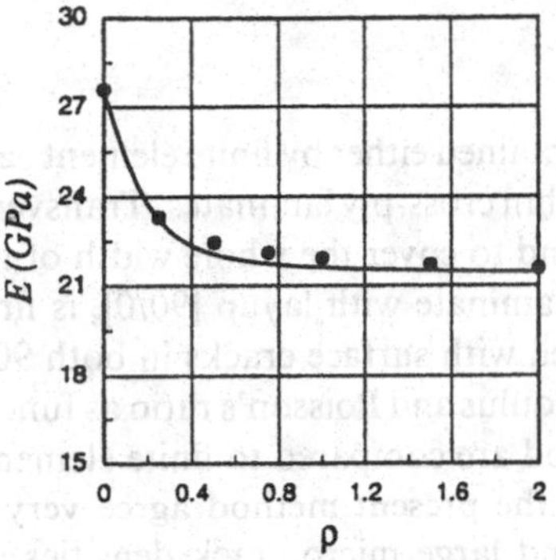

Fig. 5(a). *E*-modulus as a function of micro crack density in the 90° plies for the cross-ply laminate with surface cracks (see Fig. 4). The solid line denotes the results by the present method and the symbol the results from finite element calculations.

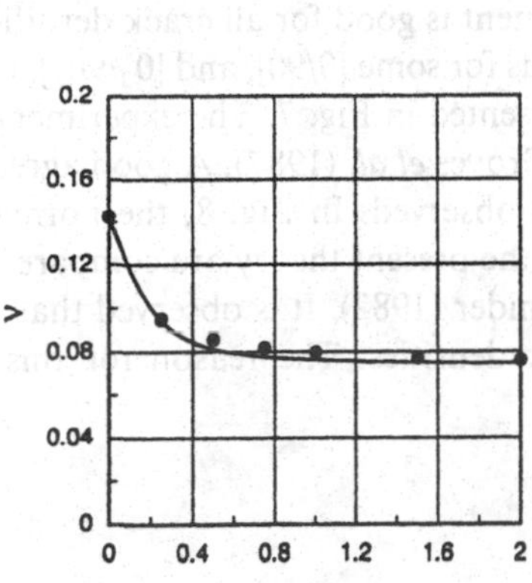

Fig. 5(b). Poisson's ratio as a function of micro crack density in the 90° plies for the cross-ply laminate with surface cracks (see Fig. 4). The solid line denotes the results by the present method and the symbol the results from finite element calculations.

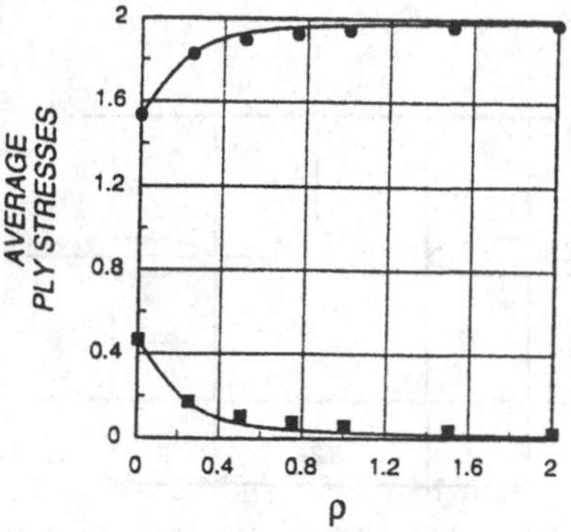

Fig. 6. Average ply stresses (in global coordinate system) as functions of micro crack density in the 90° plies for the cross-ply laminate with surface cracks (see Fig. 4). The loading is a global average stress $\bar{\sigma}_{11} = 1.0$ MPa. The solid lines denote the results by the present method and the symbols the results from finite element calculations, where ● = σ_{11} for the 0° ply and ■ = σ_{11} for the 90° plies.

3.1. *Cross-ply laminate*

Comparisons to results obtained either by finite element calculations or by experimental studies are here presented for thin cross-ply laminates. Transverse matrix cracks are assumed to exist only in the 90° plies and to cover the whole width of the plies.

A thin cross-ply GFRP laminate with layup $[90/0]_s$ is first considered. The geometry of a representative periodic cell with surface cracks in both 90° plies is shown in Fig. 4. In Figs 5(a, b), the effective *E*-modulus and Poisson's ratio as functions of micro crack densities obtained by the present method are compared to finite element calculations. It is observed that the results generated by the present method agree very well with the finite element calculations for both small and large micro crack densities. Relatively larger differences exist for intermediate matrix crack densities. However, the maximum error in this example is only about 5%. In addition, average ply stresses in the cracked laminate resulting from a unidirectional loading of $\bar{\sigma}_{11} = 1.0$ MPa has also been studied. In Fig. 6, the average ply stresses are compared to finite element calculations at discrete matrix crack densities. It is again observed that the agreement is good for all crack densities.

The normalized *E*-modulus for some $[0/90]_s$ and $[0_2/90_2]_s$ CFRP laminates as functions of micro crack densities is presented in Fig. 7. The experimental data shown in Fig. 7 are based on the results given by Groves *et al.* (1987). A good agreement between experimental and theoretical results can be observed. In Fig. 8, the normalized *E*-moduli for $[0/90_3]_s$ GFRP laminates obtained by the present theory are compared to the experimental results given by Highsmith and Reifsnider (1982). It is observed that the differences are relatively larger for larger matrix crack densities. The reason for this discrepancy could be that

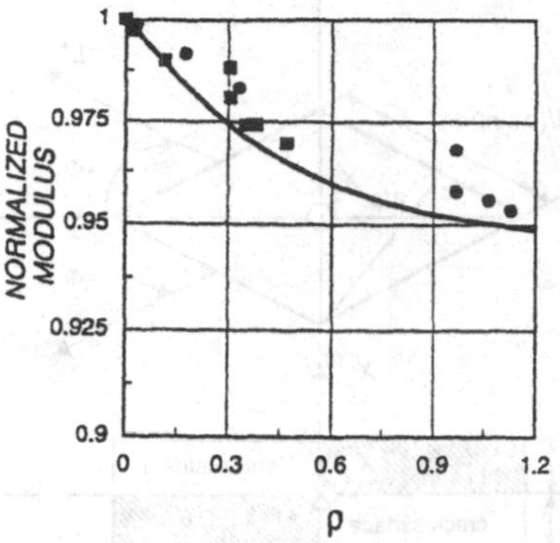

Fig. 7. Normalized E-modulus as a function of microcrack density in the 90° plies for the CFRP cross-ply laminates. The solid line denotes the prediction by the present theory and the symbols the experimental results by Groves *et al.* (1987), where ■ represents the results for $[0/90]_s$ laminates and ● for $[0_2/90_2]_s$ laminates.

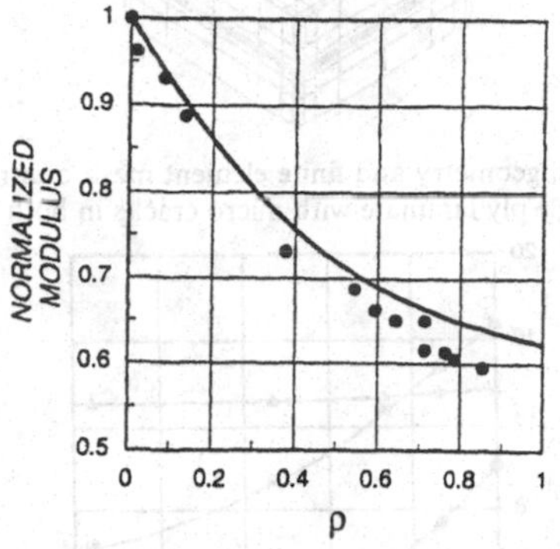

Fig. 8. Normalized E-modulus as a function of micro crack density in the 90° plies for the GFRP $[0/90_3]_s$ laminate. The solid line denotes the prediction by the present theory and the symbol the experimental results by Highsmith and Reifsnider (1982).

additional damage modes which are not included in the present theory are present in the experimental studies. However, the differences in Fig. 8 are not so large. The predictions by the present theory are still quite good.

3.2. *Thick angle ply laminate*

In this section, some thick angle-ply GFRP laminates with matrix cracks in both plies and covering the whole thickness of the laminate are investigated. In order to study the accuracy of the present method at varying layup configurations, laminates with layup $[\pm 55]_N$ and $[\pm 67.5]_N$ have been considered. Here N is assumed to be large so that three-dimensional theory can be applied. Since a laminate with layup $[\pm 45]_N$ really is a cross-ply laminate, this kind of layup is thus not included here. The finite element method has been employed to verify the efficiency and reliability of the present theory. Due to symmetry only one half of the periodic cell has been modelled by finite elements. The periodic cell geometry and the finite element mesh are illustrated in Fig. 9. In Fig. 9, the coordinate system (X_1, X_2, X_3)

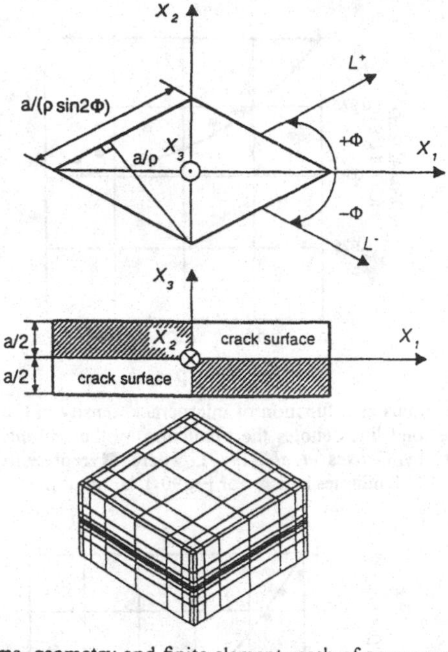

Fig. 9. Coordinate systems, geometry and finite element mesh of a representative periodic cell for a thick angle ply laminate with micro cracks in both types of plies.

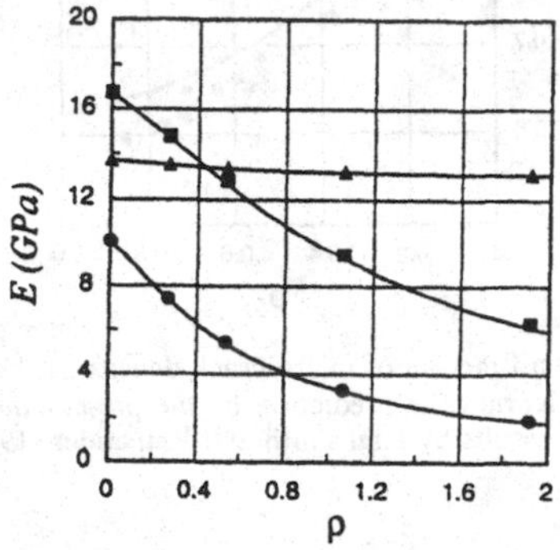

Fig. 10(a). *E*-moduli as functions of micro crack density for the angle ply ($\phi = \pm 55°$) laminate. The solid lines denote the results by the present theory, the symbols the results from finite element calculations, where ● = $\bar{E}_1$, ■ = $\bar{E}_2$ and ▲ = $\bar{E}_3$.

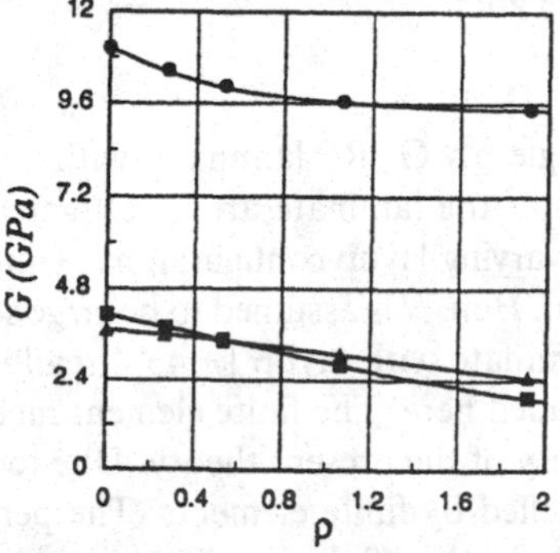

Fig. 10(b). Shear moduli as functions of micro crack density for the angle ply ($\phi = \pm 55°$) laminate. The solid lines denote the results by the present theory, the symbols the results from finite element calculations, where ● = $\bar{G}_{12}$, ■ = $\bar{G}_{13}$ and ▲ = $\bar{G}_{23}$.

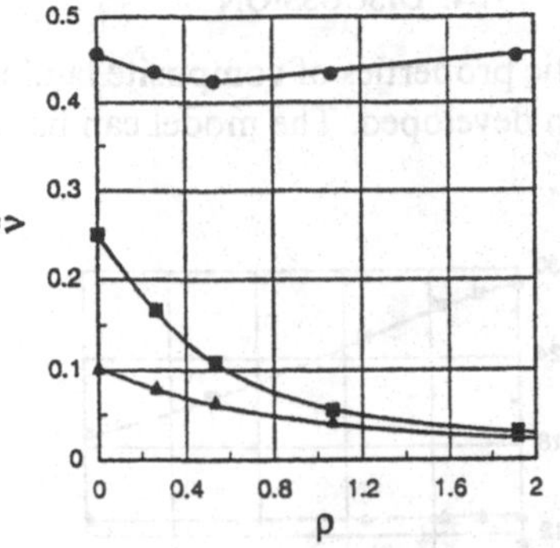

Fig. 10(c). Poisson ratios as functions of micro crack density for the angle ply ($\phi = \pm 55°$) laminate. The solid lines denote the results by the present theory, the symbols the results from finite element calculations, where ● = $\bar{\nu}_{12}$, ■ = $\bar{\nu}_{13}$ and ▲ = $\bar{\nu}_{23}$.

represents the global coordinate system. The directions L^+ and L^- are aligned to the fibre directions in the corresponding ply. The local coordinate systems are not shown in Fig. 9.

Effective global engineering constants and thermal expansion coefficients as functions of matrix crack densities are presented in Figs 10(a–d) for the $[\pm 55]_N$ laminate and in Fig. 11(a–d) for the $[\pm 67.5]_N$ laminate. In addition, the prediction of average ply stresses by the present theory has been studied for a unidirectional loading case, $\bar{\sigma}_{11} = 1.0$ MPa. Comparisons of average ply stresses to finite element calculations are presented in Figs 12(a, b). It is observed that the agreement is generally quite good for all cases. It should be pointed out that each solid line in Figs 10–12 consists of more than 150 data points. In order to illustrate the efficiency of the present method, the CPU time used by the present

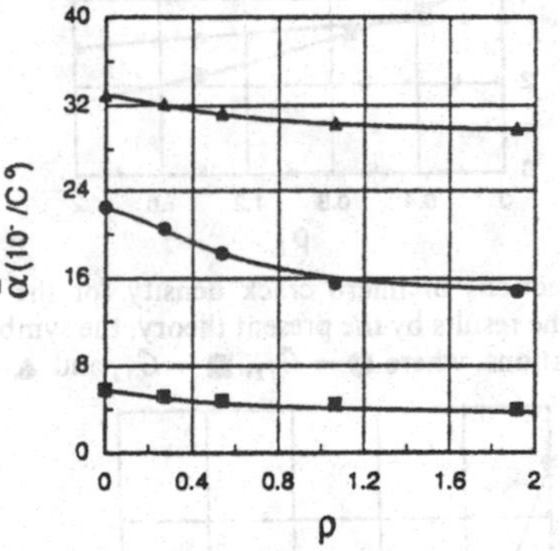

Fig. 10(d). Thermal expansion coefficients as functions of micro crack density for the angle ply ($\phi = \pm 55°$) laminate. The solid lines denote the results by the present theory, the symbols the results from finite element calculations, where ● = $\bar{\alpha}_{11}$, ■ = $\bar{\alpha}_{22}$ and ▲ = $\bar{\alpha}_{33}$.

method and by finite element calculations may be compared. It took a Macintosh SE/30 computer about 5 min to generate all theoretical results shown in Figs 10–12 in comparison to about 4 CPU hours for the finite element calculation of one single layup and one particular crack density on a DEC 3100 Work Station.

4. DISCUSSION

A model for the thermoelastic properties of composite laminates containing plies with transverse matrix cracks has been developed. The model can handle laminates of arbitrary

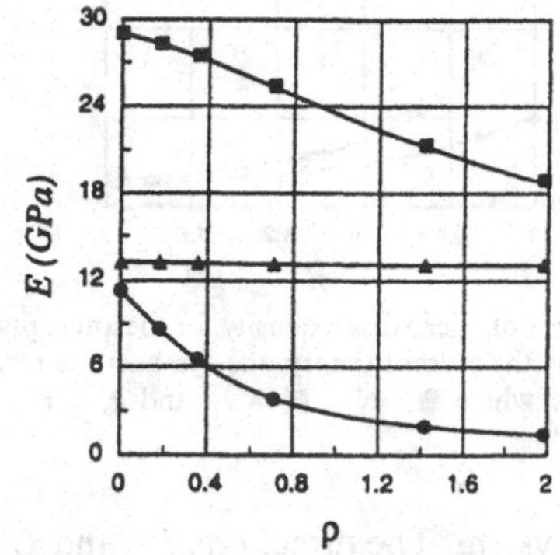

Fig. 11(a). *E*-moduli as functions of micro crack density for the angle ply ($\phi = \pm 67.5°$) laminate. The solid lines denote the results by the present theory, the symbols the results from finite element calculations, where ● = $\bar{E}_1$, ■ = $\bar{E}_2$ and ▲ = $\bar{E}_3$.

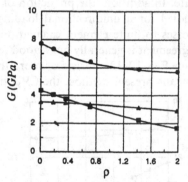

Fig. 11(b). Shear moduli as functions of micro crack density for the angle ply ($\phi = \pm 67.5°$) laminate. The solid lines denote the results by the present theory, the symbols the results from finite element calculations, where ● = $\bar{G}_{12}$, ■ = $\bar{G}_{13}$ and ▲ = $\bar{G}_{23}$.

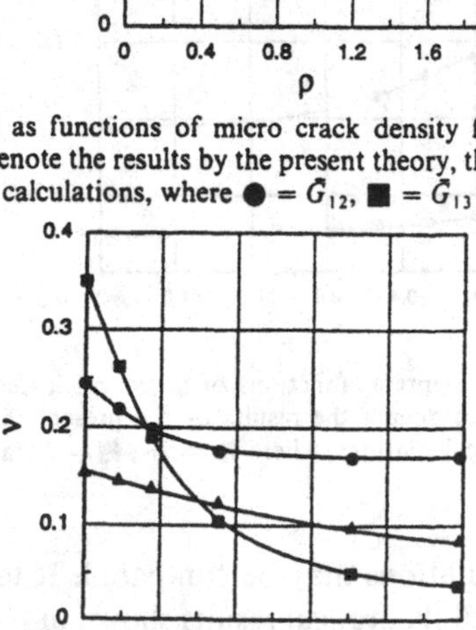

Fig. 11(c). Poisson ratios as functions of micro crack density for the angle ply ($\phi = \pm 67.5°$) laminate. The solid lines denote the results by the present theory, the symbols the results from finite element calculations, where ● = $\bar{\nu}_{12}$, ■ = $\bar{\nu}_{13}$ and ▲ = $\bar{\nu}_{23}$.

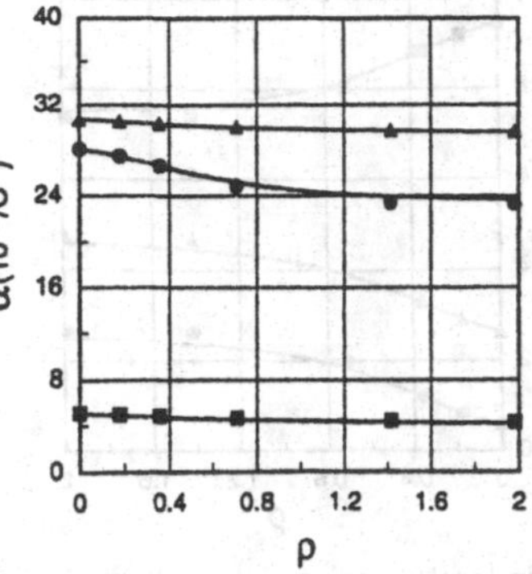

Fig. 11(d). Thermal expansion coefficients as functions of micro crack density for the angle ply ($\phi = \pm 67.5°$) laminate. The solid lines denote the results by the present theory, the symbols the results from finite element calculations, where ● = $\bar{\alpha}_{11}$, ■ = $\bar{\alpha}_{22}$ and ▲ = $\bar{\alpha}_{33}$.

layup configurations and there is no limitation on possible matrix crack densities which can be treated. The fact that the model is formulated in closed form analytical expressions is another nice feature. The model is only based on known parameters such as ply property data. In comparison to alternative models for the prediction of stress–strain relationships of matrix cracked laminates, the present model has some clear advantages. First of all, the present model is more versatile than other models, since there is no restriction concerning laminate layup nor micro crack densities. Alternative models such as the shear lag theory or the Hashin (1985, 1987, 1988) model have generally only been developed for cross-ply

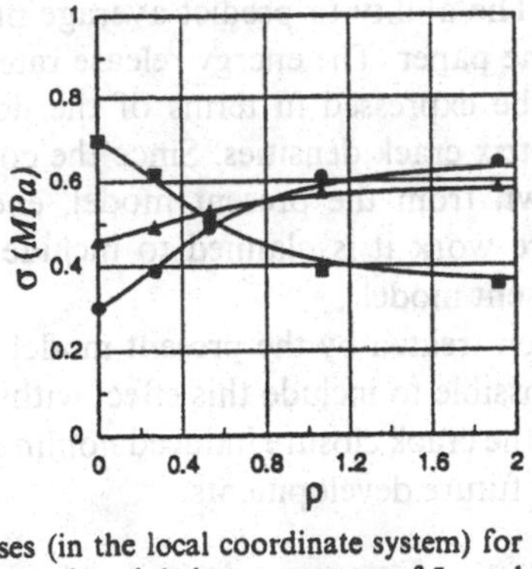

Fig. 12(a). Average ply stresses (in the local coordinate system) for the −55° ply as functions of micro crack densities. The loading is a global average stress of $\bar{\sigma}_{11} = 1.0$ MPa. The solid lines denote the results by the present theory and the symbols the results from finite element calculations where ● = σ_{11}, ■ = σ_{22} and ▲ = σ_{12}.

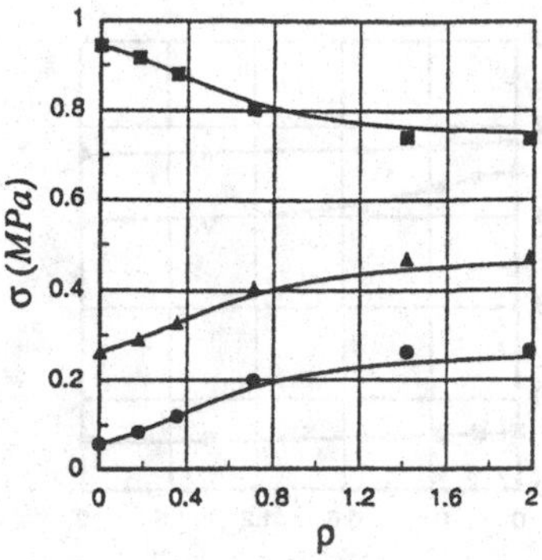

Fig. 12(b). Average ply stresses (in local coordinate system) for the $-67.5°$ ply as functions of micro crack density. The loading is a global average stress of $\bar{\sigma}_{11} = 1.0$ MPa. The solid lines denote the results by the present theory and the symbols the results from finite element calculations where ● = σ_{11}, ■ = σ_{22} and ▲ = σ_{12}.

laminates. Studies on layup configurations other than cross plies are very scarce in the literature. Secondly, the accuracy of the present model is at least as good as alternative models for cross-ply laminates. In addition, the analytical formulation of the present model makes it very easy to implement on a computer. Hence, in the authors' opinion the model presented in this paper is generally applicable and accurate enough for simulations of matrix cracked laminates.

The present paper has only considered the stress–strain relationship at given matrix crack densities. In order to simulate the behaviour of a mechanically loaded structure the model presented here must be complemented by criteria for matrix crack initiation and growth. A crack initiation and growth model could be expressed in terms of ply stresses/strains or energy release rates. The ability to predict average ply stresses and strains has already been demonstrated in the paper. The energy release rate for the creation of a new matrix crack surface area can be expressed in terms of the derivatives of stiffnesses or compliances with respect to matrix crack densities. Since the compliances as functions of micro crack densities are known from the present model, energy release rates can be accurately determined. In future work it is planned to include criteria for matrix crack initiation and growth in the present model.

An effect which has not been treated by the present model is eventual crack closures of matrix cracks. It would be possible to include this effect within the model, but it would be quite complicated because of the crack closure induced nonlinearities. This improvement of the model is therefore left for future developments.

REFERENCES

Allen, D. H., Harris, C. E. and Groves, S. E. (1987a). A thermomechanical constitutive theory for elastic composites with distributed damage—I. Theoretical development. *Int. J. Solids Structures* **23,** 1301–1318.

Allen, D. H., Harris, C. E. and Groves, S. E. (1987b). A thermomechanical constitutive theory for elastic composites with distributed damage—II. Application to matrix cracking in laminated composites. *Int. J. Solids Structures* **23,** 1319–1338.

Allen, D. H. and Lee, J.-W. (1991). Matrix cracking in laminated composites under monotonic and cyclic loadings. *Compos. Engng* **1,** 319–334.

Benthem, J. P. and Koiter, W. T. (1972). Asymptotic approximations to crack problems. In *Mechanics of Fracture 1; Methods of Analysis and Solutions of Crack Problems* (Edited by G. C. Sih), pp. 131–178. Noordhoff, Leyden.

Dvorak, G. J., Laws, N. and Hejazi, M. (1985). Analysis of progressive matrix cracking in composite laminates—I. Thermoelastic properties of a ply with cracks. *J. Compos. Mater.* **19,** 216–234.

Groves, S. E., Harris, C. E., Highsmith, A. L., Allen, D. H. and Norvell, R. G. (1987). An experimental and analytical treatment of matrix cracking in cross-ply laminates. *Expl Mech.* **27,** 73–79.

Gudmundson, P. and Östlund, S. (1992a). First order analysis of stiffness reduction due to matrix cracking. *J. Compos. Mater.* **26,** 1009–1030.

Gudmundson, P. and Östlund, S. (1992b). Prediction of thermoelastic properties of composite laminates with matrix cracks. *Compos. Sci. Technol.* **44,** 95–105.

Gudmundson, P. and Östlund, S. (1992c). Numerical verification of a procedure for calculation of elastic constants in microcracking composite laminates. *J. Compos. Mater.* **26,** 2480–2492.

Gudmundson, P., Östlund, S. and Zang, W. L. (1992). Local stresses and thermoelastic properties of composite laminates containing microcracks. In *Local Mechanics Concepts for Composite Material Systems* (Edited by J. N. Reddy and K. L. Reifsnider), pp. 283–308. Springer, Berlin.

Han, Y. M. and Hahn, H. T. (1989). Ply cracking and property degradations of symmetric balanced laminates under general in-plane loading. *Compos. Sci. Technol.* **35,** 377–397.

Hashin, Z. (1985). Analysis of cracked laminates: a variational approach. *Mech. Mater.* **4,** 121–136.

Hashin, Z. (1987). Analysis of orthogonally cracked laminates under tension. *J. Appl. Mech.* **54,** 872–879.

Hashin, Z. (1988). Thermal expansion coefficients of cracked laminates. *Compos. Sci. Technol.* **31,** 247–260.

Highsmith, A. L. and Reifsnider, K. L. (1982). Stiffness reduction mechanism in composite laminates. In *Damage in Composite Materials* (Edited by K. L. Reifsnider), pp. 103–117. ASTM STP 775, American Society for Testing and Materials, Philadelphia.

Hill, R. (1963). Elastic properties of reinforced solids: some theoretical principles. *J. Mech. Phys. Solids* **11,** 357–372.

Kachanov, M. (1992). Effective elastic properties of cracked solids: critical review of some basic concepts. *Appl. Mech. Rev.* **45,** 304–335.

Laws, N., Dvorak, G. J. and Hejazi, M. (1983). Stiffness changes in unidirectional composites caused by crack systems. *Mech. Mater.* **2,** 123–137.

Lee, J.-W. and Allen, D. H. (1989). Internal state variable approach for predicting stiffness reductions in fibrous laminated composites with matrix cracks. *J. Compos. Mater.* **23,** 1273–1291.

Lim, S. G. and Hong, C. S. (1989). Prediction of transverse cracking and stiffness reduction in cross-ply laminate composites. *J. Compos. Mater.* **23,** 695–713.

Pagano, N. J. (1974). Exact moduli of anisotropic laminates. In *Composite Materials Vol. 2, Mechanics of Composite Materials* (Edited by G. P. Sendeckyj), pp. 23–45. Academic Press.

Sun, C. T. and Li, S. (1988). Three-dimensional effective elastic constants for thick laminates. *J. Compos. Mater.* **22,** 629–639.

Tada, H., Paris, P. C. and Irwin, G. R. (1973). *The Stress Analysis of Cracks Handbook* (2nd Edn). Paris Productions Incorporated (and Del Research Corporation), 226 Woodboune Dr., St Louis, Missouri 63105.

Talreja, R. (1985). Transverse cracking and stiffness reduction in composite laminates. *J. Compos. Mater.* **19,** 355–374.

Talreja, R. (1986). Stiffness properties of composite laminates with matrix cracking and internal delaminations. *Engng Fract. Mech.* **25,** 751–762.

Tan, S. C. and Nuismer, R. J. (1989). A theory for progressive matrix cracking in composite laminates. *J. Compos. Mater.* **23,** 1029–1047.

Varna, J. and Berglund, L. (1991). Multiple transverse cracking and stiffness reduction in cross ply laminates. *J. Compos. Technol. Res.* **13,** 97–106.

CHAPTER 6
PART II

DAMAGE EVOLUTION AND THERMOELASTIC PROPERTIES OF COMPOSITE LAMINATES

W. Zang and P. Gudmundson

Swedish Institute of Composites, Piteå, Sweden

ABSTRACT: An analytic model for the prediction of the thermoelastic properties of micro cracked composite laminates is presented. The expression for the calculation of energy release rates due to growth of micro cracks is also provided. Numerical results are presented that show that the present method, to a very good accuracy, can predict thermoelastic properties of micro cracked laminates at varying crack densities and layup configurations. In addition, a resistance curve behavior of the energy release rate is observed for both carbon/epoxy and glass/epoxy composite laminates. The reasons for this *R*-curve behavior are discussed. Criteria that govern the initiation and growth of micro cracks in composite laminates are discussed and compared to experimental data.

1. INTRODUCTION

MATRIX CRACKING IS generally the first observed damage mechanism in mechanically loaded composite laminates. This kind of damage occurs at relatively low loading levels and is by itself in most cases not critical from a structural failure point of view. In order to fully utilize the load bearing capacity of composite structures, it is hence of importance to have reliable and quick tools for prediction of changes in stiffnesses, thermal expansion coefficients and redistribution of local ply stresses and strains caused by matrix cracks.

There are in general two basic aspects of a laminate theory taking micro cracks into account, i.e., criteria for micro crack initiation and growth, and prediction of all thermoelastic properties at known micro crack densities. Considerable progress has been reported concerning the later aspect. Several methods of different availability and accuracy have been proposed, such as the ply discount method, shear lag analysis [1–4], self consistent scheme [5,6], continuum damage mechanics concepts [7–11], and minimum complimentary potential energy method [12–18]. In the recent years Gudmundson and coworkers [19–22] have presented an alternative procedure for prediction of stiffnesses, thermal expansion coefficients and strain contributions from release of residual stresses. The key to their method is the use of an approximate but still quite accurate solution for average crack opening displacements.

2. THEORETICAL BASIS

2.1 Thermoelastic Properties for Laminates Containing Micro Cracks

In the following, the general three-dimensional laminate theory with micro cracking [21] will be summarized. Laminate theory is actually a homogenization process. Instead of using the properties of each ply, a set of representative (or effective) properties are employed and the laminate structure is treated as if it were made of a homogeneous material. Thus, there are two basic tasks of a laminate theory: (1) to establish the relations between ply material properties (such as compliances and thermal expansion coefficients) and the effective properties, (2) to recover the ply stresses and strains from known global average stresses and effective strains.

A general three-dimensional thick composite laminate containing micro cracks is considered (see Figure 1). The ply material properties and the number of micro cracks in each ply are known. The micro crack density in a typical ply k is denoted as ϱ^k and is in this article defined as the ratio between ply thickness and average distance between micro cracks.

$$\varrho^k = a^k/d^k \tag{1}$$

where the superscript k indicates the variables for ply k and the parameters a^k and d^k are defined in Figure 1. The global average stresses are defined as

$$\bar{\sigma}_{ij}^{(a)} = \sum_k \nu^k \sigma_{ij}^{k(a)} \tag{2}$$

where the superscript (a) indicates average variables and ν^k is the volume fraction of ply k.

The terms effective and average strains are often interchangeably used to describe the same properties. When micro cracks occur, however, these terms do have different meanings. In this article, effective strains are the strains which would be measured on a global scale and the average strains are the averages of

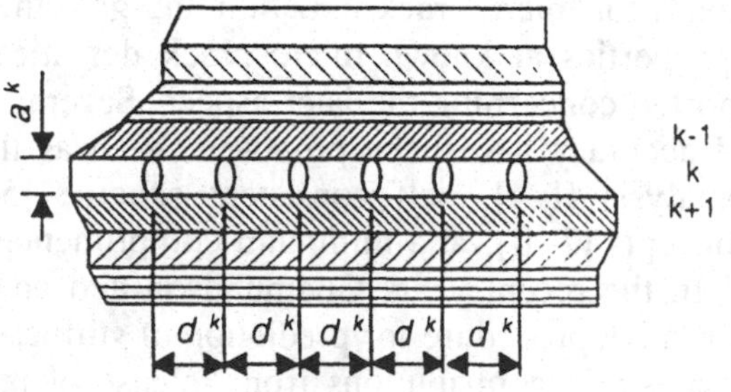

***Figure 1.** A general three-dimensional laminate with micro cracks in ply k.*

In order to apply the methods presented in References [1–22] to real problems, criteria governing the initiation and growth of micro cracks (the evolution law) must be known. There are two kinds of criteria that are often applied, either a strength criterion or an energy release rate (fracture mechanics based) criterion. The task to establish these criteria for composite laminates is not easy because none of the parameters (ply stresses, ply strains or energy release rates) can directly be measured for a micro cracked composite laminate. Therefore, it is first necessary to establish reliable procedures for evaluations of these parameters and then to verify these criteria through experiments and practical applications.

Based on the through-the-thickness-flaw concept, several models have been proposed in literature. In References [15] and [16], a constant critical value of the energy release rate was employed as a material property to predict the initiation and growth of micro cracks for composite laminates with different ply thicknesses. The employment of the constant critical value of the energy release rate implies that the behavior of micro cracks will be the same for different thicknesses of the transverse plies and at varying micro crack densities. However, it is well known that the so-called constraint effect of neighboring plies is only of importance for transverse ply thickness up to a certain limit (cf. [23]). Beyond this thickness, the strains to initiate and propagate micro cracks will be almost independent of the thickness of transverse plies. In addition, it has been found that the critical value of the energy release rate varies with transverse ply thickness and micro crack density (resistance curve) [3]. All these factors indicate that a single critical energy release rate criterion is not sufficient for modelling of initiation and growth of micro cracks in composite laminates with different ply layup and micro crack configurations.

It should be pointed out that very large experimental scatters do exist in the literature for the values of the critical energy release rates. Because of the different processes used to manufacture specimens, the initial defects in the specimens can be quite different. The method employed to count the micro crack densities is another source of scatter. As a result, the stresses to initiate and propagate micro cracks at the same micro crack densities for the same material and same layup configuration can be quite different from laboratory to laboratory. Another uncertainty comes from the methods utilized to evaluate the values of the energy release rate at a given micro crack density. It is noticed that the value of the energy release rate is a local property; hence, it is very sensitive to the accuracy of the approximations used in the theory. Therefore, caution should be exercised in applying the energy criterion to predict the behavior of micro cracks in real composite structures.

In this article, the theory in Reference [21] will first be summarized. Based on the through-the-thickness-flaw concept, the expression for the calculation of energy release rates due to growth of micro cracks will then be presented. Criteria governing the initiation and growth of micro cracks will be discussed in conjunction with experimental data. A resistance curve behavior of the energy release rate is observed for both carbon/epoxy and glass/epoxy composite laminates. Reasons for this R-curve behavior are discussed.

strains experienced by the material in different plies. The difference between effective and average strains is due to the contribution from crack opening displacements. The global effective strains are defined as,

$$\bar{\epsilon}_{ij}^{(e)} = \frac{1}{2V}\int_{\Gamma^{out}} (u_i n_j + u_j n_i) d\Gamma \tag{3}$$

where u_i denotes the displacements, n_i the unit normal vector on Γ^{out} (the outer boundary surface of a volume V which is large in comparison to distances between micro cracks and ply thicknesses) and the superscript (e) the effective variables. It is obvious that,

$$\bar{\epsilon}_{ij}^{(e)} = \sum_k \nu^k \left[\frac{1}{2V^k}\int_{\Gamma^{k,out}} (u_i^k n_j^k + u_j^k n_i^k) d\Gamma\right]$$

$$= \sum_k \nu^k \epsilon_{ij}^{k(e)} \tag{4}$$

where V^k is the volume of ply k within V and the integral is performed only on the outer boundary surfaces of ply k. By an application of the divergence theorem, the integral for the effective ply strains can be divided into two parts as,

$$\epsilon_{ij}^{k(e)} = \frac{1}{2V^k}\int_{V^k} (u_{i,j}^k + u_{j,i}^k) dV_k + \frac{\varrho^k}{2a^k}(\Delta\bar{u}_i^k n_j^k + \Delta\bar{u}_j^k n_i^k)$$

$$= \epsilon_{ij}^{k(a)} + \Delta\epsilon_{ij}^k \tag{5}$$

In Equation (5), $\Delta\bar{u}_i^k$ indicates the average crack opening displacements in ply k, n_i^k is the component of the unit normal vector on the crack surface, $\epsilon_{ij}^{k(a)}$ are the average strains and $\Delta\epsilon_{ij}^k$ are the average incremental strains due to crack opening displacements.

In the following, the stresses, strains and thermal expansion coefficients will be partitioned into in-plane parts and out-of-plane parts:

$$\sigma = \begin{pmatrix}\sigma_I\\ \sigma_O\end{pmatrix} \qquad \epsilon = \begin{pmatrix}\epsilon_I\\ \epsilon_O\end{pmatrix} \qquad \alpha = \begin{pmatrix}\alpha_I\\ \alpha_O\end{pmatrix} \tag{6}$$

where

$$\sigma_I = \begin{pmatrix}\sigma_{11}\\ \sigma_{22}\\ \sigma_{12}\end{pmatrix} \qquad \epsilon_I = \begin{pmatrix}\epsilon_{11}\\ \epsilon_{22}\\ 2\epsilon_{12}\end{pmatrix} \qquad \alpha_I = \begin{pmatrix}\alpha_{11}\\ \alpha_{22}\\ 2\alpha_{12}\end{pmatrix} \tag{7}$$

are the in-plane stresses, strains and thermal expansion coefficients and

$$\sigma_O = \begin{pmatrix} \sigma_{33} \\ \sigma_{13} \\ \sigma_{23} \end{pmatrix} \qquad \epsilon_O = \begin{pmatrix} \epsilon_{33} \\ 2\epsilon_{13} \\ 2\epsilon_{23} \end{pmatrix} \qquad \alpha_O = \begin{pmatrix} \alpha_{33} \\ 2\alpha_{13} \\ 2\alpha_{23} \end{pmatrix} \tag{8}$$

are the out-of-plane stresses, strains and thermal expansion coefficients.

The compatibility and equilibrium conditions for the laminate with micro cracks read

$$\epsilon_I^{k(e)} = \bar{\epsilon}_I^{(e)} \qquad \sigma_O^{k(a)} = \bar{\sigma}_O^{(a)} \tag{9}$$

The relation between average stresses and strains in ply k can be expressed as

$$\epsilon^{k(a)} = S^k \sigma^{k(a)} + \alpha^k \Delta T \tag{10}$$

where

$$S^k = \begin{pmatrix} S_{II}^k & S_{IO}^k \\ (S_{IO}^k)^T & S_{OO}^k \end{pmatrix} \tag{11}$$

In Equation (11), S_{II}^k, S_{IO}^k, $(S_{IO}^k)^T$ and S_{OO}^k are 3×3 sub-matrices of the ply compliance matrix. In addition, the global average stresses are related to the global effective strains by the following expression,

$$\bar{\epsilon}^{(e)} = \bar{S}_{(c)} \bar{\sigma}^{(a)} + \bar{\alpha}_{(c)} \Delta T + \bar{\epsilon}_{(c)}^{(r)} \tag{12}$$

where $\bar{S}_{(c)}$, $\bar{\alpha}_{(c)}$ and $\bar{\epsilon}_{(c)}^{(r)}$ are the 6×6 global effective compliance matrix, the 6×1 thermal expansion coefficient vector and the 6×1 effective strain vector due to release of residual stresses, respectively. Explicit expression for all of these properties can be found in Reference [21].

For a given loading condition, (either prescribed loading, prescribed displacements, or a mixture of these, including thermal loading and residual stresses) the global effective strains and average stresses can be evaluated by Equation (12). The average ply stresses and strains can then be recovered by employing Equations (5)–(11), provided that the average crack opening displacements are known. The solution for average crack opening displacements will be discussed in the next section.

2.2 Average Crack Opening Displacements and Energy Release Rate for Micro Crack Growth

It should be pointed out that all expressions in Reference [21] for the thermoelastic properties of micro cracked laminates can be exactly determined, provided that exact solutions for average crack opening displacements are available. Thus the problem is focussed on finding out the solutions for average crack opening displacements. Since a general exact solution for average crack opening displacements valid for different ply geometries and crack distribution configurations is not available, approximate solutions have to be employed. The quality of the resulting theory will therefore strongly depend on the accuracy of the approximate solution for crack opening displacements.

For any kind of loading condition, it is always possible [by employing Equation (12)] to describe it as prescribed in-plane strains $\bar{\epsilon}_I$ and prescribed out-of-plane stresses $\bar{\sigma}_O$, including prescribed thermal loading ΔT and residual stresses. If only linear elasticity is considered, the problem has been solved in Reference [21] by a superposition of two problems. In the first problem, the laminate without micro cracks loaded by prescribed global effective in-plane strains $\bar{\epsilon}_I^{(e)}$, global average out-of-plane stresses $\bar{\sigma}_O^{(a)}$, ΔT and residual stresses is considered. This problem can be solved by laminate theory without micro cracks. In particular, the solution to this problem provides the tractions on prospective crack surfaces. The tractions on the prospective crack surfaces for ply k, for example, can be expressed as

$$\tau^k = A^k\bar{\epsilon}_I^{(e)} + B^k\bar{\sigma}_O^{(e)} + (C^k - A^k\bar{\alpha}_I)\Delta T + \tau^{k(r)} \tag{13}$$

where

$$\left.\begin{aligned} A^k &= N_I^k(S_{II}^k)^{-1} \\ B^k &= -N_I^k(S_{II}^k)^{-1}S_{IO}^k + N_O^k \\ C^k &= A^k(\bar{\alpha}_I - \alpha_I^k) \\ \tau^{k(r)} &= N_I^k\sigma_I^{k(r)} \end{aligned}\right\} \tag{14}$$

and

$$N_I^k = \begin{pmatrix} n_1^k & 0 & n_2^k \\ 0 & n_2^k & n_1^k \\ 0 & 0 & 0 \end{pmatrix}$$
$$N_O^k = \begin{pmatrix} 0 & 0 & 0 \\ 0 & 0 & 0 \\ 0 & n_1^k & n_2^k \end{pmatrix} \tag{15}$$

In Equations (14) and (15), $\sigma_I^{k(r)}$ are the residual stresses and n_j^k $(j = 1,2)$ is the unit normal vector on the crack surface $(n_3^k = 0)$.

In the second problem, the tractions on crack surfaces resulting from the first problem [Equation (13)] are released under vanishing effective global in-plane strains $(\bar{\epsilon}_I^{(e)} = \mathbf{0})$ and average global out-of-plane stresses $(\bar{\sigma}_O^{(a)} = \mathbf{0})$. The solution to the second problem will enable the determination of average crack opening displacements. The average crack opening displacements in a typical ply k will depend linearly on all crack surface tractions. Thus

$$\Delta\bar{u}^k = \sum_i a^k\beta^{ki}\tau^i \tag{16}$$

where β^{ki} are 3 × 3 matrices which depend on laminate layup, ply properties and crack densities. In Reference [21], the second problem is approximately replaced by a row of parallel cracks in an infinite, homogeneous medium. This approximation implies that the effects from adjacent layers are neglected. Thus Equation (16) becomes,

$$\Delta\bar{u}^k = a^k\beta^{kk}\tau^k \tag{17}$$

where explicit expression for β^{kk} are given in Reference [21].

It is noticed that when ϱ^k tends to zero, the components of β^{kk} approach the dilute results given in Reference [19]. For ϱ^k tending to infinity, it can be shown that the results obtained by the present method are in agreement with the results from the infinite limit given in Reference [20]. The present theory is thus in agreement with the asymptotic results for small and infinite micro crack densities, respectively. For intermediate values of ϱ^k, the accuracy of the solutions obtained by the present theory has to be checked against numerical or experimental results.

The equation for energy release rate due to micro crack growth can now easily be formulated. Since the first problem does not contain any micro cracks, the attention can be focussed on the second problem. The potential energy U for the second problem can be written as,

$$U = -1/2V\sum_i v^i(\tau^i)^T\sum_k \varrho^i\beta^{ki}\tau^k \tag{18}$$

where the superscript T indicates the transpose of a matrix or a vector and τ^k can be evaluated according to Equation (13). It is noticed that the tractions on crack surfaces are prescribed (dead loading). The energy release rate due to micro crack growth in ply i can thus be expressed as,

$$G^i = -\partial U/\partial A^i$$

$$= 1/2a^i(\tau^i)^T\sum_k[\partial(\varrho^i\beta^{ki})/\partial\varrho^i]\tau^k \tag{19}$$

Equation (19) is exact provided that the exact β^{ki} (average crack opening displacements) are employed. If the approximate solution for the average crack opening displacements [Equation (17)] is utilized, the energy release rate due to micro crack growth in ply i can approximately be written as,

$$G^i = 1/2a^i(\tau^i)^T[\partial(\varrho^i\beta^{ii})/\partial\varrho^i]\tau^i \tag{20}$$

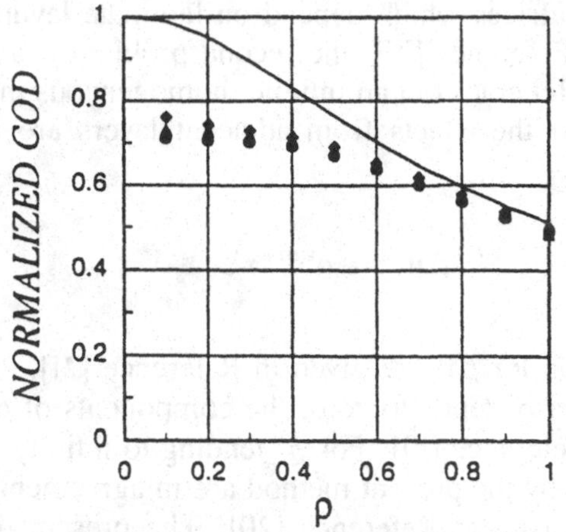

Figure 2. *Normalized crack opening displacements for carbon/epoxy cross ply $[0/90_N]_s$ laminates. The solid line indicates the results by the approximate solution and the symbols the results from finite element calculations, where ● stands for N = 1, ▲ for N = 2 and ♦ for N = 3.*

2.3 Exact Average Crack Opening Displacement for Carbon/Epoxy Cross Ply Laminates

It is known that the results for the average crack opening displacements will in general depend on micro crack densities, laminate layup configurations and ply material properties. In Reference [19], an extensive study on the accuracy of the approximate solution for the average crack opening displacements at dilute micro crack densities with varying fibre angles in the adjacent plies was performed. It was observed that in the dilute crack density case, the maximum errors of the approximate solutions for the average crack opening displacements occurred for laminate layups corresponding to cross ply laminates. They were about 10% for glass/epoxy and 30% for carbon/epoxy. In this article, some results for the average crack opening displacements of carbon/epoxy cross ply $[0/90_N]_s$ laminates under the action of a uniform normal pressure τ^o on the crack surfaces are presented as functions of micro crack densities (see Figure 2). The results in Figure 2 are normalized with respect to $\pi(1 - \nu_{LT}\nu_{TL})\tau^o a/(2E_T)$, where E_T is the ply transverse modulus, ν_{LT} and ν_{TL} are Poisson's ratios and a is the thickness of the 90° ply. The materials used in the finite element calculations are presented in Table 1. It is observed that the thickness of the 90° plies has a very small effect on the results for the normalized average crack opening displacements. In addition the maximum error between the approximate solution and the finite element results is found for dilute micro crack densities. The error decreases with increasing micro crack densities. A curve fitting of the finite element results for $N = 1$ yields,

$$\Delta u = a\tau^o\beta$$

$$= a\tau^o\{[\pi(1 - \nu_{LT}\nu_{TL})/(2E_T)](0.7140 + 0.0353\varrho - 0.2690\varrho^2)\} \quad (21)$$

Equation (21) will be employed in the next section for the evaluation of experimental results.

3. NUMERICAL RESULTS AND DISCUSSION

It should be pointed out that all formulations presented in the previous section are exact, provided that exact solutions for the crack opening displacements (β^{k_i}) are employed. For cases when exact expressions for the crack opening displacements (β^{k_i}) are not available, the approximate solutions in Reference [21] can be utilized to obtain all required thermoelastic properties. The solution accuracy will of course be influenced by the use of the approximate solution for the crack opening displacements (β^{k_i}). In order to demonstrate the accuracy of the present theory employing approximate solutions for crack opening displacements, a few two- and three-dimensional problems have been investigated by the present theory and compared to finite element calculations. Two kinds of laminate systems have been considered, thin cross ply laminates with micro cracks in one set of plies and thick angle ply laminates with micro cracks in both sets of plies. In the finite element calculations, periodic cells with appropriate periodic boundary conditions were employed and the finite element program ABAQUS was used. A detailed description of the finite element modelling is given in Reference [24].

It is noticed that there are two sets of results that are of interest for a micro cracked laminate structure: first, global properties such as effective stiffnesses and global stresses and strains; and second, local properties such as ply average stresses and strains are of importance. An extensive investigation of the accuracy of the present theory using approximate solution for the average crack opening displacements has been performed in Reference [21] for glass/epoxy laminates and in Reference [22] for carbon/epoxy laminates. For the global laminate properties, excellent agreements with the results from finite element calculations and experiments have been obtained. The attention in the present investigation will focus on the local laminate properties such as energy release rates due to micro crack growth. In addition, criteria that govern the initiation and growth of micro cracks will also be discussed.

Table 1. Material properties for unidirectional composites.

Properties	Glass/Epoxy	Carbon/Epoxy	AS4/3501-6 [9]	AS4/3501-6 [15,16]
E_L (GPa)	44.00	130.00	140.10	130.00
E_T (GPa)	15.80	9.72	8.36	9.70
G_{LT} (GPa)	3.40	5.36	4.31	5.00
ν_{LT}	0.26	0.31	0.25	0.30
ν_{TT}	0.42	0.49	0.30	0.50
α_L (ppm/°C)	6.72	−0.77	−0.43	−0.09
α_T (ppm/°C)	29.32	24.50	27.30	28.80
T_{eff} (°C)	−150.00	—	−150.00	−125.00
Ply thickness (mm)	0.13	0.13	0.15	0.13

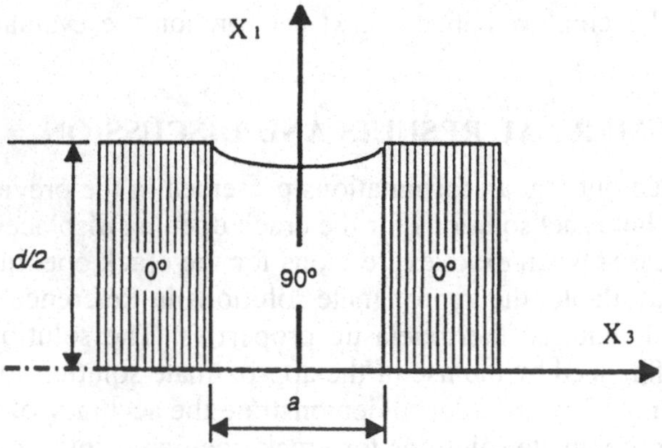

Figure 3. A representative periodic cell for a cross ply laminate with micro cracks in the 90° ply.

3.1 Effective Laminate Properties

Thin cross ply carbon/epoxy laminates with $[0/90_N]_s$ ($N = 1,2,3$) layup are first considered. The geometry of a representative periodic cell with micro cracks in the 90° ply is shown in Figure 3. In Figures 4a and 4b, the effective Young's modulus and Poisson's ratio (normalized by the initial values) as functions of micro crack densities obtained by the present method are compared to finite element calculations. Thick angle ply carbon/epoxy laminates $[\pm 55]_N$ with micro cracks in both plies and covering the whole thickness of the laminate are then investigated. The finite element method has been employed to verify the efficiency and reliability of the present theory. Due to symmetry only one-half of the

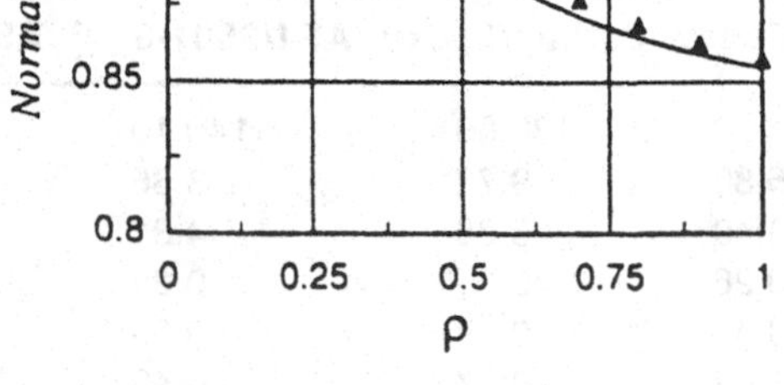

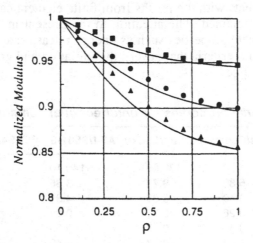

Figure 4a. Normalized Young's moduli as functions of micro crack densities in the 90° plies for the carbon/epoxy cross ply $[0/90_N]_s$ laminates (see Figure 3). The solid lines denote the results by the present method and the symbols the results from finite element calculations, where ■ stands for N = 1, ● for N = 2 and ▲ for N = 3.

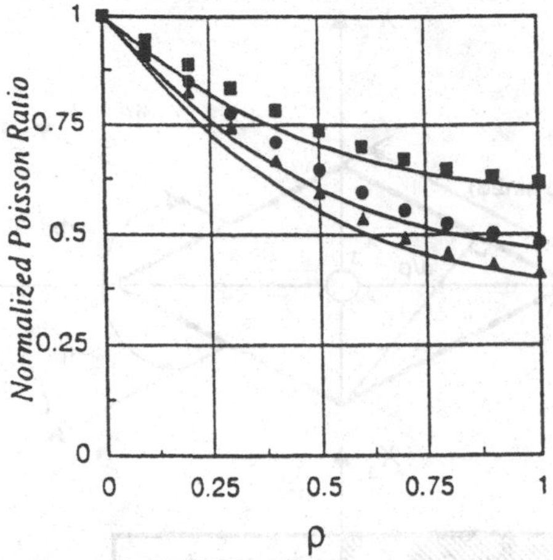

Figure 4b. *Normalized Poisson's ratios as functions of micro crack densities in the 90° plies for the carbon/epoxy cross ply $[0/90_N]_s$ laminates (see Figure 3). The solid lines denote the results by the present method and the symbols the results from finite element calculations, where ■ stands for N = 1, ● for N = 2 and ▲ for N = 3.*

periodic cell has been modelled by finite elements. The periodic cell geometry and the finite element mesh are illustrated in Figure 5. Effective global engineering constants (normalized by the initial values) as functions of micro crack densities are presented in Figures 6a and 6b. From the results in Figures 4 and 6 it is observed that quite good agreements between the present theory and finite element calculations are obtained.

Since the global laminate properties will in general depend on both average crack opening displacements and the ply properties (ply stiffnesses, layup, etc.) in the whole laminate, it could be expected that the solution for the global laminate properties will be rather insensitive to the accuracy of the approximate solution for the average crack opening displacements. This conclusion has been confirmed by the results presented in References [21] and [22] and the results in Figures 4 and 6. As far as the global laminate properties are concerned, the accuracy of the present theory appears to be sufficient.

3.2 Energy Release Rate

Even though the error of the approximate solution for the average crack opening displacements could in some cases be quite large, numerical results presented in References [21] and [22] and in the previous section indicate that the results for the global laminate properties obtained by the present theory are still quite accurate. For the local laminate properties, this conclusion may not be true. Referring to Equations (19) and (20), the solution for the energy release rate due to micro crack growth is directly related to the average crack opening displacements. The accuracy of the solution for the energy release rate due to micro crack growth is thus more sensitive to the accuracy of the solutions for the average crack opening displacements employed in the theory.

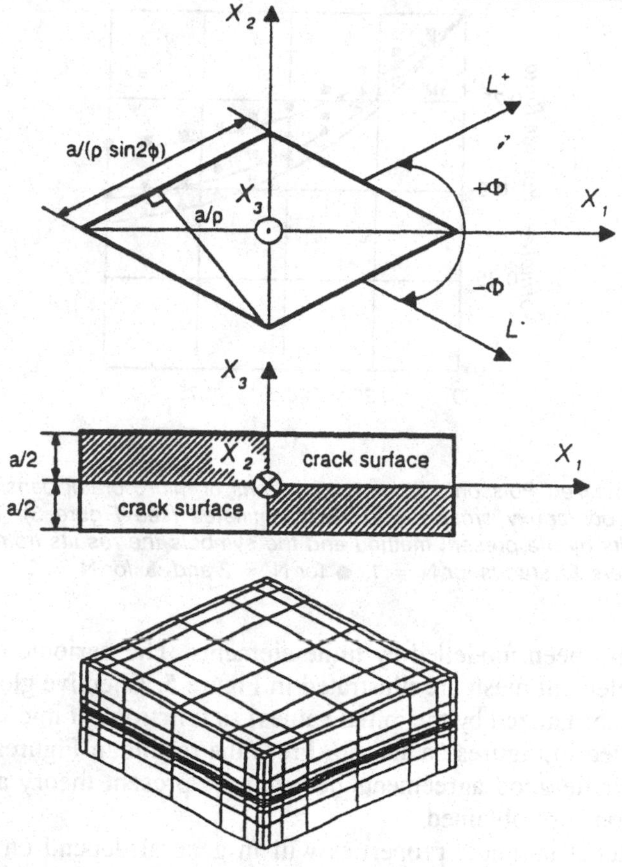

Figure 5. Coordinate systems, geometry and finite element mesh of a representative periodic cell for a thick angle ply laminate with micro cracks in both types of plies.

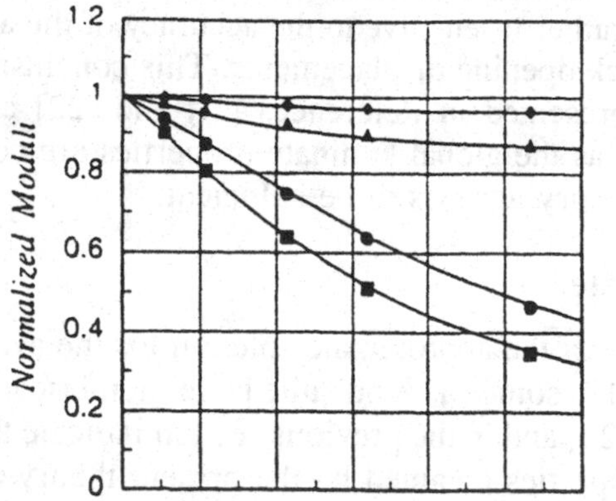

Figure 6a. Normalized moduli as functions of micro crack densities for the carbon/epoxy thick angle ply ($\phi = \pm 55°$) laminate. The solid lines denote the results by the present theory, the symbols the results from finite element calculations, where ■ = $\bar{E}_1$, ● = $\bar{E}_2$, ▲ = $\bar{E}_3$ and ♦ = $\bar{G}_{12}$.

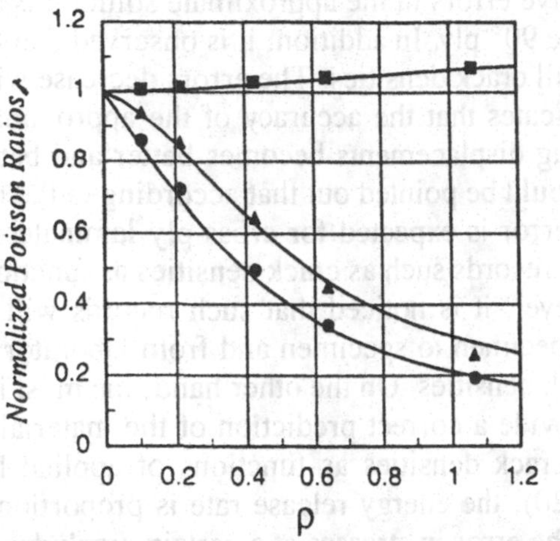

Figure 6b. *Normalized Poisson's ratios as functions of micro crack densities for the carbon/epoxy thick angle ply ($\phi = \pm 55°$) laminate. The solid lines denote the results by the present theory, the symbols the results from finite element calculations, where* ■ $= \bar{\nu}_{12}$, ● $= \bar{\nu}_{13}$, *and* ▲ $= \bar{\nu}_{23}$.

In Figure 7, results for energy release rates as functions of micro crack densities for carbon/epoxy laminates with layup $[0/90_N]_s$ ($N = 1,2,3$) under the action of a unidirectional loading $\bar{\sigma}_{11}^{(a)}$ are compared to results by finite element calculations. The results in Figure 7 are normalized with respect to C^*, given by

$$C^* = a(\bar{\sigma}_{11}^{(a)})^2/\bar{E}_1^{(o)} \tag{22}$$

where $\bar{E}_l^{(o)}$ is the initial E-modulus and a is the thickness of the 90° ply. It is

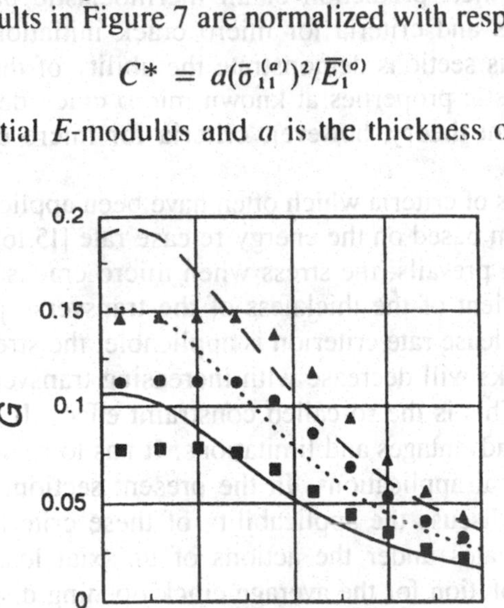

Figure 7. *Normalized energy release rates as functions of micro crack densities in 90° plies for the carbon/epoxy cross ply $[0/90_N]_s$ laminates (see Figure 3). The smooth lines denote the results by the present theory, the symbols the results from finite element calculations, where the solid line and* ■ *stand for* N = *1, the dotted line and* ● *for* N = *2, and the dashed line and* ▲ *for* N = *3.*

noticed that the relative errors in the approximate solution is almost independent of the thickness of the 90° ply. In addition, it is observed that the maximum error is about 30% for small crack densities. The errors decrease with increasing crack densities, which indicates that the accuracy of the approximate solution for the average crack opening displacements becomes better and better with increasing crack densities. It should be pointed out that according to the results in Reference [19], the maximum error is expected for cross ply laminates.

In laboratory tests, records such as crack densities as functions of applied loads are registered. However, it is noticed that such records will always show quite large scatters from specimen to specimen and from laboratory to laboratory, especially at small crack densities. On the other hand, the most important objective for a theory is to provide a correct prediction of the material and structural behavior, such as the crack densities as functions of applied loads. Referring to Equations (19) and (20), the energy release rate is proportional to the square of applied loads. Thus the error in stresses at a certain crack density will be limited to 15% for a maximum error of 30% in energy release rate. Hence, errors due to the application of the approximate solutions for the average crack opening displacements could be within the scatters of tests.

3.3 Criteria for Micro Crack Initiation and Growth

In general, there are two basic aspects for a laminate theory taking micro cracks into account, i.e., prediction of all thermoelastic properties at known micro crack densities and criteria for micro crack initiation and growth. The results in the previous sections demonstrate the ability of the present theory to predict all thermoelastic properties at known micro crack densities. In order to fully apply the present theory, however, criteria for micro crack initiation and growth are necessary.

There are two kinds of criteria which often have been applied: either a strength criterion or a criterion based on the energy release rate [15,16]. It is obvious that if a strength criterion prevails, the stress when micro cracks initiate and propagate will be independent of the thickness of the transverse plies. On the other hand, if the energy release rate criterion is applicable, the stresses to initiate and propagate micro cracks will decrease with increasing transverse ply thicknesses [cf. Equation (19)]. This is the so-called constraint effect. It is known that each criterion has its own advantages and limitations. It has to be verified through experiments and practical applications. In the present section, experimental data will be employed to discuss the applicability of these criteria. Since only data from cross ply laminates under the actions of uniaxial loading condition are available, the exact solution for the average crack opening displacements [Equation (21)] will be employed in the present discussion. Thus all results should be realized as exact solutions.

In the literature, experimental data are very often presented in the form of micro crack densities as functions of applied average stresses. However, it would be more instructive if the micro crack densities as functions of applied effective strains were reported. In Figures 8 and 9, the micro crack densities are plotted

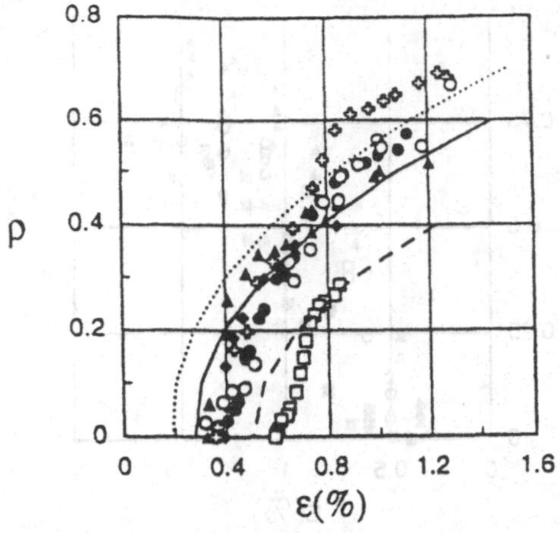

***Figure 8.** Micro crack densities as functions of applied global effective strains for the AS4/3501-6 cross ply laminates. The smooth lines denote the prediction by the present theory and symbols the results from experiments [9,15,16], where the dashed line stands for $[0/90]_s$, the solid line for $[0/90_2]_s$, the dotted line for $[0/90_4]_s$, □ for $[0/90]_s$, ● and ▲ for $[0/90_2]_s$, ♦ and ○ for $[0_2/90_2]_s$, and ✢ for $[0/90_4]_s$.*

as functions of applied effective strains or energy release rates for carbon/epoxy (AS4/3501-6) cross ply laminates. The results in Figure 8 are based on the experimental data from References [9], [15] and [16]. The same results based on the experiment data from Reference [18] are shown in Figures 10 and 11 for glass/epoxy cross ply laminates.

Referring to Figure 8, it is observed that the strains to initiate and propagate micro cracks for a thin transverse ply ($N = 1$) are larger than the strains for other cases. This factor clearly demonstrates the constraint effect for thinner transverse

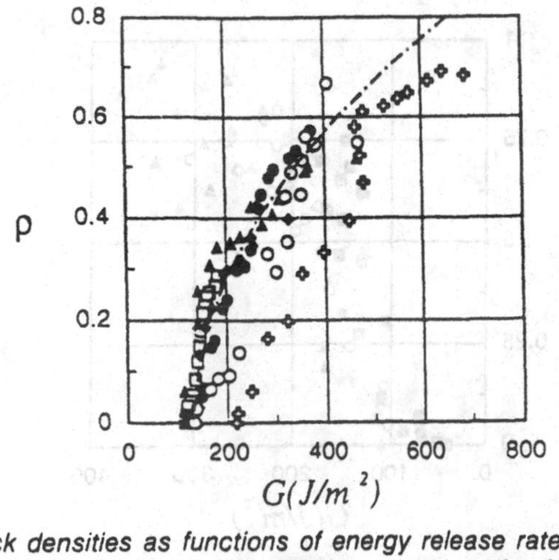

***Figure 9.** Micro crack densities as functions of energy release rates for the AS4/3501-6 cross ply laminates, where □ stands for $[0/90]_s$, ● and ▲ for $[0/90_2]_s$, ♦ and ○ for $[0_2/90_2]_s$, ✢ for $[0/90_4]_s$ and the smooth line is the curve fit based on the solid symbols.*

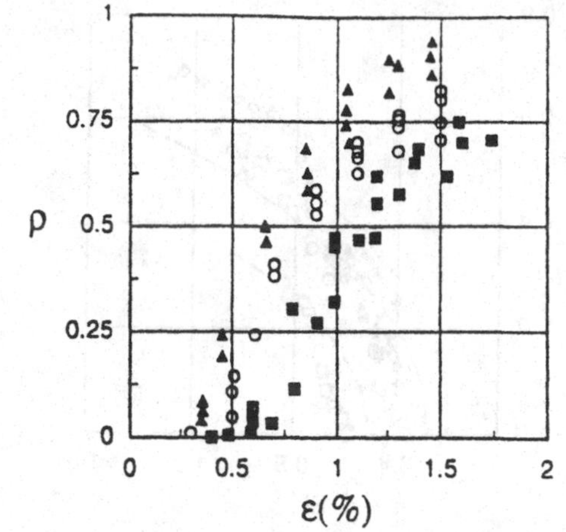

***Figure 10.** Micro crack densities as functions of applied global effective strains for the glass/epoxy cross ply $[0/90_N]_s$ laminates. The symbols indicate the results from experiments [18], where ■ stands for N = 1/2, ○ for N = 1 and ▲ for N = 2.*

plies. The constraint effect diminishes and vanishes with increasing transverse ply thicknesses, since the effective strains are almost the same for $N = 2$ and $N = 4$. In addition, the results in Figure 9 indicate that the single G (energy release rate) criterion cannot govern the behavior of micro crack initiation and growth with different transverse ply thicknesses, since the G-values are obviously different for different ply layup configurations. The G-values are in general larger for thicker transverse plies. Combining the results in Figures 8 and 9, it seems that the G criterion is valid for micro cracks in thinner transverse plies and the strength criterion prevails for micro cracks in thicker transverse plies. A similar trend can be observed from the results shown in Figures 10 and 11.

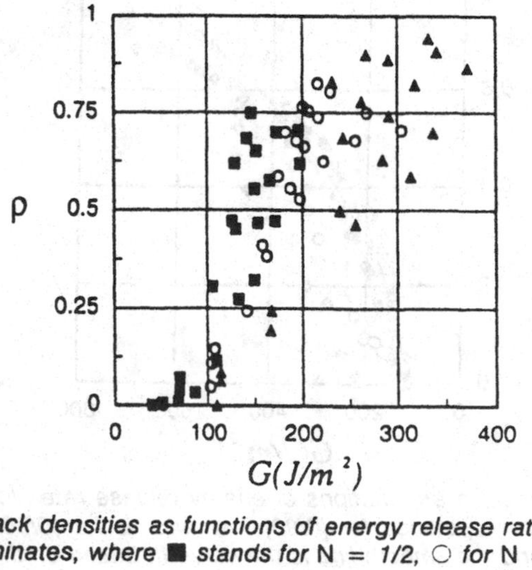

***Figure 11.** Micro crack densities as functions of energy release rates for the glass/epoxy cross ply $[0/90_N]_s$ laminates, where ■ stands for N = 1/2, ○ for N = 1 and ▲ for N = 2.*

Referring to Figures 9 and 11, it is noticed that all materials studied in this article show a R (resistance) curve behavior. A similar phenomenon has been observed in Reference [3]. This phenomenon can be explained in different ways. It is known that material and structural defects will be introduced during manufacturing. Micro cracks will primarily initiate from the largest defects, since the energy release rate is larger for larger defects. The average size of remaining defects will therefore decrease with increasing micro crack density. Thus, it can be expected that the critical energy release rate will increase with increasing micro crack densities.

It will be interesting to see how the present theory (employing the approximate solution for the average crack opening displacements) works. Based on the results from the smooth curve in Figure 9 (fitted from the data for $N = 2$), the present theory (employing approximate solutions and G criterion) predicts the micro crack densities as functions of applied effective strains. These results are presented as the smooth curves in Figure 8. It is noticed that the experimental results usually show quite a large scatter. The prediction by the present theory employing the approximate solutions for the average crack opening displacements are fairly reasonable for thinner plies. The prediction for the $N = 4$ case fails because of the invalidity of the G criterion for thicker plies.

4. CONCLUSION

The prediction of the evolution of micro cracks in composite laminates and the responses of composite laminates containing micro cracks have been discussed. The presented theory for thermoelastic properties is exact, except for the solution for average crack opening displacements. The price for employing the approximate solution for average crack opening displacements is paid by the loss of accuracy in some of the results. However, all formulations provided by the present theory are explicit and valid for all kinds of laminate layup and micro crack distributions. The simplicity and flexibility are the main advantages of the present theory in comparison with alternative methods.

Based on experimental data, energy release rates have been computed by the present theory for different materials with different transverse ply thicknesses. It is observed that all materials investigated in this article show a R (resistance) curve behavior. In addition, the energy release rate curves for the same material are different for different transverse ply thicknesses. This factor indicates that a single G criterion cannot govern the initiation and growth of micro cracks in transverse plies of any thickness.

ACKNOWLEDGEMENTS

The authors gratefully acknowledge Dr. J. Varna from the Department of Materials and Production Engineering, Lulea University of Technology, Sweden and Dr. P. W. M. Peters from DLR, Köln, Germany for providing their experimental data.

REFERENCES

1. Highsmith, A. L. and K. L. Reifsnider. 1982. "Stiffness-Reduction Mechanisms in Composite Laminates," in *Damage in Composite Materials, ASTM STP 775*, K. L. Reifsnider, ed., Philadelphia, PA: American Society for Testing and Materials, pp. 103–117.
2. Lim, S. G. and C. S. Hong. 1989. "Prediction of Transverse Cracking and Stiffness Reduction in Cross-Ply Laminated Composites," *Journal of Composite Materials*, 23:695–713.
3. Han, Y. M., H. T. Hahn and R. B. Croman. 1988. "A Simplified Analysis of Transverse Ply Cracking in Cross-Ply Laminates," *Composites Science and Technology*, 31:165–177.
4. Tan, S. C. and R. J. Nuismer. 1989. "A Theory for Progressive Matrix Cracking in Composite Laminates," *Journal of Composite Materials*, 23:1029–1047.
5. Laws, N., G. J. Dvorak and M. Hejazi. 1983. "Stiffness Changes in Unidirectional Composites Caused by Crack Systems," *Mechanics of Materials*, 2:123–137.
6. Laws, N. and G. J. Dvorak. 1985. "The Loss of Stiffness of Cracked Laminates," *Proceedings IUTAM Eshelby Memorial Symposium*. London: Cambridge University Press, pp. 119–127.
7. Talreja, R. 1985. "Transverse Cracking and Stiffness Reduction in Composite Laminates," *Journal of Composite Materials*, 19:355–375.
8. Talreja, R. 1986. "Stiffness Properties of Composite Laminates with Matrix Cracking and Interior Delamination," *Engineering Fracture Mechanics*, 25:751–762.
9. Talreja, R., S. Yalvac, L. D. Yats and D. G. Wetters. 1992. "Transverse Cracking and Stiffness Reduction in Cross Ply Laminates of Different Matrix Toughness," *Journal of Composite Materials*, 26:1644–1663.
10. Allen, D. H., C. E. Harris and S. E. Groves. 1987. "A Thermomechanical Constitutive Theory for Elastic Composites with Distributed Damage – I. Theoretical Development," *International Journal of Solids and Structures*, 23:1301–1318.
11. Allen, D. H., C. E. Harris and S. E. Groves. 1987. "A Thermomechanical Constitutive Theory for Elastic Composites with Distributed Damage – II. Application to Matrix Cracking in Laminated Composites," *International Journal of Solids and Structures*, 23:1319–1338.
12. Hashin, Z. 1985. "Analysis of Cracked Laminates: A Variational Approach," *Mechanics of Materials*, 4:121–136.
13. Hashin, Z. 1985. "Analysis of Orthogonally Cracked Laminates under Tension," *Journal of Applied Mechanics*, 54:872–879.
14. Hashin, Z. 1988. "Thermal Expansion Coefficients of Cracked Laminates," *Composite Science and Technology*, 31:247–260.
15. Liu, S. and J. A. Nairn. 1992. "The Formation and Propagation of Matrix Microcracks in Cross-Ply Laminates during Static Loading" *Journal of Reinforced Plastics and Composites*, 2:158–178.
16. Nairn, J. A. and S. F. Hu. "Micromechanics of Damage: A Case Study of Matrix Microcracking," in *Damage Mechanics of Composite Materials*, R. Talreja, ed.
17. Varna, J. and L. Berglund. 1992. "Thermo-Elastic Properties of Composite Laminates with Transverse Cracks," *Journal of Composite Technology and Research*, 13:97.
18. Peters, P. W. M., L. Buligin, J. Varna and L. A. Berglund. 1993. Submitted to *Composite*.
19. Gudmundson, P. and S. Östlund. 1992. "First Order Analysis of Stiffness Reduction Due to Matrix Cracking," *Journal of Composite Materials*, 26:1009–1030.
20. Gudmundson, P. and S. Östlund. 1992. "Thermoelastic Properties of Composite Laminates with Matrix Cracks," *Composite Science and Technology*, 44:95–105.
21. Gudmundson, P. and W. Zang. 1992. "A Universal Model for Thermoelastic Properties of Micro Cracked Composite Laminates," SICOMP Technical Report 92-007, Swedish Institute of Composites, Box 271, S-941 26 Piteå, Sweden. To appear in *International Journal of Solids and Structures*.
22. Gudmundson, P. and W. Zang. 1992. "Thermoelastic Properties of Micro Cracked Composite Laminates," SICOMP Technical Note 92-06, Swedish Institute of Composites, Box 271, S-941 26

Piteå, Sweden. *The Eighth International Conference on Mechanics of Composite Materials*, April 20–22, 1993, Riga, Latvia.

23. Parvizi, A., K. W. Garrett and J. E. Bailey. 1978. "Constrained Cracking in Glass Fiber-Reinforced Epoxy Cross-Ply Laminates," *Journal of Materials Science*, 13:195–201.
24. Gudmundson, P. and S. Östlund. 1992. "Numerical Verification of a Procedure for Calculation of Elastic Constants in Micro Cracking Composite Laminates," *Journal of Composite Materials*, 26:2480–2492.

CHAPTER 7

FATIGUE OF COMPOSITES: A CDM PERSPECTIVE

M. Chrzanowski
Cracow University of Technology, Crakow, Poland

ABSTRACT

The paper gives an overview of the Continuum Damage Mechanics (CDM) capability to describe fatigue failure in composites. As both phenomena: fatigue failure and composites deterioration are of high complexity, the general approach offered by Continuum Damage Mechanics is presented, rather than detailed description of both.

The main features of both processes are highlighted and a qualitative description of different aspects of fatigue failure is given. Two stages of the failure process, namely macrocrack nucleation and its propagation are distinguished and shown to be described by means of CDM. High and low cycle fatigue are distinguished, as well as the influence of time dependent behaviour of a material upon its fatigue resistance. A simple damage law for anisotropic materials is proposed, too.

The paper is meant as a source of references for further study of this complex phenomenon which still deserves systematic and fundamental investigations.

1. INTRODUCTION

The description of fatigue in composites sets an extreme challenge to study by a research community, and to its application in engineering practice. Both, the fatigue phenomenon and the mechanical response of composites form separate branches of applied mechanics, and both have been studied extensively during the last few decades, none of these studies reaching a state of saturation. This becomes obvious after we name some problems connected with each.

Fatigue by itself is a very complex phenomenon, as it reflects variable loading effects upon material and structure integrity. Its description falls into the frame of fracture mechanics, involving both elastic and inelastic material response, and also often time dependent behaviour. Moreover, its essential feature is that it is a process which develops with number of cycles leading ultimately to final failure. Typically it consists of two - at least - distinct stages: nucleation of one (or more) predominant crack and its propagation throughout the material which undergoes cyclic loading. Therefore, the classical fracture mechanics, for which the existence of a crack - understood as displacement discontinuity - is "sine qua non" condition of existence, is unable to describe both stages; description of the nucleation period requires special methods to be developed.

There are numerous factors influencing the fatigue failure process:

- mean stress effect,
- effect of compression,
- frequency effect,
- wave shape effect,
- prescribed stress or strain (which distinction is necessary when inelastic deformations are involved),
- stress multiaxiality effect (with the distinction between proportional and nonproportional loading),
- material hardening,
- cyclic hardening,
- influence of loading sequences,
- shake-down phenomenon,
- creep-fatigue interaction including the ratcheting phenomenon, etc.

Fatigue is also very sensitive to the material microstructure, even for isotropic materials, so all the above factors should be reconsidered for composites. On a macroscopic level the mechanical behaviour of composites is characterised by anisotropy, which in a very broad sense is a violation of stress and strain tensor co-axiality. In fact, instead of relating deviatoric parts of the strain and stress tensors

$$e_{ij} = \varepsilon_{ij} - \varepsilon_{kk}\delta_{ij}/3 \tag{1.1}$$

and

$$s_{ij} = \sigma_{ij} - \sigma_{kk}\delta_{ij}/3 \tag{1.2}$$

respectively, by multiplication by a scalar (shear modulus G) for isotropic materials:

$$s_{ij} = 2G\varepsilon_{ij}, \tag{1.3}$$

a tensorial law has to be used for anisotropic composites:

$$\varepsilon_{ij} = S_{ijkl}\sigma_{kl} \tag{1.4}$$

where S_{ijkl} is the compliance matrix.

Thus, it is not only the matter of 21 material constants to be determined in the case of general anisotropy (or, perhaps 9 or 5 material constants, for a FRC which exhibits orthotropy or transverse isotropy), but also problems of off-axis loading, which yield the well known shear coupling effect. If we consider a very broad class of materials which are used for both matrix and fibres, then it becomes apparent that the description of fatigue for anisotropic materials encounters numerous difficulties.

The complexity of the phenomenon is reflected - among others - in a vast number of ASTM STP publications related to fatigue of composites exclusively [1-4] or to fatigue in a broader context of composite material behaviour [5-19]. The proceedings of the ASME Winter Annual Meeting 1990 [20] and materials of the recent school on Fracture and Damage Mechanics of Composite Materials [21] can be used as good sources of reference. However, a coherent and applicable theory of fatigue of composites is non-existent, even if some books have been published covering [22] or relating to this topic [23,24].

In the course of this lecture a potential capability of Continuum Damage Mechanics (termed CDM after proposition by Jan Hult [25]) , which is a branch of fracture mechanics originated from creep applications rather than fatigue problems [26,27], is investigated. Since its quite satisfactory application for isotropic materials will be demonstrated, some conclusions will bear only qualitative character, pointing out the still dormant capabilities of CDM application in the field of fatigue failure.

2. ENGINEERING VERSUS MICROMECHANICAL APPROACHES

There are two extreme approaches to fatigue description (which are also valid for any mechanical behaviour description): microstructural and phenomenological. The first is based on a detailed study of the material structure (which, of course is especially important for man-made materials, but by no means easier than that for natural polycrystalline materials like metals), and leads again to two extreme situations: a very complex quantitative description of a simple phenomenon (like fatigue under constant stress amplitude), or to only qualitative explanation of some special effects, listed above. None of these results satisfies engineers. Therefore they turn towards a phenomenological description based on simple observations. Let us begin from this end.

The typical S-N curves (stress versus number of cycles to fatigue failure) for two FRC are shown in Fig.1 (after reference [32]).

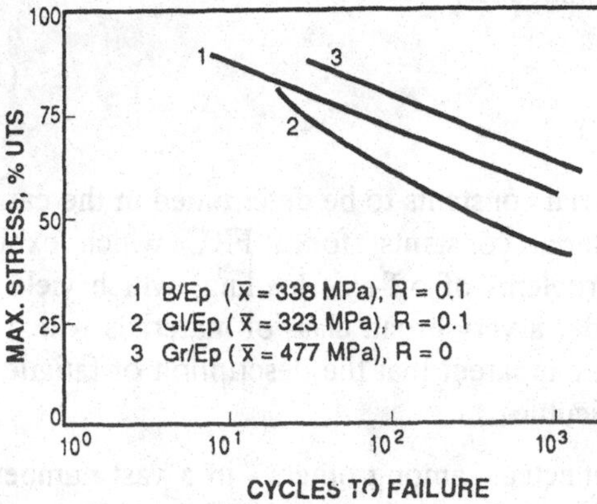

Fig.1. S-N relations for B/Ep, Gr/Ep and Gl/Ep laminates ($\bar{x}$ stands for UTS)

Thus, the simplest attempt to describe fatigue of composites is to approximate experimental results by means of a straight line in a *log*S - *log*N diagram, which yields the following formula for cycling with a constant amplitude S and stress ratio R= -1:

$$S = a + b/N^x, \tag{2.1}$$

where a, b and x are material constants [28], or by a formula proposed in [29] for representing linear grouping of data in an S-*log*N plot:

$$S = UTS - B\, log N,$$

where UTS is ultimate tensile strength and B is a material constant (of order about 0.1 UTS).

Data for cycling with different mean stress can be represented in a similar manner by the Goodman diagram shown in Fig.2 [23], and consequently by a simple formula:

$$\frac{S_a}{S} = 1 - \frac{S_m}{UTS}, \tag{2.2}$$

where S_a and S_m are applied stress amplitude and mean stress, whereas S is stress amplitude to cause failure at a given number of cycles, as previously.

These simple formulae are not satisfactory when a variable stress amplitude is applied. If it is of a step-wise nature (i.e. the stress amplitude in each time interval S_i remains constant) then the Palmgren-Miner formula can be applied:

$$\sum_{i=1}^{m} \frac{n_i}{n_{*i}} = 1, \tag{2.3}$$

where n_i is the number of cycles with amplitude S_i, and n_{*i} is the number of cycles to failure for this stress amplitude, and m is number of applied blocks.

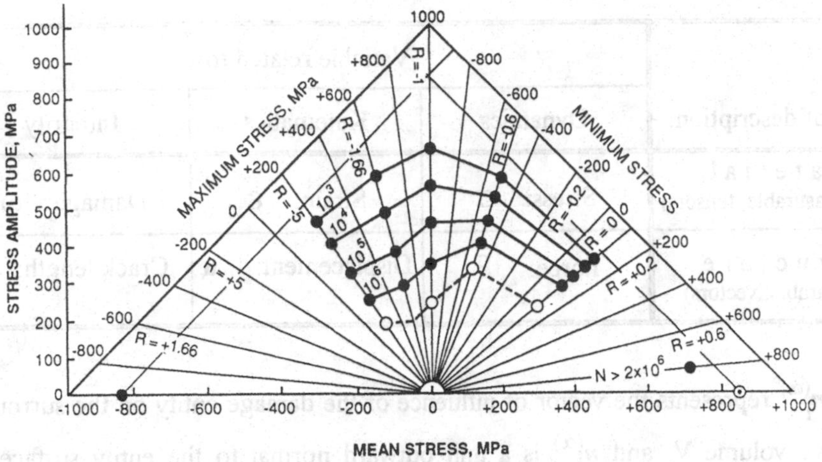

Fig.2. Constant life diagram of Gr/Ep

Between these engineering approaches and micromechanics lies a vast area of continuum mechanics which make use of useful, though arbitrarily defined, tensorial quantities representing the intensity of inner forces and deformation in a material point, i.e. stress and strain tensors. These are common tools for engineers analysing structural behaviour, but incapable to describe gradual material deterioration which is a characteristic process for fatigue or creep rupture. It is then quite natural to add a new tensorial quantity on the material point level to describe this process. Table 1 systematises the use of engineering variables on a material level (non-measurable tensorial quantities) and on a structural level (measurable vectorial quantities).

Like stress and strain, damage requires two questions to be answered:

1. How it is defined, and
2. How it is related to its corresponding vectorial quantity (similarly to the equilibrium equation for stress-force relation $\sigma_{ij,j} + F = 0$,
   or to the strain - displacement relation: $\varepsilon_{ij} = (u_{i,j} + u_{j,i})/2$).

Separate studies have to be carried out to answer both questions, referred to micromechanics to answer question #1, and to thermodynamics of irreversible processes to answer question #2. As it is out of the scope of this lecture, let us confine ourselves to the simplest definition of a damage tensor given by Vakulenko and Kachanov,Jr. [30]:

$$\omega_{ij} = \frac{1}{V}\sum_{\alpha} a^{(\alpha)} n_i^{(\alpha)} \otimes n_j^{(\alpha)} dA^{(\alpha)} , \tag{2.4}$$

Table 1. State variable definitions

Level of description:	Variable related to:		
	Dynamics	Kinematics	Integrity
Material (non-measurable, tensors)	Stress: σ	Strain: ε	Damage: ω
Structure (measurable, vectors)	Force: **F**	Displacement: **u**	Crack length: **l**

where $a^{(\alpha)} n_i^{(\alpha)}$ represents the vector of influence of the damage entity on the surrounding representative volume V, and $n_j^{(\alpha)}$ is a unit outward normal to the entity surface $A^{(\alpha)}$ (Fig.3). The summation is to be taken over all damage entities in V. More details can be found elsewhere [20,22].

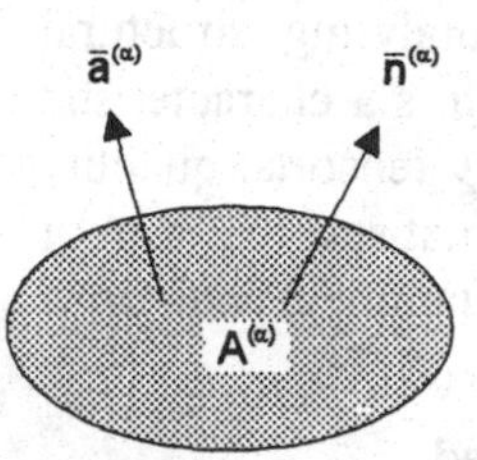

Fig. 3. Damage entity definition

An evolution law for damage was originally proposed by L.M.Kachanov [26] for creep conditions:

$$\dot{\omega} = C\left(\frac{\sigma}{1-\omega}\right)^m, \tag{2.5}$$

where C and m are material constants to be evaluated from a *log* σ - *log t** diagram, *t** being time to failure in creep conditions, and σ - main principal stress in a given point; a dot over a quantity represents its time derivative. The scalar damage representation ω used here can be understood as a special case of damage tensor (uniaxial damage), and therefore equation (2.5) becomes a special case of the more general nonlinear (with respect to both damage and stress) evolution law:

$$\dot{\omega}_{ij} = g(\sigma_{ij}, \omega_{ij})\, \sigma_{ij} \tag{2.6}$$

which is analogous to the viscoplasticity of isotropic materials (cf. e.g. Zienkiewicz-Cormeau [31]):

$$\dot{e}_{ij} = \Gamma(\sigma_{ij})\, \sigma_{ij}, \tag{2.7}$$

where $\dot{e}_{ij}$ stands for the rate of the deviatoric part of visco-plastic strain , extended to visco-plasticity coupled with damage [32]:

$$\dot{e}_{ij} = \Gamma(\sigma_{ij}, \omega_{ij})\, \sigma_{ij} \,. \tag{2.8}$$

Before we will propose a corresponding law for fatigue of composites we have to examine the ability of the above equation to be used for description of fatigue of isotropic materials (e.g. metals). First, let us study the development of fatigue deterioration due to load cycling.

3. FATIGUE AS A PROCESS

As the CDM approach to fatigue description originates from the description of creep material deterioration, let us make a brief comparison of both processes, which are summarised in Fig.4.

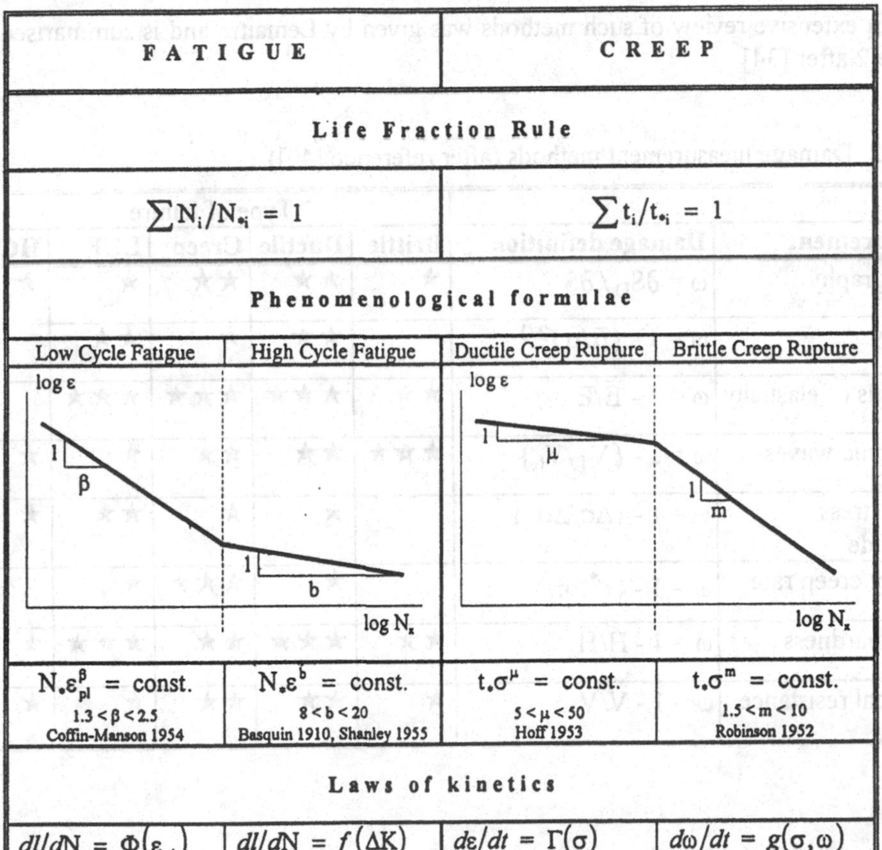

Fig.4. Fatigue versus creep comparison

The similarities between fatigue and creep failure are obvious. The main difference can be observed when examining laws of kinetics: these for fatigue are given in terms of crack growth, whereas for creep they are expressed in terms of state variables. Thus, the kinetics of damage growth which precedes the crack propagation cannot be described in terms of classical fracture mechanics. In chapter 4 the extension of damage kinetics law (2.5) will be discussed, and the results of its application for fatigue and creep-fatigue failure will be shown. The general conclusion which can be drawn from Fig.4 is that the use of a damage variable should be possible in the range of cycling loading to reproduce typical fatigue behaviour.

The problem of damage measurement and identification remains still open. However, according to the classification given in the Table 1 damage is a non-measurable quantity, and the only way to evaluate it is to perform indirect measurement of these material characteristics whose change can be attributed to damage development (called damage monitoring, cf. e.g. [33]).

An extensive review of such methods was given by Lemaitre and is summarised here in Table 2 after [34].

Table 2. Damage measurement methods (after reference [40])

		Type of failure				
Measurement	**Damage definition**	**Brittle**	**Ductile**	**Creep**	**LCF**	**HCF**
Micrography	$\omega = \partial S_D / \partial S$	★	★★	★★	★	★
Density	$\omega = 1 - \{\bar{\rho}/\rho\}^{2/3}$		★★	★	★★	
Modulus of elasticity	$\omega = 1 - \bar{E}/E$	★★	★★★	★★★	★★★	
Ultrasonic waves	$\omega = 1 - \{\bar{V}_L/V_L\}$	★★★	★★	★★	★	★
Cyclic stress amplitude	$\omega = 1 - \{\Delta\sigma/\Delta\sigma^*\}$		★	★	★★	★
Tertiary creep rate	$\omega = 1 - \{\dot{\varepsilon}^*/\dot{\varepsilon}_t\}^n$		★	★★★	★	
Micro-hardness	$\omega = 1 - H/\bar{H}$	★★	★★★	★★	★★★	★
Electrical resistance	$\omega = 1 - V/\bar{V}$	★	★★	★★	★	★

Notations:

∂S_D - fractured area of studied surface

∂S - initial area of studied surface,

ρ - density,

E - modulus of elasticity,
V_L - longitudinal wave velocity,
$\Delta\sigma$ - stress amplitude under strain cycling,
$\Delta\sigma^*$ - saturated stress amplitude under strain cycling,
$\dot{\varepsilon}_t$ - tertiary creep rate,
$\dot{\varepsilon}^*$ - minimum creep rate,
H - micro-hardness,
V - potential difference.
and all quantities with a bar denote its value for a damaged material.

Grades in Table 2 are given here by stars from very good (★★★) to poor (★). These are mostly referred to metals and some of them are irrelevant for composites (e.g. hardness or electrical resistance). Some of them, especially for fatigue, are very convincing, at least for metals for low cycle fatigue (c.f. Fig.5 from [35] showing a drop in stress

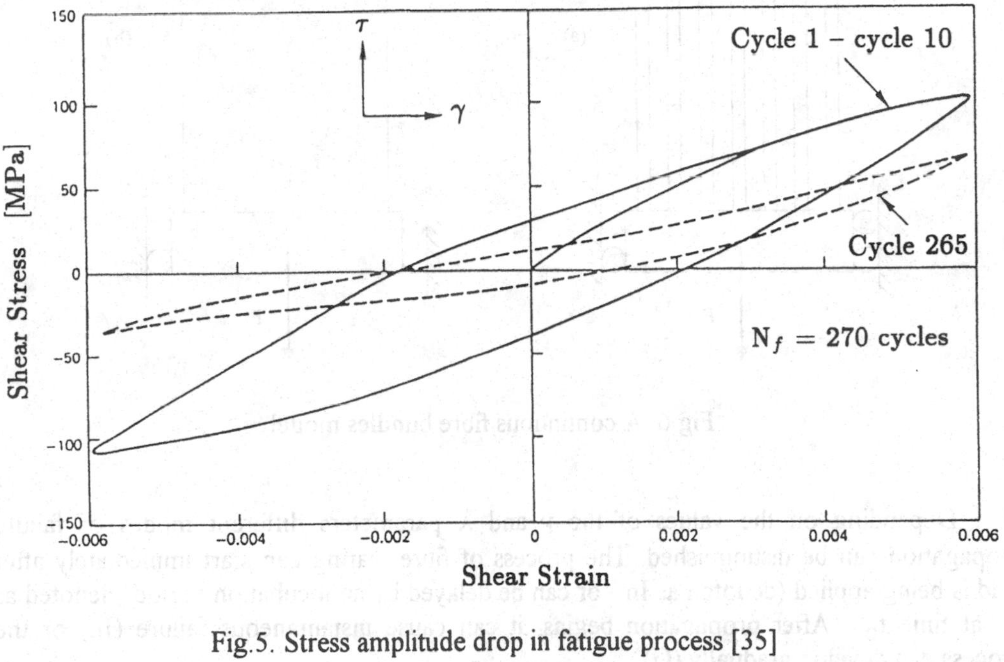

Fig.5. Stress amplitude drop in fatigue process [35]

amplitude under prescribed strain cycling).

Although the most characteristic for composites is dispersed damage connected with matrix cracking, gradual failure may cause the process to be extended over an essential part of the structure's life. This is the case of weak interface, which may cause the composite to behave like a dry fibre bundle. Such a study of fibre bundles (in creep conditions, but we have already seen the similarities between creep and fatigue processes) was carried out in[36] and had demonstrated different behaviour of a structure, depending on a variation in creep modulus and ultimate strength:

$$M = \overline{M}\ \{ 1 + \lambda\ (2x\text{-}1)\}\ , \tag{3.1a}$$

$$U = \overline{U}\ \{ 1 + \nu\ (2x\text{-}1)\}\ , \tag{3.1b}$$

where $\overline{M}$ and $\overline{U}$ are mean values of creep modulus (in linear viscoelasticity), and ultimate strength, respectively, and x is a dimensionless coordinate across bundle width (0<x<1), cf. Fig.6.

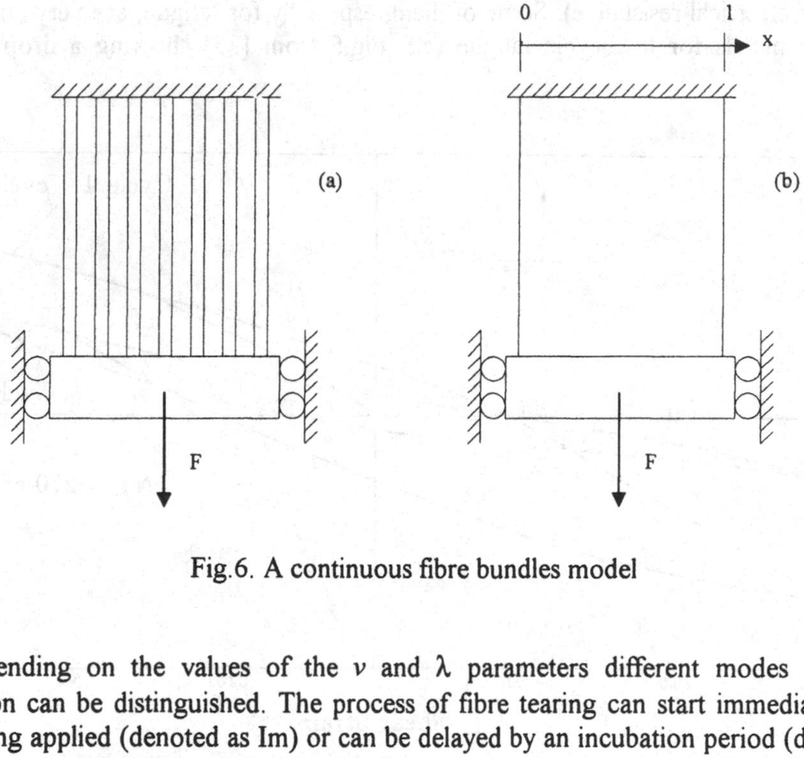

Fig.6. A continuous fibre bundles model

Depending on the values of the ν and λ parameters different modes of failure propagation can be distinguished. The process of fibre tearing can start immediately after load is being applied (denoted as Im) or can be delayed by an incubation period (denoted as D) at time t_o . After propagation begins, it can cause instantaneous failure (In) or the process can develop gradually (G).

All these situations are indicated by corresponding notations in a diagram (Fig.7) on the ν,λ plane , and summarised in Fig.8, with c denoting advancement of failure front across fibre bundles, and t_r standing for time to rupture.

The above study becomes, of course, much more complicated when the fibre bundle is immersed in a matrix allowing fibre-matrix interaction. Nevertheless, it demonstrates that failure is a process which can take different courses depending on material properties.

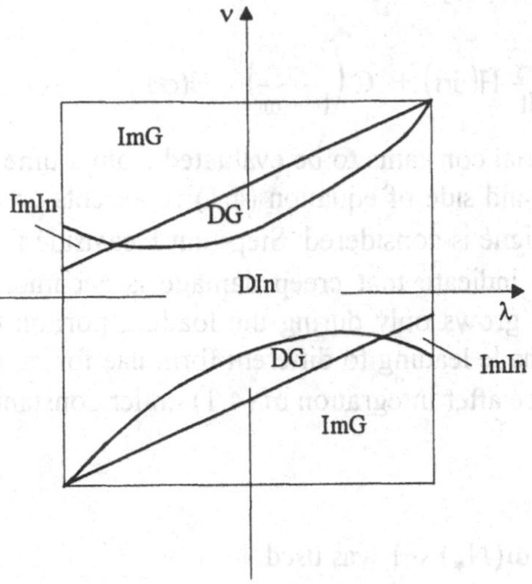

Fig.7. Different failure modes in the ν,λ plane

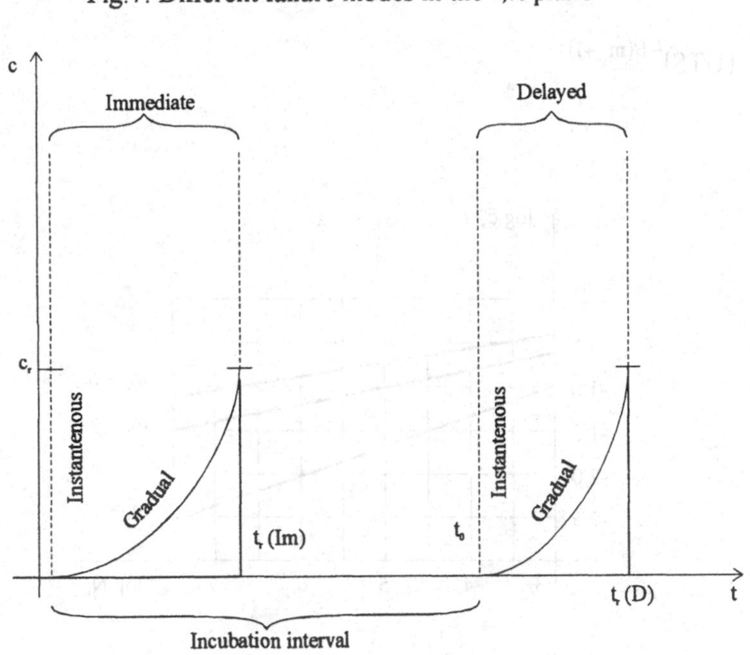

Fig.8. Different qualitative behaviour of fibre bundle model

4. CONTINUUM DAMAGE MECHANICS APPLIED TO FATIGUE DESCRIPTION

To describe fatigue failure, which essentially is a time-independent process (at least when creep effects are not taken into account), the evolution law for damage (2.5) has to be extended, e.g. as proposed in [37]:

$$\dot{\omega} = C_0\left(\frac{\sigma}{1-\omega}\right)^{m_0} \frac{d\sigma}{dt} H(d\sigma) + C\left(\frac{\sigma}{1-\omega}\right)^{m} H(\sigma)\,, \tag{4.1}$$

where C_0 and m_0 are material constants to be evaluated from a time - independent test. The second term on the right hand side of equation (4.1) represents creep damage, so it has to be neglected when only fatigue is considered. Step-unit Heaviside functions $H(d\sigma)$ and $H(\sigma)$ were introduced here to indicate that creep damage is accumulated only under tensile stress, and fatigue damage grows only during the loading portion of a cycling. However, other assumptions can be made leading to different formulae for number of cycles to failure N_*, than these obtained here after integration of (4.1) under constant stress amplitude σ_a :

$$N_* = 1/C_0\sigma_a^{(m_0+1)}\,, \tag{4.2}$$

where the failure condition $\omega(N_*) = 1$ was used.

If in the first cycle the ultimate strength UTS will be reached, then by integration of the first term in (4.1) one can find that:

$$C_0 = (\mathrm{UTS})^{-1/(m_0+1)} \tag{4.3}$$

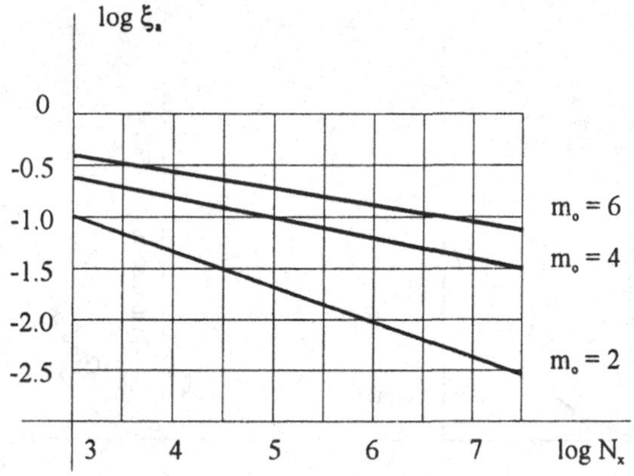

Fig.9. Representation of S-N diagram by means of a CDM equation (4.1)

The value of the material constant m_0 determines the slope of $log\sigma_a$ - $logN_*$ as shown in Fig.9. [38], where the dimensionless stress $\xi_a = \sigma_a$ /UTS is used.

The effect of mean stress can be also reproduced by means of the first term of Eq. (4.1), as shown in Fig.10.

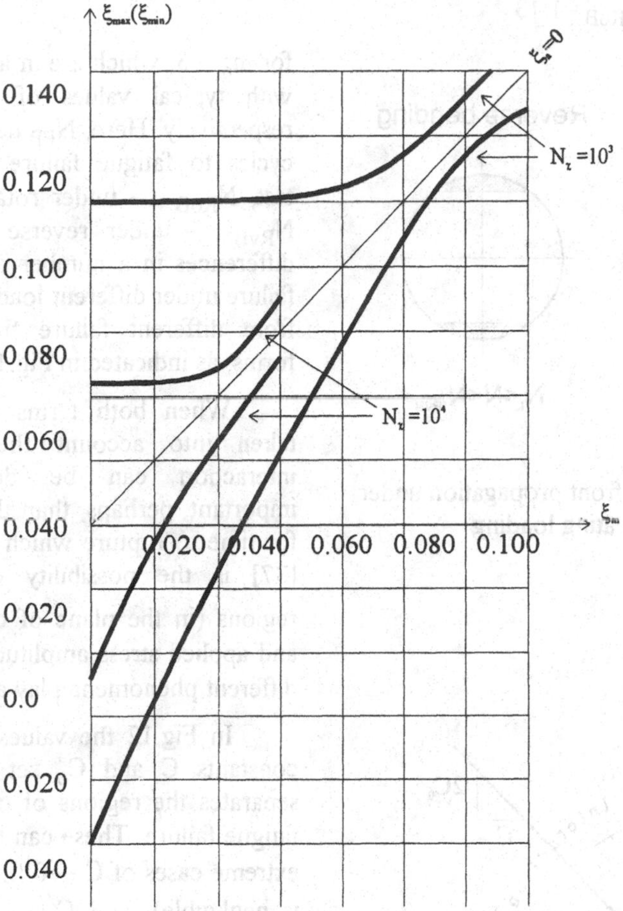

Fig.10. Mean stress influence on fatigue life (Smith diagram)

The above diagrams correspond to the time of first macroscopic crack incubation. In the case of uniaxial state of stress this time coincides with the time of total failure as damage is uniformly distributed over the volume of a specimen and reaches its critical value of ω=1 at the same instant of time in a given cross-section.

The most interesting results were, however, obtained for the second stage of fatigue failure propagation, i.e. failure front movement in the case of uniaxial but non-uniformly distributed stresses. Comparison of rotary bending (a typical fatigue test) and reverse bending fatigue, with that of uniaxial push-pull test has given the values:

$$N_{*PP} / N_{*RoB} = 0.91$$

$$N_{*RvB} / N_{*RoB} = 1.13 \tag{4.4}$$

for $m_0 = 3$, which are in a good agreement with typical values of 0.85 and 1.05, respectively. Here, N_{PP} denotes number of cycles to fatigue failure under push-pull test, N_{RoB} - under rotary bending, and N_{RvB} - under reverse bending. These differences in a number of cycles to final failure under different loading modes result from different failure front propagation forms, as indicated in Fig.11.

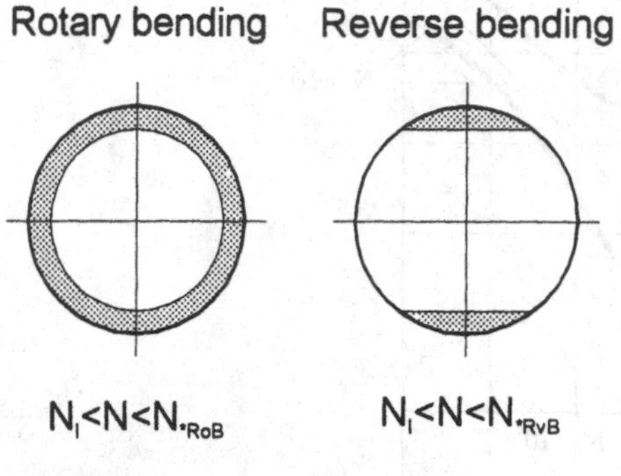

Fig.11. Failure front propagation under alternating loading

When both terms of Eq.(4.1) are taken into account then creep-fatigue interaction can be described. More important, perhaps, than the exact formula for time to rupture which can be found in [37] is the possibility of defining the regions (in the plane of cycling period T and applied stress amplitude σ_a) in which different phenomena play a crucial role.

In Fig.12 the values of the material constants C and C_0 set the line which separates the regions of creep failure and fatigue failure. These can be viewed as the extreme cases of $C \rightarrow 0$ (i.e. creep damage is negligible), and $C_0 \rightarrow 0$ (no fatigue damage case) as indicated in Fig.13a and 13b, respectively.

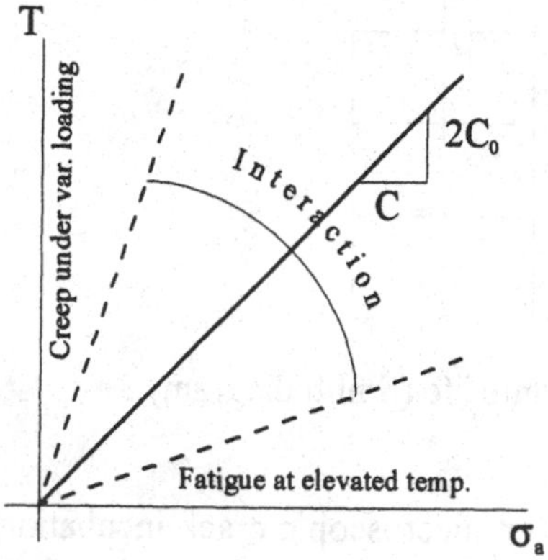

Fig.12. Creep-fatigue dominant region definition

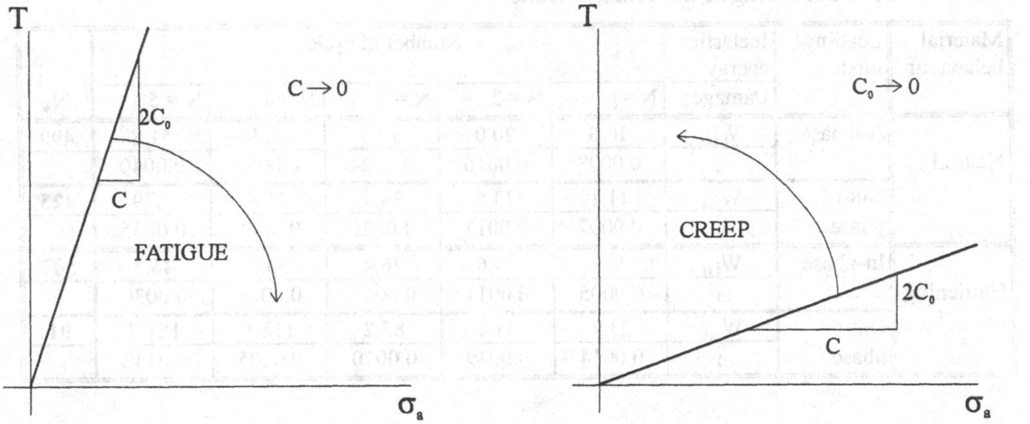

Fig.13. Extreme cases of pure creep and fatigue failure

All the above description holds when stress is prescribed, and therefore is suitable for high cycle fatigue rather then for low cycle fatigue, when inelastic deformations come to play a decisive role. Because of the complexity of the fatigue process in these circumstances a totally different approach was proposed in [39] to take into account numerous effects listed in section 1.

An assumption was made that damage growth is controlled by the inelastic energy dissipated in the course of the fatigue process. The evolution law takes the form:

$$(1-\omega)^n \frac{d\omega}{dN} = T^k B\, W_h^n\,, \tag{4.5}$$

where B and n are material constants, and W_h is a hysteresis energy which has to be measured or calculated by an appropriate constitutive law. In the paper [39] the Viscoplasticity Based on Overstress (VBO) by Krempl [40] was used to evaluate W_h which varies with number of cycles and depends on the cyclic material behaviour (neutral or hardening). The results of calculations for in- and out- of phase cycling are given in Table 3 demonstrating the ability of the proposed theory to be used for life prediction in low cycle fatigue circumstances.

Table 3. Low cycle fatigue modelling results

Material behaviour	Loading mode	Inelastic energy / Damage	Number of cycles					
			N=1	N=2	N=3	N=4	N=5	N_*
Neutral	In-phase	W_{in}	10.0	20.0	30.3	40.9	51.8	**499**
		ω	0.0008	0.0016	0.0024	0.0032	0.0040	
	Out-of-phase	W_{in}	11.1	23.6	34.7	45.8	56.9	**425**
		ω	0.0007	0.0017	0.0026	0.0035	0.0045	
Hardening	In-phase	W_{in}	9.3	17.6	26.4	35.3	44.5	**507**
		ω	0.0005	0.0011	0.0017	0.0023	0.0030	
	Out-of-phase	W_{in}	23.7	53.4	85.2	118.1	151.7	**91**
		ω	0.0014	0.0039	0.0070	0.0105	0.0142	

The results discussed above were obtained for isotropic laws of damage growth, and therefore are not suitable for composites which exhibit strong anisotropy. However an appropriate constitutive equation can be proposed following the way in which a coupled theory for isotropic materials was formulated.

Let us begin from viscoplasticity theory coupled with damage given by the Eqs. (2.6) and (2.8):

$$\dot{e}_{ij} = \Gamma\left(\sigma_{ij}, \omega_{ij}\right) \sigma_{ij}, \tag{4.6a}$$

$$\dot{\omega}_{ij} = g\left(\sigma_{ij}, \omega_{ij}\right) \sigma_{ij}, \tag{4.6b}$$

where g and Γ are scalar functions of stress and damage tensor invariants.

For anisotropic materials these functions have to be replaced by compliance matrices W_{ij} and B_{ij}, respectively:

$$\dot{e}_{ij} = B_{ijkl}\, \sigma_{kl}, \tag{4.7a}$$

$$\dot{\omega}_{ij} = W_{ijkl}\, \sigma_{kl}, \tag{4.7b}$$

whose elements have to be non-increasing functions of the damage tensor invariants (if the influence of the damage fatigue process on its rate can be identified), and , possibly, the functions of stress if non-linearity with respect to stress is observed. Similarly to the extension of viscoplasticity theory to incorporate elastic strains:

$$\dot{e}_{ij} = S_{ijkl}\left(\sigma_{kl}, \omega_{kl}\right)\dot{\sigma}_{kl} + B_{ijkl}\left(\sigma_{kl}, \omega_{kl}\right) \sigma_{kl} \tag{4.8a}$$

and making use of the analogy with Eq. (4.1), one can postulate an extended damage law for anisotropic materials:

$$\dot{\omega}_{ij} = V_{ijkl}(\sigma_{kl}, \omega_{kl})\dot{\sigma}_{kl} + W_{ijkl}(\sigma_{kl}, \omega_{kl})\,\sigma_{kl}\,, \tag{4.8b}$$

where V_{ijkl} is a new compliance matrix characterising the ability of a material to undergo time-independent deterioration. The matrix S_{ijkl} is a classical compliance matrix, modified for nonlinear elasticity. Though one can assume that this matrix should be stress independent (as classical an-isotropic elasticity is), the V_{ijkl} matrix has to be both stress and damage dependent, as static load causes nonlinearity of the material deterioration process.

Particular forms of all the above matrices may be hard to determine, and perhaps some simplification should be made, e.g. assuming that these matrices are influenced only by damage tensor invariants. Further, if we assume that all components of all matrices are multiplied by a scalar function of damage invariants (e.g. damage tensor trace), then an initial anisotropy will remain unaffected by the damage process, though the elements of all matrices will change in course of deterioration development.

It is worthwhile to notice that the above equations give:
for both purely time independent processes:

$$d\varepsilon = S\,d\sigma\,, \tag{4.9a}$$

$$d\omega = V\,d\sigma\,, \tag{4.9b}$$

and solely time-dependent ones:

$$d\varepsilon/dt = B\,d\sigma\,, \tag{4.10a}$$

$$d\omega/dt = W\,d\sigma\,, \tag{4.10b}$$

respectively (an absolute notation was used here for a clarity) Stress and time can be eliminated from these special cases yielding:

$$d\omega = \mathbf{V}\,\mathbf{S}^{-1}d\varepsilon\,, \tag{4.11a}$$

$$d\omega = \mathbf{W}\,\mathbf{B}^{-1}d\varepsilon\,, \tag{4.11b}$$

i.e. damage growth in both cases is governed by the growth of deformation only, provided the matrices **B**, **S**, **V**, **W** are not directly influenced by the state of stress (however they are indirectly influenced by stress via the damage tensor).

Finally, the set of governing equations (4.8) has to be completed by the failure criterion, which in a general form can be written as:

$$\Phi(\omega_{ij}) = 1\,, \tag{4.12}$$

where a scalar function $\mathbf{\Phi}$ of damage tensor invariants has to be chosen to fit experimental data. In its simplest form it can be set to:

$$\omega_1 = 1\,, \tag{4.13}$$

where ω_1 is main principal value of the damage tensor, which is a classical assumption for scalar representation of damage.

The sketched theory is capable of describing not only time-independent and time-dependent phenomena, but also the influence of the damage process upon initial anisotropy and take into account a nonlinear dependence upon state of stress.

Acknowledgement

This work was partially supported by the KBN grant 3 P404 043 06.

REFERENCES

1. Fatigue of Composite Materials, ASTM STP 569, 1975.
2. Fatigue of Filamentary Composite Materials, ASTM STP 636, 1977.
3. Fatigue of Fibrous Composite Materials, ASTM STP 723, 1981.
4. Composite Materials: Fatigue and Fracture, ASTM STP 907, 1986.
5. Composite Materials: Testing and Design (Second Conference), ASTM STP 497, 1972.
6. The Test Methods for High Modulus Fibers and Composites, ASTM STP 521, 1973.
7. Composite Materials: Testing and Design (Third Conference), ASTM STP 546, 1974.
8. Composite Reliability, ASTM STP 580, 1975.
9. Fracture Mechanics of Composites, ASTM STP 593, 1975.
10. Composite Materials: Testing and Design (Fourth Conference), ASTM STP 617, 1977.
11. Advanced Composite Materials - Environmental Effects, ASTM STP 658, 1978.
12. Composite Materials: Testing and Design (Fifth Conference), ASTM STP 674, 1979.
13. Fatigue Mechanisms, ASTM STP 676, 1979.
14. Nondestructive Evaluation and Flow Criticality for Composite Materials, ASTM STP 696, 1979.
15. Test Methods and Design Allowables for Fibrous Composites, ASTM STP 734, 1981.
16. Damage in Composite Materials, ASTM STP 775, 1982
17. High Modulus Fiber Composites in Ground Transportation and High Volume Applications, ASTM STP 873,1983.
18. Delamination and Debonding of Materials, ASTM STP 876, 1985.

19. Composite Materials: Testing and Design (Ninth Volume), ASTM STP 1059, 1990.

20. Dvorak, G.J., Lagoudas, D.C. (eds): Microcracking-Induced Damage in Composites, AMD-Vol.111, MD-Vol.22, ASME, 1990.

21. Beaumont, W.R, Crane, R.L., Ryder, J.T.: Fracture and Damage Mechanics of Composite Materials, Reference Materials for the seminar held 1-3 April, 1992, San Antonio, Texas, USA, Technomic Publishing Co., Lancaster, PA 1992.

22. Talreja, R.: Fatigue of Composite Materials, Technomic Publishing Co., Lancaster, PA 1987.

23. Zwben,C., Hahn.H.T., Chou,T.-W.: Mechanical Behavior and Properties of Composite Materials, in: Delaware Composites Design Encyclopaedia, vol.1., Technomic Publishing Co., Lancaster, PA, 73-127, 1989.

24. Datoo, M.H.: Mechanics of Fibrous Composites, Elsevier, 1991.

25. Janson, J., Hult, J.: Fracture Mechanics and Damage Mechanics a Combined Approach, J.mécanique appliquée, 1,1(1977),69-84.

26. Kachanov, L.M.: On time to rupture in creep conditions (in Russian), Izv. An SSSR, OTN (1958), 26-31.

27. Rabotnov, Yu.N., On a Mechanism of Delayed Fracture (in Russian), in: Vopr. Prochn. Mat. Konstr., Izd. AN SSSR (1959), 5-7.

28. Sims, D.F., Brogdon, V.H.: Fatigue Behavior of Composites Under Different Loading Modes, in: ASTM STP 636 (1977), 185-205.

29. Mandel, J.F., Huang, D.D., McGarry, F.J.: In: Proc. 36th Ann.Tech. Conf. RP/C Inst., SPI, Inc., 10-A, 1981.

30. Vakulenko, A.A., Kachanov, M.L.: Continuum Theory of Media with Cracks (in Russian), Mekh. Tv. Tela, 4 (1971), 159-166.

31. Zienkiewicz, O.C., Cormeau, I.C.: Visco-Plasticity, Plasticity, and Creep in Elastic Solids - A Unified Numerical Solution Approach, Int.J.Num.Meth. in Eng., 8 (1974), 821-853.

32. Chrzanowski, M., Bodnar, A., Latus, P.: Lifetime Evaluation of Creeping Structures, in: Proc. 5th Int.Conf. on Creep, Lake Buena Vista, Florida, USA, 18-21 May, 1992, ASM, 1992, 461-469.

33. Coffin L.F.: Notch Fatigue Crack Initiation Studies in a High Strength Nickel Base Superalloy, Eng. Fract. Mech., 28 (1987), 485-503.

34. Lemaitre, J.: Damage Measurements, Eng. Fract. Mech., 28 (1987), 643-661.

35. Skelton, R.P.: High Strain Fatigue Testing at Elevated Temperature: a Review, High Temp.Techn., 3 (1985), 179-194.

36. Chrzanowski, M., Hult, J.: Ductile Creep Rupture of Fibre Bundles, Eng. Fract. Mech., 28 (1987), 681-688.

37. Chrzanowski, M., Kolczuga, M.: Continuous Damage Mechanics Applied to Fatigue, Mech. Res. Comm., 7 (1980), 41-46.

38. Chrzanowski, M.: The Use of the Damage Concept in Describing Creep-Fatigue Interaction Under Prescribed Stress, Int.J.Mech.Sc., 18 (1976), 69-73.

39. Chrzanowski, M.: A Strain Energy Governed Damage Law For High Temperature Low Cycle Fatigue, Eng. Trans., 39 (1991), 389-418.

40. Krempl, E., McMahon, J.J.,Yao, D.: Viscoplasticity Based on Overstress with a Different Growth Law for the Equilibrium Stress, Mech. Mat., 5 (1986), 35-48.

CHAPTER 8

COMPUTATIONAL METHODS IN COMPOSITE ANALYSIS AND DESIGN

F.G. Rammerstorfer
Vienna Technical University, Vienna, Austria
and
K. Dorninger
SATURN Corp., Troy, MI, USA
and
A. Starlinger
AIREX Composite Engineering, Sins, Switzerland
and
I.C. Skrna-Jakl
Vienna Technical University, Vienna, Austria

ABSTRACT

In this Chapter formulations of special finite shell elements for the analysis of composite shell structures (layered fiber-composite and sandwich shells) as well as computational considereations of local effects are described. Nonlinearities due to large deformations and progressive damage as well as local and global loss of stability are treated. Finally, some practical applications of the computational procedures are described.

List of Variables:

The following notations (with or without sub- or superscripts, respectively) are used for describing the material behavior:

E: Young's modulus
G: Shear modulus
ν: Poisson's ratio
α: Linear coefficient of thermal expansion (CTE)
σ_{lCu}: Ultimate uniaxial compressive stress in fiber direction
σ_{lTu}: Ultimate tensile stress in fiber direction
σ_{qCu}: Ultimate compressive stress normal to fiber direction
σ_{qTu}: Ultimate tensile stress normal to fiber direction
τ_{lqu}: Ultimate shear stress
$\underset{\approx}{\mathbf{C}}$: Material (tangent) stiffness matrix
$\underset{\approx}{\mathbf{K}}$: (Tangent) stiffness matrix (or contributions)
$\underset{\sim}{\tau}$: Cauchy stresses
$\underset{\approx}{T}$: Matrix containing Cauchy stress components
$\underset{\sim}{\varepsilon}$: Almansi strains
r, s, t: Natural, i.e. normalized coordinates in isoparametric formulation
$\underset{\sim}{u}$: Vector of nodal displacements
$\underset{\sim}{\mathbf{r}}$: Vector of external nodal forces
$\underset{\sim}{\mathbf{f}}$: Vector of internal nodal forces
λ: Load multiplier
$\bar{\eta}$: Eigenvalue, i.e. factor for estimating critical load multiplier

Sub- and Superscripts:

$_l$: Longitudinal
$_q$: Transverse, mostly used in-plane
$_t$: Transverse, mostly used out-of-plane
: Overall effective – no superscript
$_f$: Face layers (of sandwich shell)
$_c$: Core (of sandwich shell)
$_{th}$: Temperature effect
$^{(e)}$: Of the element
^{m} : Increment number
* : Critical state
$\underline{\ }$: At the reference surface
$\hat{\ }$: Difference between upper and lower surface

Operators:

$(\ )^T$: Transposed of matrix ( )
$(\ )^{-1}$: Inverse of matrix ( )

Introduction

In many technical applications of composite materials the geometry or the boundary conditions are rather complicated, and analytical methods do not have the capability of calculating the mechanical behavior with sufficient accuracy. In such cases numerical, i.e. computational, methods are used. As a typical representative of these methods the finite element method is discussed in this Chapter. In order to allow the treatment of large deformations and progressive damage nonlinear formulations are presented for composite shells.

If deformations, stresses, vibrations or the stability behavior of composite lightweight structures are analyzed by computational methods like the finite element method, it generally will not be possible by far to consider micromechanical and macromechanical effects in the same model by simply using a sufficiently fine discretization for the structure. Hence, it makes sense to formulate the micromechanical (or local) phenomena in a way (e.g. analytically, like in other Chapters of this book) which allows their treatment on the level of integration points of the finite elements of the macroscopic, i.e. structural, model. Figure 1 schematically shows the principal ideas of the concept for the semi-analytical treatment of local effects, compare [1].

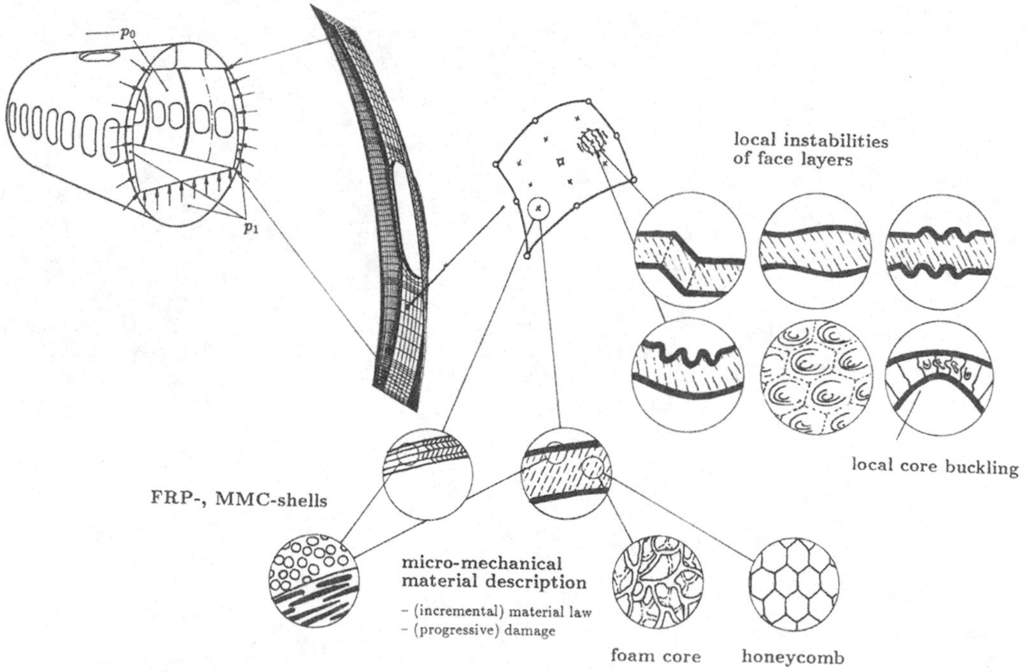

Fig. 1 Concept of semi-analytical treatment of local effects

In this Chapter the above concept is described for some typical local phenomena like progressive matrix or fiber failure in layered fiber reinforded composite shells or face layer wrinkling in sandwich shells, see other Chapters of this book.

On the other hand, some local effects cannot be treated analytically, and numerical methods and models can give insight into the local behavior. The results obtained this way can lead to formulations which can, normally on an approximative level, be used in analytical methods. In the Chapter "Micromechanics for Macroscopic Material Description" [2] this approach has been discussed in connection with micromechanical considerations, and the classical analytical approximation of stress fields near free edges according to e.g. [3,4,5], as discussed in another Chapter of this book [6], is also an application of this concept. In the present Section the numerical investigation of interlaminar stresses in integrally stiffened plates is described as a typical representative of this kind of problems.

1 The LFC-Element for FRP- or Sandwich-Shells

In this subsection an isoparameteric nonlinear finite shell element for layered composite shells is briefly described using Mindlin–Reissner kinematics. LFC means **L**ayered **F**iber reinforced **C**omposite shell.

1.1 FRP Shells

In the case of layered fiber reinforced polymer (FRP) shells many types of finite shell elements for analyzing layered composite shell structures have been developed; see for example the review article [7]. Some of these elements allow the consideration of micromechanical effects like progressive damage due to failure on the micro-level, e.g. fiber or matrix cracking or fiber buckling, in the form of a nonlinear overall material description; see for example [8], where these local effects are taken into account by stiffness degradation of the individual layers on the integration point level.

A detailed description of the theoretical background of the LFC-element along with a number of illustrative examples can be found in [8]. Here, only the outline of the element formulation is presented with special emphasis on the failure analysis modelling for FRP shells.

The element formulation is based on the degeneration principle, see [9]. By using objective strain and stress measures (here Cauchy stresses and Almansi strains are used), geometrically nonlinear behavior in terms of large deformations is included.

The incremental finite element equation has the well known form:

$$[{}^{m}\underset{\approx}{\mathbf{K}}_{e} + {}^{m}\underset{\approx}{\mathbf{K}}_{g}](\Delta \underset{\sim}{u})^{1} = {}^{m+1}\underset{\sim}{r} - ({}^{m}\underset{\sim}{f} - \Delta \underset{\sim}{f}_{th}) \qquad (1).$$

The usual iterative application of eqn. (1) in each increment improves the result up to a given accuracy.

For the updated Lagrange formulation the stiffness matrices and the nodal force vectors for one element (e) are given by the following integrals over the element volume (^{m}V):

$$ {}^m\underset{\approx}{\mathbf{K}}_e^{(e)} = \int\limits_{{}^mV} {}^m\underset{\approx}{\mathbf{B}}_l^T \, {}^m\underset{\approx}{\mathbf{C}} \, {}^m\underset{\approx}{\mathbf{B}}_l \, d^mV \qquad (2), $$

$$ {}^m\underset{\approx}{\mathbf{K}}_g^{(e)} = \int\limits_{{}^mV} {}^m\underset{\approx}{\mathbf{B}}_{nl}^T \, {}^m\underset{\approx}{\mathcal{T}} \, {}^m\underset{\approx}{\mathbf{B}}_{nl} \, d^mV \qquad (3), $$

$$ {}^m\underset{\sim}{\mathbf{f}}^{(e)} = \int\limits_{{}^mV} {}^m\underset{\approx}{\mathbf{B}}_l^T \, {}^m\underset{\sim}{\tau} \, d^mV \qquad (4), $$

$$ \Delta\underset{\sim}{\mathbf{f}}_{th}^{(e)} = \int\limits_{{}^mV} {}^m\underset{\approx}{\mathbf{B}}_l^T \, {}^m\underset{\approx}{\mathbf{C}} \, \underset{\sim}{\alpha} \Delta\vartheta \, d^mV \qquad (5). $$

The current material matrix ${}^m\underset{\approx}{\mathbf{C}}$ (in the Chapter "Lamination Theory" [6] this matrix is denoted by $\underset{\approx}{\hat{\mathbf{E}}}$ for the linear elastic material) depends on the local state of damage. $\Delta\underset{\sim}{\mathbf{f}}_{th}$ is the vector of internal nodal forces equivalent to stress increments resulting from a temperature increment $\Delta\vartheta$ and is computed by using the direction dependent coefficients of linear thermal expansion (vector $\underset{\sim}{\alpha}$). (Moisture effects can be treated in analogy with temperature effects, compare [6]. They are not included in the formulation presented here.)

The computation of the element stiffness matrix requires a three-dimensional integration (see eqns. (2–5)) which usually is performed by some type of numerical integration technique. Since we are dealing with multi-layer shells (with a very large number of layers allowed), where each layer requires at least one (even better two or more) integration points over its thickness, the numerical effort would considerably increase with the number of layers. An effective way for overcoming this difficulty is the use of a quasi-analytical thickness integration; for details see [8].

If the temperature field is assumed to be linearly distributed over the thickness of the shell, with $\overline{\vartheta}$ being the temperature load of the midsurface and $\widehat{\vartheta}$ the temperature difference between opposite points on the two surfaces of the shell, the vector of Cauchy stress components becomes:

$$ \begin{aligned} {}^m\underset{\sim}{\tau}(r,s,t) = {}^m\underset{\approx}{\mathbf{C}}(r,s,t)\Big(& \big[{}^m\underset{\sim}{\overline{\varepsilon}}(r,s) - {}^m\underset{\sim}{\alpha}(r,s,t)\, {}^m\overline{\vartheta}(r,s)\big] + \\ & + t\Big[{}^m\underset{\sim}{\widehat{\varepsilon}}(r,s) - {}^m\underset{\sim}{\alpha}(r,s,t)\, \frac{{}^m\widehat{\vartheta}(r,s)}{2}\Big] + t^2\, {}^m\underset{\sim}{\widetilde{\varepsilon}}(r,s)\Big) \end{aligned} \qquad (6), $$

where $\underset{\sim}{\overline{\varepsilon}}$, $\underset{\sim}{\widehat{\varepsilon}}$ and $\underset{\sim}{\widetilde{\varepsilon}}$ are portions of $\underset{\sim}{\varepsilon}$, which are not, linearly and quadratically depending on t, and

$$^m\underset{\approx}{C}(r,s,t) = {}^m\underset{\approx}{G}^T(r,s)\,\underset{\approx}{C}'(r,s,t)\,{}^m\underset{\approx}{G}(r,s) \tag{7}$$

$$^m\underset{\sim}{\alpha}(r,s,t) = {}^m\underset{\approx}{G}^{-1}(r,s)\,\underset{\sim}{\alpha}'(r,s,t) \tag{8}$$

where quantities $(\ )'$ represent properties with respect to the (r,s,t)-system. The matrix $^m\underset{\approx}{G}$ performs the transformation from the global $\underset{\sim}{x}$ system to the local $\underset{\sim}{x}'$ system, corresponding to the r,s,t directions, compare the Chapter on lamination theory.

The material matrix and the vector of coefficients of linear thermal expansion of the individual layers are derived from analytical micromechanical considerations, see other Chapters of this book [2,10]. They are defined individually with respect to the principal material axes of the layer, i.e. (l,q).

A rotation transformation serves for computing $\underset{\approx}{C}'$ and $\underset{\sim}{\alpha}'$ from $\underset{\approx}{C}_L$ (which corresponds to $\hat{\underset{\approx}{E}}_L$ in the Chapter on lamination theory.) and $\underset{\sim}{\alpha}_L$.

Improvements of such shell elements with respect to their transverse shear behavior can be found e.g. in [1], [11,12] and [13].

In other Chapters of this book the failure behavior of FRP-laminae is described. As long as rather small rotations are involved the relations obtained in [6] can be approximatively applied to Cauchy stresses, too; for problems with large rotations 2nd Piola Kirchhoff stresses should be used instead.

As mentioned in the above Chapter, eqns. (47,48) in [6] represent "secant"-stiffnesses of the material and, therefore, stresses can be computed directly from the total strains. The following procedure, applied at each load step m at each stiffness sampling point r_i, s_j (= 2/D integration points) for all N layers of the laminate, accounts for stiffness changes due to material cracking: The local inplane strains $^n\underset{\sim}{\varepsilon}_L$ at the midsurface of layer n with respect to the layer's local axis, i.e. l,q, are derived from the global strain vector $^m\underset{\sim}{\varepsilon}$; then, with $^n\underset{\sim}{\varepsilon}_L$ and the corresponding local elasticity matrix $^n\underset{\approx}{C}_L$ of the previous step an estimate of the stresses is computed, and in the next step the combined failure criterion is examined. If failure is indicated by a violation of eqn. (42) or eqn. (46) of [6] the local stiffness matrix is changed according to eqn. (47) or eqn. (48), respectively, of [6].

The above approach shows a certain mesh dependency of the finite element results. This behavior can also be used as a criterion for the quality of the finite element mesh.

The simulation of progressive failure is demonstrated in the following example.

Example: Three point bending specimen

For testing the ultimate flexural strength and modulus of an FRP laminate of the type typically used in aircraft structures a simple flexural specimen (i.e. a three point bending bar) is used. This bar is loaded up to complete failure, and the load–

displacement path is recorded, see also [14] and [1]. A comparison of the measured results with the finite element investigations shows the applicability of the strength calculations included in the LFC formulation.

Lay-up: 30 layers of Kevlar29 fabric, effective layer thickness = 0.09 mm
For the FE model each fabric layer is subdivided into two sublayers in order to approximate the woven reinforcement by UD-layers.
→ assumed lay-up: $[(0/90)_{15s}]$... cross-ply, layer thickness = 0.045 mm
Model: one quarter of the bar has been modeled by eight 16-node elements
Material: Kevlar29 (UD)

$E_l = 57.2$ kN/mm²	$E_q = 3.9$ kN/mm²	$\nu_{ql} = 0.35$
$G_{lq} = 2.3$ kN/mm²	$G_{lt} = 2.3$ kN/mm²	$G_{qt} = 2.3$ kN/mm²
$\sigma_{lTu} = 1300$ N/mm²	$\sigma_{lCu} = 227$ N/mm²	$\tau_{lqu} = 34$ N/mm²
$\sigma_{qTu} = 12$ N/mm²	$\sigma_{qCu} = 53$ N/mm²	$\beta_E = 0.2$

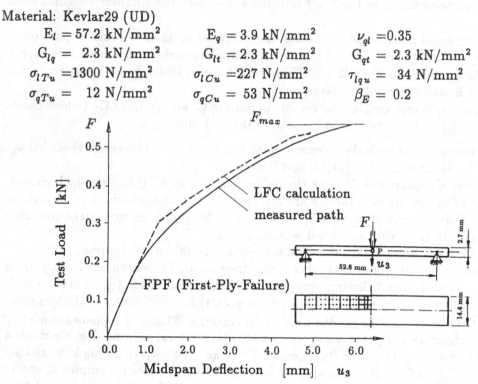

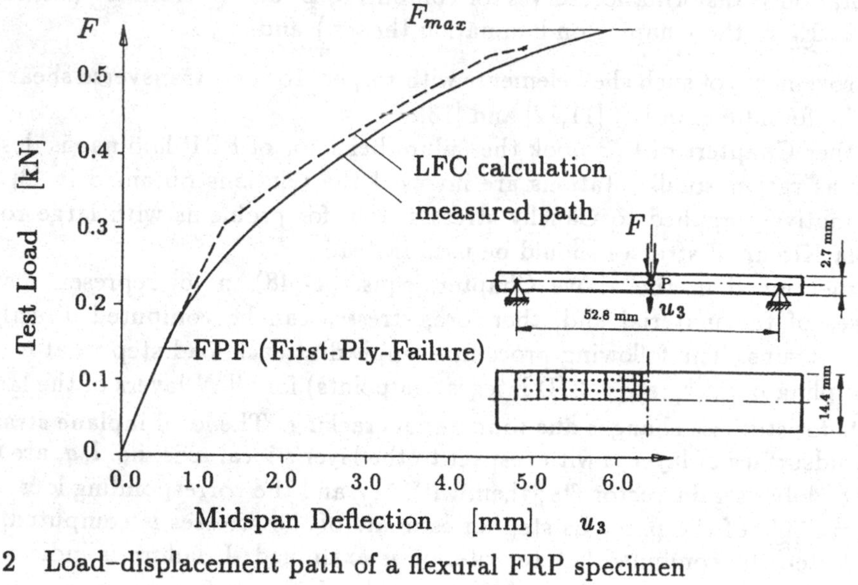

Fig. 2 Load–displacement path of a flexural FRP specimen

Figure 2 shows the calculated and the measured load–displacement path of the flexural specimen. Reasonably good agreement is obtained. First-ply-failure occurs at a relatively low load level and might be interpreted as the beginning of the nonlinear deformations of the experimentally obtained curve.

1.2 Sandwich Shells

The above LFC-element can also be used for sandwich shells with thin face layers and thick core. Under the assumption of antiplane core conditions the specific material matrices and the treatment of local instabilities are discussed in [6]. Using these modifications the LFC-element becomes the LM-element (Layer Model element) [15].

As the stiffness of the sandwich structure may be changed considerably by local buckling phenomena, the influence of regions with buckled face layers on the global deformation and stability behavior has to be considered for further postcritical loading. The following steps describe the procedure in accounting for local instability phenomena; see also [15]:

- Calculation of membrane forces in the face layers,
- Determination of critical loads inducing local buckling, i.e. buckling of the face layers at short wavelengths including shear buckling and intracell buckling in the case of honeycomb cores,
- Check for the occurence of local instability: The smaller of the two critical loads is taken for checking the safety against local buckling:

$$P_{crit}^{buckl} = Min\ \left(P_{crit}^{wrinkl}, P_{crit}^{icb}\right) \qquad (9),$$

- Update of the vector of out-of-balance loads and of the tangential stiffness matrices in case of appearance of local buckling.

Since the determination of local stability limits is influenced by the actual loading state (biaxial loading, bending moments, ...), the critical loads have to be calculated for each state m.

If local buckling occurs, the stiffness matrix of the locally buckled shell element has to be updated. The postbuckling behavior of the wrinkled face layer is approximated by the assumption of a constant post-critical membrane compression force perpendicular to the wrinkles, as described in the above Chapter on lamination theory, [6], leading to a modified ${}^{m}\underset{\approx}{C}$-matrix. Thus, in effect, a nonlinear material law is introduced for the shell element.

This way the stiffness reduction and stress redistribution due to the growth of the locally buckled areas are taken into account in an approximative manner, and the global analysis can be continued up to complete collapse.

Example: Wrinkling in a compressed sandwich plate with a central hole

As an example for the application of this sandwich shell element a square sandwich plate with a central hole, see Fig. 3, is considered, see also [1]. The plate is uniaxially loaded by compressive in-plane loads.

Due to the central cylindrical hole, a stress concentration is induced near the hole, where the local instability will occur first. The material for the test specimen and the material data were provided by AIREX AG, Sins, Switzerland. The specimens were produced from sandwich plates with the following material properties: core material AIREX R63.50 (isotropic foam, $E_c = 28N/mm^2$, $\nu_c = 0.166$, $c = 35mm$), and face layer material Ac 110 (Anticorodal, AlSiMgMn, $E_f = 70kN/mm^2$, $\nu_f = 0.3$, $\sigma_{Y,0.2} = 240N/mm^2$, $t_f = 0.5mm$). The reference compressive load $q_{l,ref} = 2N/mm$ acts along two opposite edges of the plate; $a = 500$ mm, $R = 100$ mm.

Due to the double symmetry conditions, only one quarter of the sandwich shell is modelled. The load displacement development for the edge at which the compressive loads are applied is plotted in Fig. 3 together with figures showing the growth of the wrinkled areas. Wrinkling starts at the location of maximum compressive stress, i.e. at the apex of the hole at a load level $\lambda = 35$, corresponding to a critical distributed load $q_l^* = 70N/mm$. The wrinkled zone grows towards the outer edge, and just before reaching the collapse load at $\lambda^{coll} = 67$, wrinkling is observed at the corners, too. Experimental results verify the computated behavior, see [1].

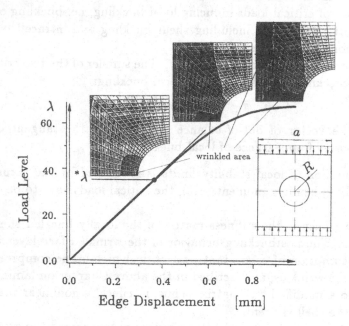

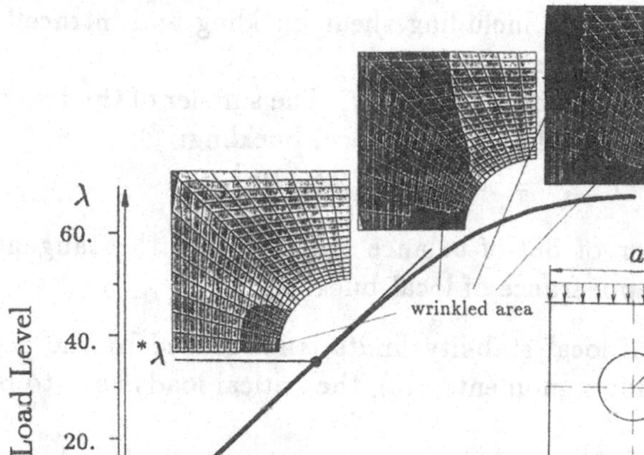

Fig. 3 Behavior of the compressed sandwich plate with central hole

2 Algorithms for Stability Analysis

Frequently composite shell structures are thin shells which may tend to buckle.

In many cases, especially if limit points appear, the observation of the load–displacement path indicates critical mechanical or thermal load configurations. However, bifurcation buckling requires a more detailed consideration.

Based on the static stability criterion [16], bifurcation or limit points, i.e. bifurcation buckling or snap-through buckling, appear if at least one further equilibrium state exists adjacent to the original configuration without any load variation. Hence, in the critical state the equation

$$^*\underset{\approx}{\mathbf{K}}\,\delta\underset{\sim}{u} = \underset{\sim}{0} \tag{10}$$

has a non-trivial solution $\delta\underset{\sim}{u} \neq \underset{\sim}{0}$. If it is assumed that the material behavior is linear elastic and temperature-independent and if, in addition, proportional mechanical and thermal loading

$$^{m}\underset{\sim}{\mathrm{r}} = {}^{m}\lambda\, \underset{\sim}{\mathrm{r}}_{ref} \qquad (11),$$

$$^{m}\vartheta = {}^{m}\lambda\, \vartheta_{ref} \qquad (12)$$

is applied with $\underset{\sim}{\mathrm{r}}_{ref}$ denoting the reference mechanical loading and ϑ_{ref} representing the reference thermal loading, then condition (10) can be rewritten as

$$({}^{*}\underset{\approx}{\mathbf{K}}_{e} + \underset{\approx}{\mathbf{K}}_{g}({}^{*}\lambda))\delta\underset{\sim}{u} = \underset{\sim}{0} \qquad (13).$$

The relation $\underset{\approx}{\mathbf{K}}_g(\lambda)$ is nonlinear. Having reached the state m we can ask for the multiplier $^{m}\eta$ which leads to

$$^{*}\lambda = {}^{m}\eta\, {}^{m}\lambda \qquad (14).$$

After linearizing the nonlinear relation between $\underset{\approx}{\mathbf{K}}_g$ and λ and neglecting, for the time being, the difference between $^{m}\underset{\approx}{\mathbf{K}}_e$ and $^{*}\underset{\approx}{\mathbf{K}}_e$ we obtain a linear eigenvalue problem

$$({}^{m}\underset{\approx}{\mathbf{K}}_{e} + {}^{m}\overline{\eta}\, {}^{m}\underset{\approx}{\mathbf{K}}_{g})\delta\underset{\sim}{\overline{u}} = \underset{\sim}{0} \qquad (15)$$

leading to an estimate for the critical load multiplier

$$^{*}\lambda \approx {}^{*}\overline{\lambda} = {}^{m}\overline{\eta}_1\, {}^{m}\lambda \qquad (16)$$

and an approximation for the fundamental buckling mode

$$\delta\underset{\sim}{u} \approx \delta\underset{\sim}{\overline{u}}_1 \qquad (17).$$

$^{m}\overline{\eta}_1$ and $\delta\underset{\sim}{\overline{u}}_1$ are the smallest eigenvalue and the corresponding eigenvector of the eigenvalue problem eqn. (15), respectively.

The error caused by the linearization of $\underset{\approx}{\mathbf{K}}_g(\lambda)$ and the neglection of the difference between $^{m}\underset{\approx}{\mathbf{K}}_e$ and $^{*}\underset{\approx}{\mathbf{K}}_e$ vanishes and, hence, the estimate (eqn (16)) becomes accurate when the lowest eigenvalue $^{m}\overline{\eta}_1$ approaches 1, i.e. the current configuration m approaches the critical one ($^{m}\lambda \to {}^{*}\lambda$). This leads to the strategy of "accompanying eigenvalue analyses" in which the eigenvalue problem eqns. (15–17) is formulated and solved at several load levels during the incremental prebuckling analysis. $^{*}\overline{\lambda}(u_{ref})$, with u_{ref} being a properly chosen component of the displacement vector $^{m}\underset{\sim}{u}$, represents an "estimate curve" crossing the $^{m}\lambda(u_{ref})$ curve, i.e. the load–displacement path, whenever the critical load situation is reached.

Very often a stability analysis for thermal (or mechanical) loading only is needed with the mechanical loads (or the thermal load) being held constant. A procedure which accounts for this is implemented in the LFC-element:

The $\underset{\approx}{\mathbf{K}}_g^{(e)}$-matrix (eqn (3)) can be split into a part which depends on mechanical loads and a part related to thermal loading:

$$^m\underset{\approx}{\mathbf{K}}_g^{(e)} = {}^m\underset{\approx}{\mathbf{K}}_g^{(e)}{}_{mech} + {}^m\underset{\approx}{\mathbf{K}}_g^{(e)}{}_{th} \tag{18}$$

After assembling the global stiffness matrices, the procedure for detection of buckling loads, described above, can be applied either for constant thermal loading or constant mechanical loads by rewriting eqn. (15):

- for buckling due to mechanical loads only:

$$({}^m\underset{\approx}{\mathbf{K}}_e + {}^m\underset{\approx}{\mathbf{K}}_{g\,th} + {}^m\overline{\eta}\,{}^m\underset{\approx}{\mathbf{K}}_{g\,mech})\delta\underset{\sim}{\overline{u}} = \underset{\sim}{0} \tag{19}$$

with a resulting estimate for the critical load (${}^m\vartheta$ being held constant)

$$^*\underset{\sim}{r} \approx {}^m\overline{\eta}\,{}^m\underset{\sim}{r} \tag{20}$$

- for buckling due to thermal loading only:

$$({}^m\underset{\approx}{\mathbf{K}}_e + {}^m\underset{\approx}{\mathbf{K}}_{g\,mech} + {}^m\overline{\eta}\,{}^m\underset{\approx}{\mathbf{K}}_{g\,th})\delta\underset{\sim}{\overline{u}} = \underset{\sim}{0} \tag{21}$$

with a resulting estimate for the critical temperature load (${}^m\underset{\sim}{r}$ being held constant)

$$\begin{aligned} {}^*\overline{\vartheta} &\approx {}^m\overline{\eta}\,{}^m\overline{\vartheta} \\ {}^*\widehat{\vartheta} &\approx {}^m\overline{\eta}\,{}^m\widehat{\vartheta} \end{aligned} \tag{22}$$

Another approach, which is even more efficient in some respects, is presented in [17].

The following examples demonstrate the application of the stability algorithm.

Example: Thermal buckling of a square layered FRP-plate

The thermal buckling behavior, analyzed by nonlinear approaches, sometimes shows strange effects. Most of the papers dealing with thermal buckling of layered composite plates or shells use classical buckling theory, see e.g. [18]. Classical buckling theory has, of course, a limited range of application, especially in the case of unsymmetrically layered composites. A very demonstrative example is a cross-ply square plate under uniform temperature rise, which for simply supported, straight edges (as investigated by [18]) is also considered in [1]. The plate under consideration has the following properties:

Lay-up: [0/90]... two layer cross-ply, layer thickness = 0.2 mm, width of the square = 300 mm.

Material: Graphite/Epoxy

$E_l = 127.5\,\mathrm{kN/mm^2}$ $E_q = 11.0\,\mathrm{kN/mm^2}$ $\nu_{lq} = 0.35$

$G_{lq} = 5.5\,\mathrm{kN/mm^2}$ $G_{lt} = 5.5\,\mathrm{kN/mm^2}$ $G_{qt} = 4.6\,\mathrm{kN/mm^2}$

$\alpha_l = -0.08\times10^{-5}\,{}^\circ\mathrm{C}^{-1}$ $\alpha_q = 2.90\times10^{-5}\,{}^\circ\mathrm{C}^{-1}$

Five different sets of boundary conditions, which are equal for all edges, are considered: a) hinged, with in-plane movement restricted to retain straight edges; b) hinged and free to move in-plane; c) fixed rotations and normal displacements, but free to move in-plane; d) fixed rotations and normal displacements, with in-plane movement restricted to retain staight edges; and e) hinged and no inplane motions.

The corresponding load (i.e. temperature) displacement paths, prebuckling deformations and estimate curves for critical temperature rises are shown in Fig. 4. The estimate curves ${}^*\overline{\vartheta}_{appr}(u)$ or ${}^*\overline{\vartheta}_{appr}(w)$, respectively, result from accompanying linearized eigenvalue analyses as described above.

Figure 4 shows some interesting results. Cases (c) and (d) lead to identical results, namely no out-of-plane deformations w and no stability loss. This can be explained by considering the clamped layered plate with a constant temperature rise as an analogy to a homogeneous plate with a constant temperature moment, which would also show no out-of-plane deformations or buckling under the same boundary conditions. With boundary conditions (a), (b) and (e) out-of-plane pre-buckling deformations and bucking can be observed, however, at significantly different critical temperatures and with different buckling modes. Only in case (e) classical buckling analysis renders correct buckling estimates (the estimate curve is horizontal), whereas in the case (a) classical buckling analysis would considerably underestimate the critical temperature and would lead to the wrong buckling mode, compare also [1]. In contrast to (a), in the case (b) classical buckling analysis would overestimate the critical temperature. The significant difference between the buckling behavior of (a) and (b) comes from the small difference in the in-plane boundary conditions. In the case (b) warping of the edges can be observed: The in-plane displacements u of the corner points of the plate are equal in cases (a) and (b), whereas the mid-edge in-plane displacements are much smaller in the case (b). The case (b) shows significantly larger out-of-plane displacements than case (a).

Example: Stability of a clamped sandwich beam

The analysis of the stability behavior of a clamped sandwich beam under compression shows the applicability of the algorithms derived in the present work to the case of global as well as local instabilities. Since there exist formulas for the analytical approximative determination of the global buckling load accounting for the shear effects, the results of the numerical analysis can be checked, provided the sandwich parameters fulfill the conditions assumed for the derivation of the analytical method.

The sandwich beam (as considered in [15]) modeled for the FE analysis is shown schematically in Fig. 5.

In order to analyze the influence of an isotropic low-density core on the stability

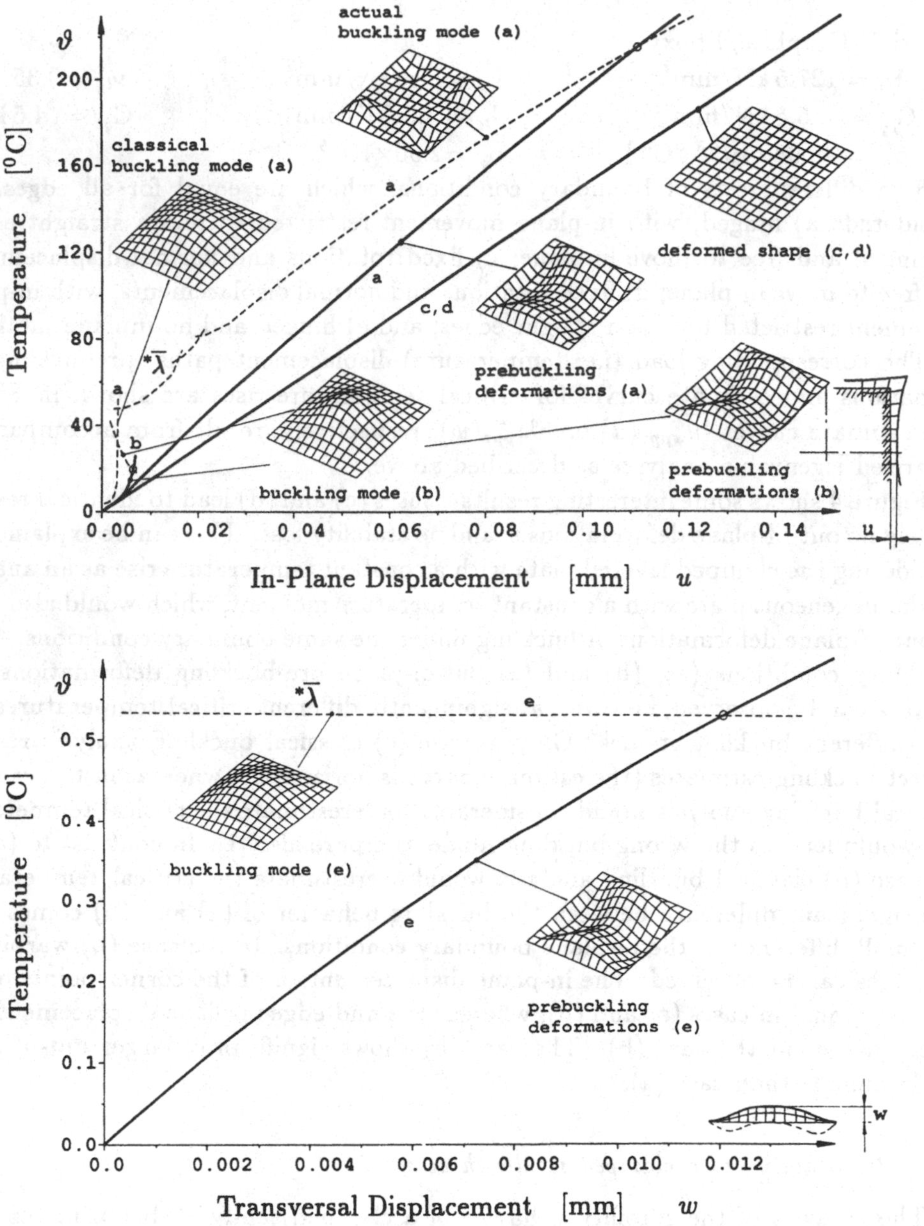

Fig. 4 Thermally loaded [0/90] laminate plate

behavior, the Young's modulus of the core is varied in a series of computations. In Fig. 6 the critical loads corresponding to the individual kinds of stability loss are plotted in dependence on the ratio of the Young's moduli E_c/E_f.

Figure 6 shows how increased core stiffness, i.e. increased E_c, leads to transitions from shear buckling of the core (point E to D) to antisymmetrical wrinkling of the

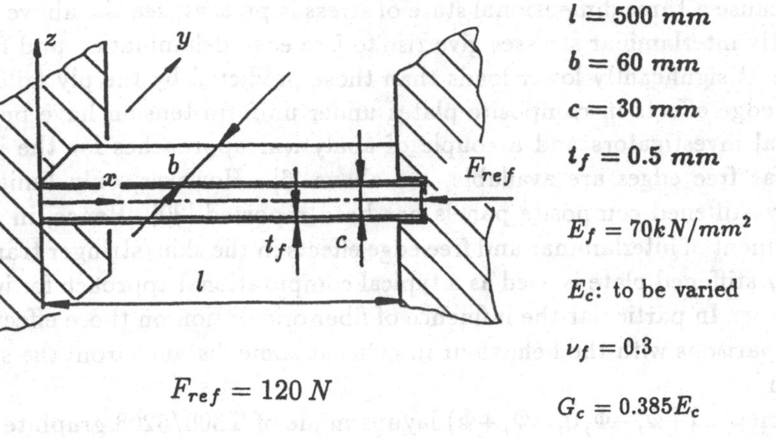

Fig. 5 Clamped sandwich beam

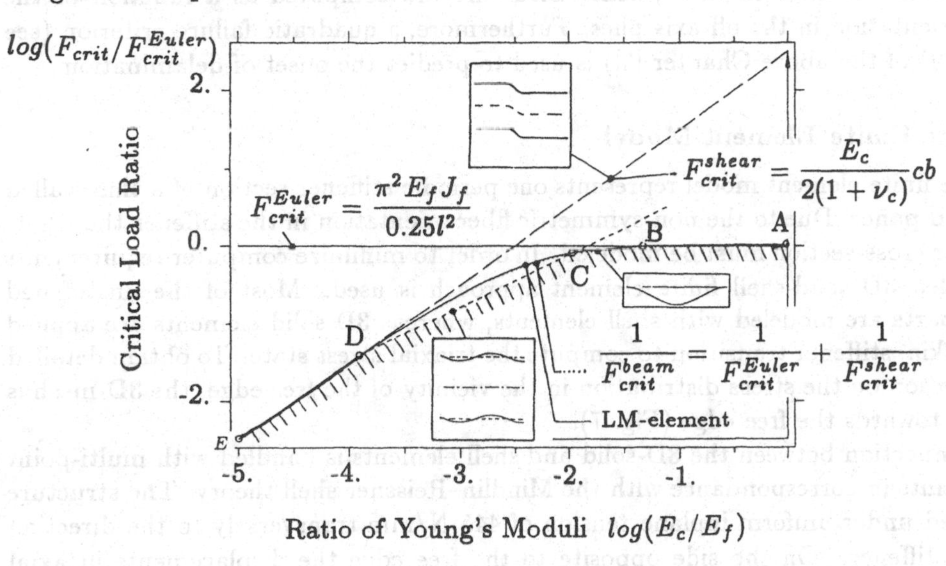

Fig. 6 Critical load ratio $F_{crit}/F_{crit}^{euler}$ in dependence on E_c/E_f

face layers (D to C), global buckling influenced by the small shear stiffness of the core (C to B) and, finally, to pure Euler buckling (B to A). The good correspondance with some approximative analytical formulae taken from [19] becomes obvious.

3 Local Investigations – Integrally Stiffened Plates

It is well known that at free edges of layered anisotropic materials interlaminar stresses may occur due to the mismatch in material properties between adjacent layers. In the region near the free edge the classical laminated plate theory is not valid because a three dimensional state of stress is present; see the above Chapter [6]. Frequently interlaminar stresses give rise to free edge delamination and failure of the laminate at signficantly lower loads than those predicted by the ply failure criteria.

Free edge effects in composite plates under uniform tension have been analyzed by several investigators and a couple of analytical approaches for the interlaminar stresses at free edges are available, see above [6]. However, only limited work on integrally stiffened composite panels has been reported [20]. Hence, in this Section the treatment of interlaminar and free edge effects in the skin/stringer transition of an integrally stiffened plate is used as a typical computational approach for investigating local effects. In particular the influence of fiber orientation on those effects is studied and comparisons with the behaviour in areas at some distance from the stiffeners are discussed.

Symmetric $(+\Phi, -\Phi, 0, -\Phi, +\Phi)$ layups made of T300/5208 graphite epoxy laminates [21] with various fiber orientations in the off-axis plies are considered. The interlaminar stresses in the stiffened structure are computed as a function of the fiber orientation in the off-axis plies. Furthermore, a quadratic failure criterion (see eqn. (49) of the above Chapter [6]) is used to predict the onset of delamination.

3.1 The Finite Element Model

The finite element model represents one periodic stiffener section of a thin walled stiffened panel. Due to the non-symmetric fiber orientation in the stiffener the whole stiffener cross-section must be modeled. In order to minimize computer requirements a coupled 3D solid-shell finite element approach is used. Most of the unstiffened panel parts are modeled with shell elements, whereas 3D solid elements are applied in the skin–stiffener transition to compute the triaxial stress state. To obtain detailed information on the stress distribution in the vicinity of the free edge, the 3D-mesh is refined towards the free edge (Fig. 7).
The connection between the 3D-solid and shell elements is handled with multi-point constraints in correspondance with the Mindlin–Reissner shell theory. The structure is loaded under uniform inplane tension of 445 N/mm transversely to the direction of the stiffener. On the side opposite to the free edge the displacements in axial direction of the stiffener are fixed whereas the displacements in the vertical direction, except on a horizontal nodeline, and the displacements in the load direction, except for one single node, are free.

3.2 Results and Discussion

Figure 8 shows results of the linear finite element analysis in terms of the interlaminar stresses at the $+\Phi/-\Phi$ interface (for $\Phi = 45°$). In the skin/stringer transition, i.e. position A, these interlaminar stresses are higher than at some distance from the stiffener.
By comparing the influence of the free edge effects on the interlaminar stresses in the unstiffened parts (position B) of the panel with their behavior in the skin/stringer

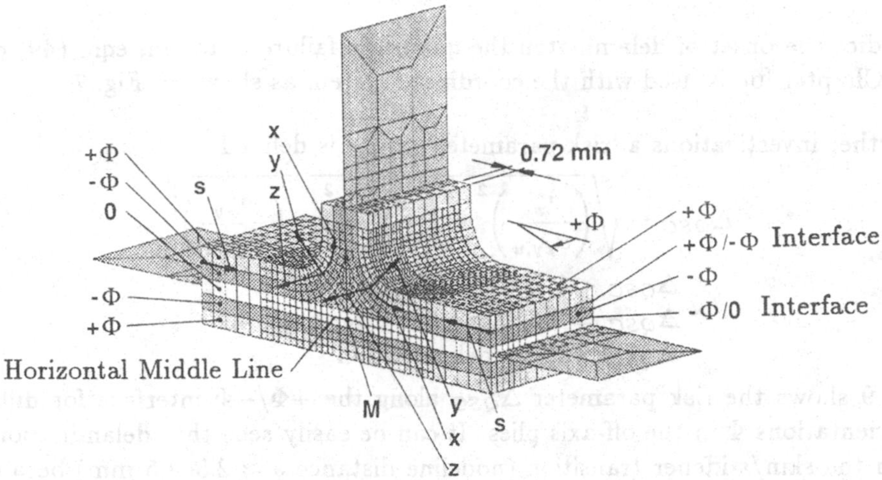

Fig. 7 The combined 3D/shell element model

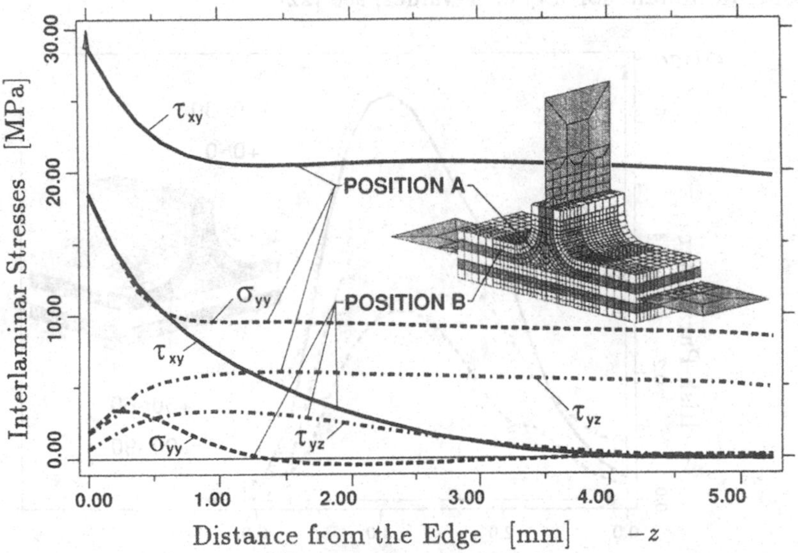

Fig. 8 Interlaminar stresses at the free edge

transition the following observations can be made: The absolute as well as the relative increase of the interlaminar shear stresses, σ_{xy}, σ_{yz}, when approaching the free edge is much more pronounced in the unstiffened parts of the panel than in the skin/stiffener transition, whereas the absolute increase of the inerlaminar normal stresses, σ_{yy}, is much higher in the skin/stringer transition than in the unstiffened parts of the panel.

To predict the onset of delaminaton the quadratic failure criterion, eqn. (49) of the above Chapter [6], is used with the coordinate system as shown in Fig. 7.

For further investigations a 'risk parameter' Δ_{QSC} is defined:

$$\Delta_{QSC} = \sqrt{\left(\frac{\sigma'_{xy}}{\sigma_{xy,u}}\right)^2 + \left(\frac{\sigma'_{yz}}{\sigma_{yz,u}}\right)^2 + \left(\frac{\sigma'_{yy}}{\sigma_{yy,u}}\right)^2} \qquad (23).$$

$\Delta_{QSC} < 1$ no delamination
$\Delta_{QSC} = 1$ onset of delamination
$\Delta_{QSC} > 1$ delamination

Figure 9 shows the risk parameter Δ_{QSC} along the $+\Phi/-\Phi$ interface for different fiber orientations Φ in the off-axis plies. It can be easily seen that delamination will start in the skin/stiffener transition (nodeline distance $s \approx 2.5$–4.5 mm) because of the high values of the risk parameter. Due to the fact that the magnitude of the risk parameter in the skin/stiffener transition is lowered by increasing the fiber angle beyond 30° in the off-axis plies, the onset of delamination can be prevented by an appropriate choice of the fiber orientation. However, one has to notice that ply failure may become dominant for higher Φ-values; see [22].

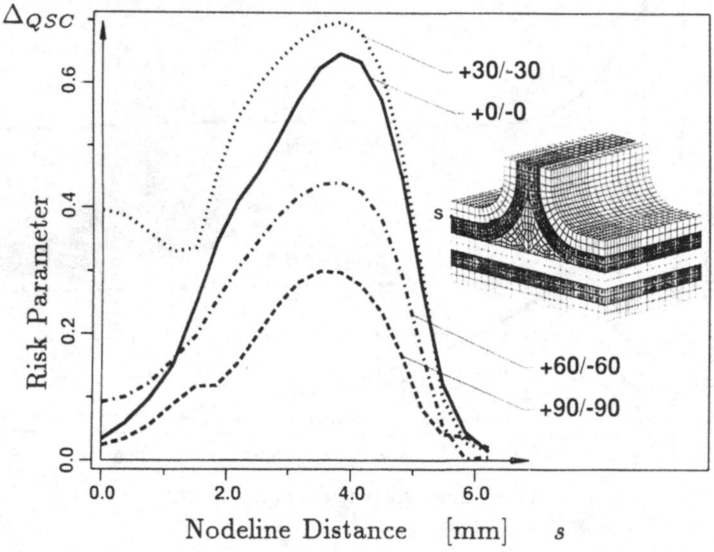

Fig. 9 Value of the risk paramter Δ_{QSC} along the $+\Phi/-\Phi$ interface

The magnitude of the risk parameter as a function of the off-axis ply angle shows a behavior which is significantly different from that in the skin/stringer transition.

While in the skin/stringer transition the risk parameter shows a strictly monotonic dependence on Φ, in the skin area the following behavior is found: when the fiber angle is increased from $0°$ to $30°$, the risk parameter rises, but further increases of the off-axis ply angle causes it to decrease again which agrees with observations described in [23].

4 Industrial Applications

The use of the LFC-element in practical applications is shown exemplarily by the analysis of a composite leaf spring and of a side wall panel for the MD–11 airliner.

Material: T300/5208 Graphite/Epoxy

$E_l = 132.38 \text{ kN/mm}^2$
$E_q = 10.75 \text{ kN/mm}^2$
$\nu_{lq} = 0.24$
$G_{lq} = 5.65 \text{ kN/mm}^2$
$G_{lt} = 5.65 \text{ kN/mm}^2$
$G_{qt} = 5.65 \text{ kN/mm}^2$

Festigkeitswerte

$\sigma_{lTu} = 1513. \text{ N/mm}^2$
$\sigma_{lCu} = 1696. \text{ N/mm}^2$
$\sigma_{qTu} = 44. \text{ N/mm}^2$
$\sigma_{qCu} = 44. \text{ N/mm}^2$
$\tau_{lqu} = 87. \text{ N/mm}^2$
$\beta_E = 0.9$

Laminate lay-up:
region A: $[0_{11}/45_{58}/0_{11}]$
region B: $[0_{11}/45_{38}/0_{11}]$

region C: $[0_{11}/45_{28}/0_{11}]$
region D: $[0_{11}/45_{18}/0_{11}]$
ply thickness = 0.25 mm

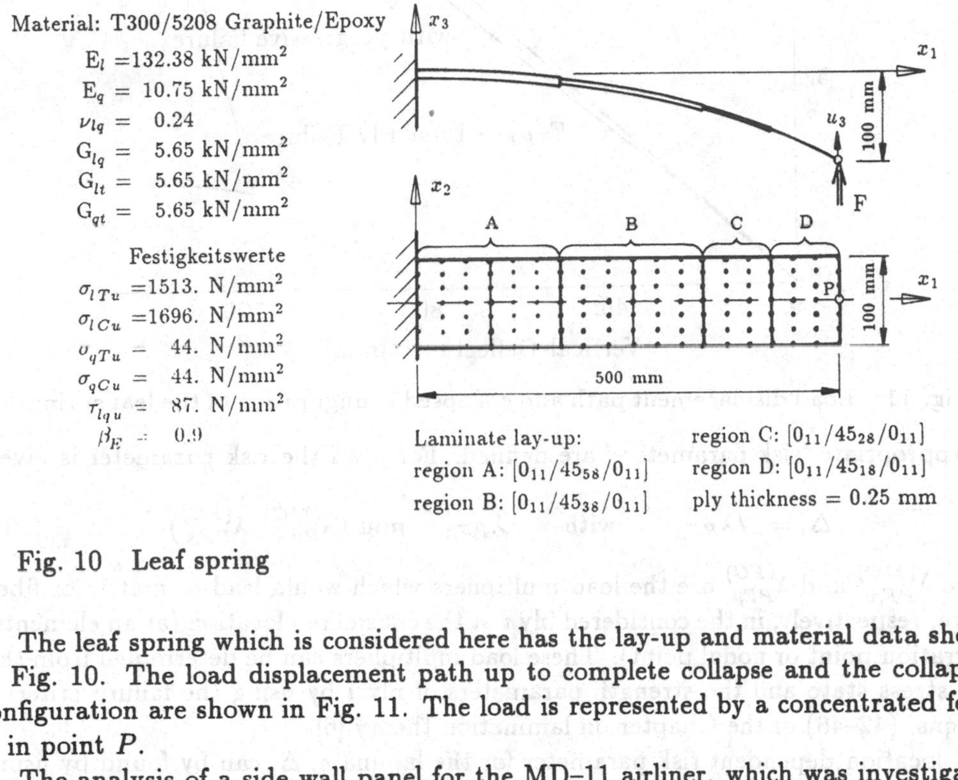

Fig. 10   Leaf spring

The leaf spring which is considered here has the lay-up and material data shown in Fig. 10. The load displacement path up to complete collapse and the collapsed configuration are shown in Fig. 11. The load is represented by a concentrated force F in point P.

The analysis of a side wall panel for the MD–11 airliner, which was investigated in cooperation with Fischer Advanced Composite Components (FACC), Ried i.I., Austria, is described briefly without giving details in material data and structural dimensions. Figure 12 shows the finite element model. The laminate lay-up varies over the structure in a rather complicated way, some portions of the shell are sandwiches. Many different stacking sequences appear in one panel. More details are given in [24].

One of the load cases which has been studied in the course of the design process was external pressure loading, see Fig. 12.

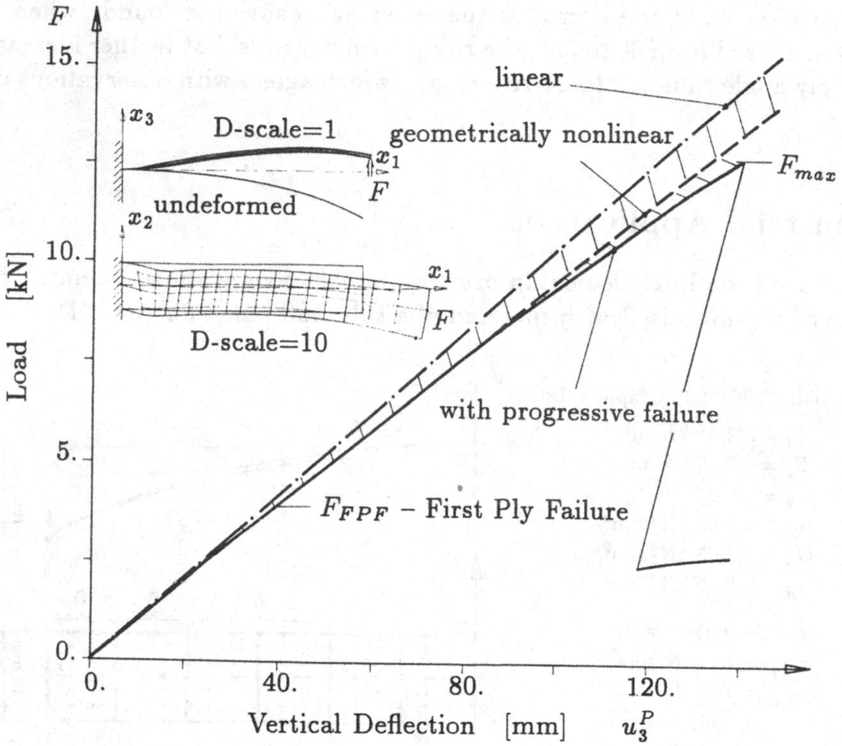

Fig. 11 Load displacement path and collapsed configuration of the leaf spring

Appropriate 'risk parameters' are defined. For ply i the risk parameter is given by

$$\Delta_i = 1/\lambda_{PF,i} \qquad \text{with} \qquad \lambda_{PF,i} = \min\left(\lambda_{PF,i}^{(MC)}, \lambda_{PF,i}^{(FC)}\right) \tag{23}$$

where $\lambda_{PF,i}^{(MC)}$ and $\lambda_{PF,i}^{(FC)}$ are the load multipliers which would lead to matrix or fiber failure, respectively, in the considered ply i at the considered location (at an element's integration point or nodal point). These load multipliers can be determined from the local stress state and the strength parameters of ply i by using the failure criteria, i.e. eqns. (42–46) of the Chapter on lamination theory [6].

A location dependent risk parameter for the laminate, $\bar{\Delta}$, can by found by using the maximum value of the layerwise Δ-values at the position under consideration. Using such parameters, the results of the finite element analysis can be interpreted in terms of finding areas of the panel or plies with critical stress states and those which are not properly utilized, i.e. zones where the lay-up shoud be improved or material, and hence weight, could be saved. This way a quasi-optimization can be carried out iteratively. Figure 13 shows the distribution of the risk parameter for one individual layer and for the complete panel, respectively, at an intermediate stage of the design process.

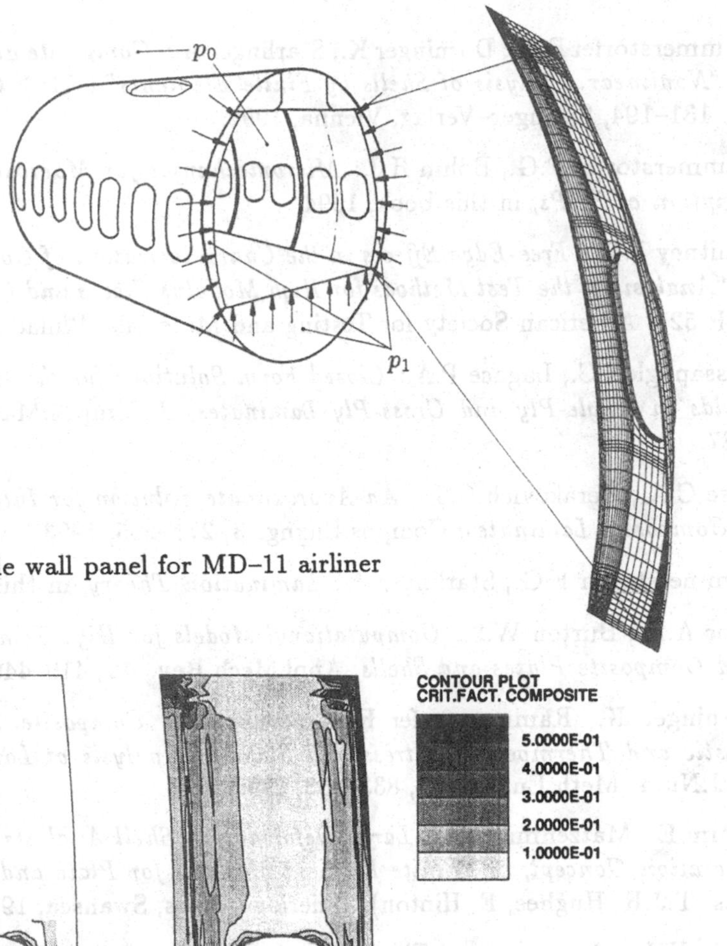

Fig. 12 Side wall panel for MD–11 airliner

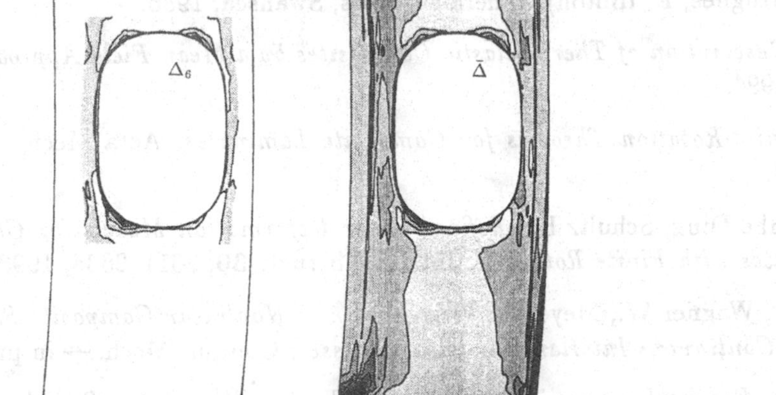

Fig. 13 Risk parameter, a) at an individual layer, b) over the complete panel

References

1. Rammerstorfer F.G., Dorninger K., Starlinger A.: *Composite and Sandwich Shells;* in *"Nonlinear Analysis of Shells by Finite Elements"* (Ed. F.G.Rammerstorfer), pp. 131–194, Springer–Verlag, Vienna, 1992.

2. Rammerstorfer F.G., Böhm H.J.: *Micromechanics for Macroscopic Material Description of FRPs;* in this book, 1994.

3. Whitney J.M.: *Free-Edge Effects in the Characterization of Composite Materials;* in *"Analysis of the Test Methods for High Modulus Fibers and Composites"* ASTM STP 521, American Society for Testing and Materials, Philadelphia, PA, 1973.

4. Kassapoglou C., Lagace P.A.: *Closed Form Solutions for the Interlaminar Stress Fields in Angle-Ply and Cross-Ply Laminates;* J.Compos.Mater. **21**, 292–308, 1987.

5. Rose C.A., Herakovich C.T.: *An Approximate Solution for Interlaminar Stresses in Composite Laminates;* Compos.Engng. **3**, 271–285, 1993.

6. Rammerstorfer F.G., Starlinger A.: *Lamination Theory;* in this book, 1994.

7. Noor A.K., Burton W.S.: *Computational Models for High-Temperature Multilayered Composite Plates and Shells;* Appl.Mech.Rev. **45**, 419–446, 1992.

8. Dorninger K., Rammerstorfer F.G.: *A Layered Composite Shell Element for Elastic and Thermoelastic Stress and Stability Analysis at Large Deformations;* Int.J.Num. Meth.Engng. **30**, 833–858, 1990.

9. Ramm E., Matzenmiller A.: *Large Deformation Shell Analysis Based on the Degeneration Concept;* in *"Finite Element Methods for Plate and Shell Structures"* (Eds. T.J.R. Hughes, E. Hinton), Pineridge Press, Swansea, 1986.

10. Böhm H.J.: *Description of Thermoelastic Composites by a Mean Field Approach;* in this book,1994.

11. Başar Y.: *Finite-Rotation Theories for Composite Laminates;* Acta Mech. **98**, 159–176, 1993.

12. Başar Y., Yunhe Ding, Schultz R.: *Refined Shear Deformation Models for Composite Laminates with Finite Rotations;* Int.J.Sol.Struct. **30**, 2611–2638, 1993.

13. Gruttmann F., Wagner W., Meyer L., Wriggers P.: *A Nonlinear Composite Shell Element with Continuous Interlaminar Shear Stresses;* Comput.Mech. — in press

14. Dorninger K.: *Entwicklung von nichtlinearen FE–Algorithmen zur Berechnung von Schalenkonstruktionen aus Faserverbundschalen.* VDI–Fortschrittsberichte **18/65**, VDI–Verlag, Düsseldorf, 1989.

15. Starlinger A.: *Development of Efficient Finite Shell Elements for the Analysis of Sandwich Structures Under Large Deformations and Global as Well as Local Instabilities.* VDI–Fortschrittsberichte **18/93**, VDI–Verlag, Düsseldorf, 1991.

16. Ramm E.: *Geometrisch nichtlineare Elastostatik und finite Elemente.* Habilitationsschrift, Universität Stuttgart, Stuttgart, 1975.

17. Wagner W.: *Nonlinear Stability Analysis of Shells with the Finite Element Method;* in *"Nonlinear Analysis of Shells by Finite Elements"* (Ed. F.G.Rammerstorfer), Springer–Verlag, Vienna, 1992.

18. Tauchert T.R.: *Thermal Stresses in Plates — Statical Problems;* in *"Thermal Stresses I"* (Ed. R.B.Hetnarski), North–Holland, Amsterdam, 1986.

19. Rammerstorfer F.G.: *Repetitorium Leichtbau.* Oldenbourg Verlag, Vienna, 1992.

20. Wang J.T., Lotts C.G., Davis D.D., Krishnamurthy T.: *Coupled 2D–3D Finite Element Method for Analysis of a Skin Panel with a Discontinous Stiffener;* Proc. AIAA/ASME/ASCE/AHS/ASC 33rd Structures, Structural Dynamics and Materials Conference, Dallas, TX, Part 2, Paper No. 92–2474–CP, pp. 818–827, 1992.

21. Reddy J.N., Pandey A.K.: *A First-Ply Failure Analysis of Composite Laminates;* Comput.Struct. **25**, 371–393, 1987.

22. Skrna–Jakl I., Rammerstorfer F.G.: *Numerical Investigation of the Free Edge Effects in Integrally Stiffened Layered Composite Panels;* Int.J.Comput.Struct. **25**, 129–137, 1993.

23. Christensen R.M., DeTeresa S.J.: *Elimination/Minimization of Edge-Induced Stress Singularities in Fiber Composite Laminates;* Int.J.Sol.Struct. **29**, 1221–1231, 1992.

24. Jakl I.: *Strukturanalyse des Side Wall Panels des MD-11 Flugzeuges.* Diploma Thesis, TU Wien, Vienna, 1989.

Acknowledgement

Some protions of this Chapter are taken from the paper "Combined micro- and macromechanical considerations of layered composite shells"by F.G.Rammerstorfer, A.Starlinger and K.Dorninger, to appear in Int.J.Numer.Meth.Engng, Vol.37 (1994) and from the Chapter "Composite and Sandwich Shells"by F.G.Rammerstorfer, K.Dorninger and A.Starlinger, published in *"Nonlinear Analysis of Shells by Finite Elements"* (Ed. F.G.Rammerstorfer), CISM/Springer–Verlag, Vienna, 1992. The permission by John Wiley & Sons, Ltd. and by the International Centre for Mechanical Sciences (CISM) is gratefully acknowledged.

CHAPTER 9

HYBRID MECHANICS

J. Hult
Chalmers University of Technology, Göteborg, Sweden
and
H.L. Bjarnehed
Prosolvia Konsult AB, Göteborg, Sweden

CHAPTER 9
PART I

HYBRID MECHANICS

J. Hult
Chalmers University of Technology, Göteborg, Sweden
and
H.L. Bjarnehed
Prosolvia Konsult AB, Göteborg, Sweden

ABSTRACT

In engineering structures load carrying FRP structural elements are often joined to metal structural elements. Three main types of joints, all with different load transfer properties, may be used in this connection: *bolted*, *adhesive*, or *clamped*. Bolted and clamped joints may be disassembled, in contrast to adhesive joints. Design of such *hybrid systems* involves considerations of load transfer across the common boundary between FRP and metal. Oscillating, singular stress concentrations may occur at sharp corners or cracks between adjacent materials with different mechanical properties. The severity of such stress concentrations may be diminished by suitable shaping of the adhering structural parts.

1. INTRODUCTION

Basic composite material mechanics deals with certain model concepts, viz.:

- Standard material: one type of fibres embedded in one type of matrix
- Standard basic element: a thin lamina with parallel fibres
- Standard structural element: a stack of thin laminæ, in adhesive contact, forming a laminate

It is sometimes advantageous to use several different types of fibre or different matrix materials or both in laminate. One then talks about a *hybrid material* or a *hybrid structure.* Fig. 1:1 shows a hybrid lamina with two types of parallel fibres (A and B) in random distribution. Fig. 1:2 shows a hybrid laminate with the same two fibre types located in separate laminæ. Fig. 1:3 shows a common type of hybrid construction, viz. a sandwich assembly with a "honeycomb" layer between the two faces.

The advantage of having more than one fibre type in a laminate may be demonstrated by considering the cases shown in Fig:s. 1:4a and 1:4b. Fibre type A is here assumed to be stiffer than fibre type B. With the stacking sequence ABBA, as in Fig. 1:4a, the bending stiffness of the laminate beam is larger than that in Fig. 1:4b with the stacking sequence BAAB. The tensile stiffness, on the other hand, is identical in the two cases. Suitable rearrangement of fibres may thus lead to more optimum designs.

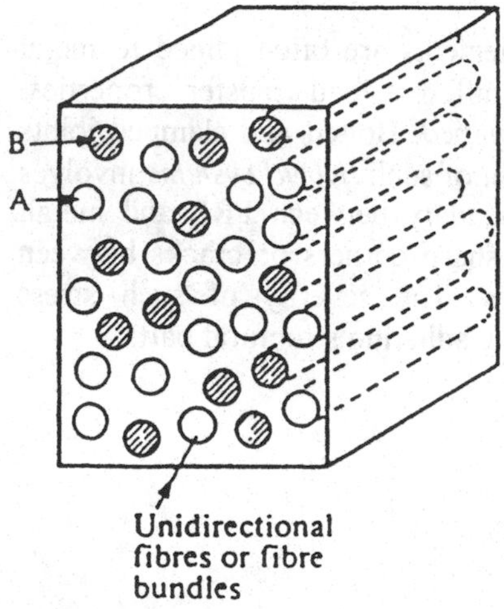

Fig. 1:1.

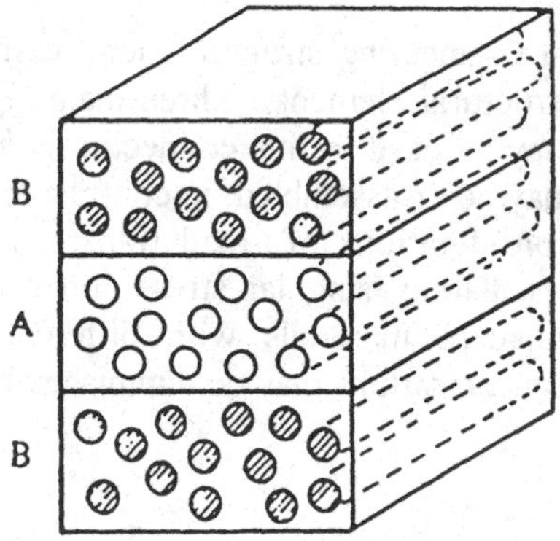

Fig. 1:2.

Fibre composite with two types of fibre in random (Fig. 1:1) and in layered distribution (Fig. 1:2). (from D. Hull, *An introduction to composite materials.* Cambridge University Press 1981, p. 235).

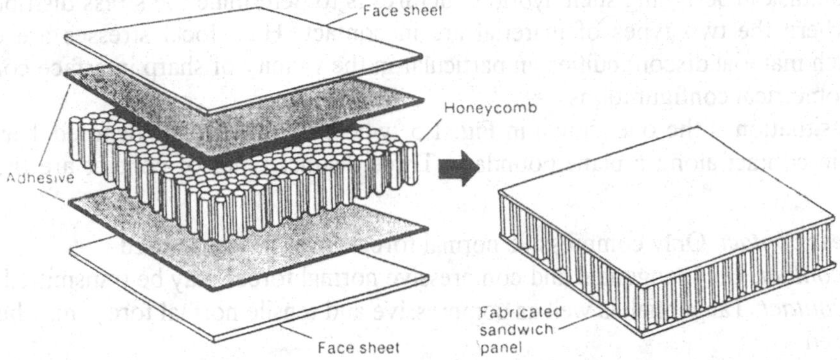

Fig. 1:3. Sandwich assembly (from ASM *Engineered Materials Handbook*, Vol. 1, 1987, p. 721).

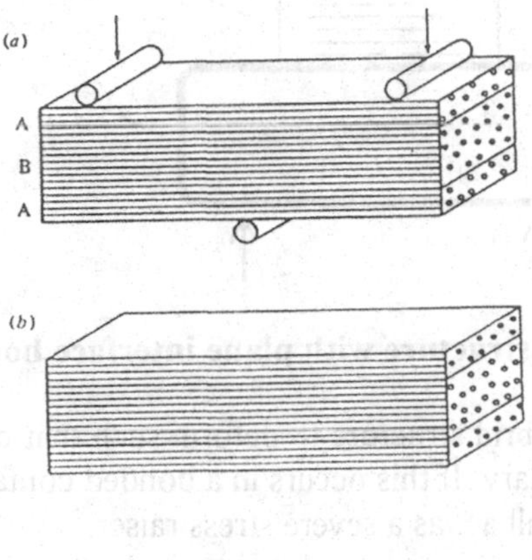

Fig. 1:4. Demonstration of effect of stacking sequence in hybrid beam (from D. Hull, *loc. cit.*, p. 236).

FRP elements are often integrated into structures made of conventional materials such as steel, aluminium or some other metal or alloy. The design of such a hybrid structure does often involve problems of load transfer between the metal parts and the FRP parts. Conventional methods of joining, such as soldering or welding, cannot be used. *Riveted or bolted joints* are possible in certain limited cases, but must often be avoided because of the local load effects created by the rivets or bolts (cf. section 2.1). *Adhesive joints or shrink fits* are often to be preferred, because of the less severe stress concentration that can then be achieved (cf. section 2.2). *Clamped joints* may easily be disassembled, and have therefore found use in certain applications (cf. section 2.3).

A common task in designing such hybrid structures is to determine the stress distribution in regions where the two types of material are in contact. High local stresses are often present at such material discontinuities, in particular in the vicinity of sharp interface corners or similar geometrical configurations.

A typical situation is the one shown in Fig. 1:5, where two materials, denoted 1 and 2, are brought in contact along a plane boundary. Three distinct types of contact are then of interest:

(a) *frictionless contact.* Only compressive normal forces may be transmitted
(b) *friction contact*. Only tangential and compressive normal forces may be transmitted
(c) *bonded contact*. Tangential as well as compressive and tensile normal forces may be transmitted.

Case (b) occurs in *bolted and clamped* joints, where sliding may occur, causing friction to arise, case (c) occurs in *adhesive* joints. The analysis is simplest in case (a) and the most complicated in case (b).

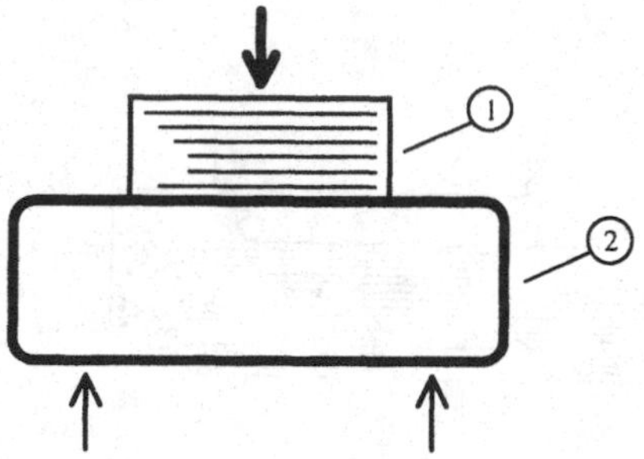

Fig. 1:5. Hybrid structure with plane interface boundary.

The loading may cause the hybrid structure to deform such that contact ceases in certain regions along the common boundary. If this occurs in a bonded contact, an *interphase* crack is being formed, cf. Fig. 1:6. It will act as a severe stress raiser.

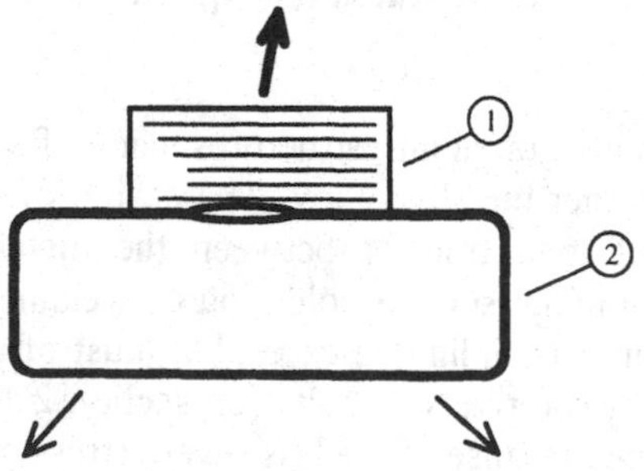

Fig. 1:6. Interphase crack in hybrid structure.

The local stress fields in the vicinity of interphase discontinuities, such as sharp corners or cracks, have been analyzed in a number of investigations, cf. Ref:s [1-9]. Most of these studies refer to different isotropic materials in contact, but the results are qualitatively valid also for contact between isotropic and anisotropic materials, as may be inferred from the analysis of the edge loaded anisotropic half plane given below.

The greatest attention has been devoted to the stress field near the tip of an interphase crack. Here, as with the case of a sharp crack in a uniform medium, the stress field is singular at the crack tip, but the singularity is of a different, oscillating character.

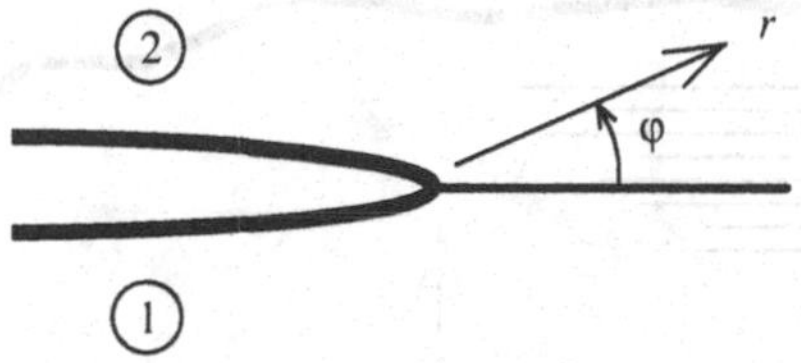

Fig. 1:7. Interphase crack tip.

With the notation of Fig. 1:7, where the two dissimilar materials are isotropic and linearly elastic, the leading term in the stress fields takes the form

$$\sigma \propto r^{-1/2} \begin{pmatrix} \sin \\ \cos \end{pmatrix} \varepsilon \ln r \tag{1.1}$$

Here the parameter ε depends on the properties of the two materials as follows:

$$\varepsilon = \frac{1}{2\pi} \cdot \ln \frac{\eta_1 / G_1 + 1/G_2}{\eta_2 / G_2 + 1/G_1} \tag{1.2}$$

with G denoting the shear modulus modulus. The parameter η is related to Poisson's ratio as follows:

$\eta_i = 3 - 4\nu_i$ for plane strain (1.3)

$\eta_i = (3 - \nu_i)/(1 + \nu_i)$ for plane stress (1.4)

If the two materials have identical mechanical properties, the classical square root singularity of linear fracture mechanics is regained. The stress oscillations which arise for dissimilar materials, and which are confined to a very small region near the crack tip, have not been found to have any significance in relation to crack growth conditions.

For a straight bond line the standard J-integral of fracture mechanics may be extended also to an interphase crack in a bi-material, cf. Ref. 10.

Edge load on orthotropic half plane

An edge loaded orthotropic half plane may be used to illustrate the effect of anisotropy on the contact stresses in various kinds of mechanical joints. The orthotropic, linearly elastic half plane in Fig. 1:8 is subjected to the harmonic edge traction

$$q = q_0 \cos\frac{2\pi x}{\lambda} \tag{1.5}$$

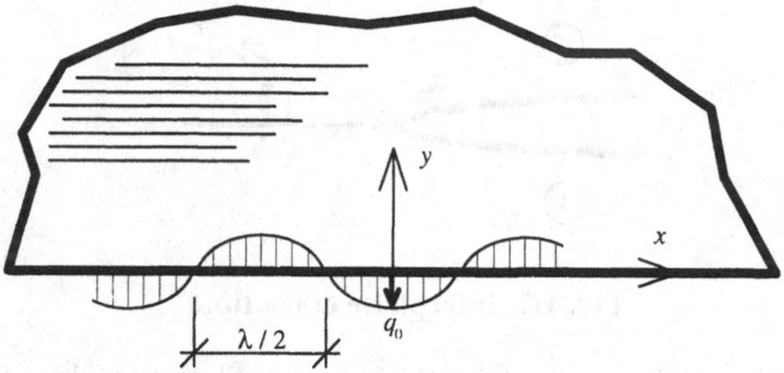

Fig. 1:8. Orthotropic half plane with harmonic edge traction.

The corresponding edge deformation may then be derived as follows. With standard notation for the constitutive equations

$$\varepsilon_x = \frac{1}{E_x}\sigma_x - \frac{\nu_{yx}}{E_y}\sigma_y \tag{1.6}$$

$$\varepsilon_y = -\frac{\nu_{xy}}{E_x}\sigma_x + \frac{1}{E_y}\sigma_y \tag{1.7}$$

$$\gamma_x = \frac{1}{G_{xy}}\tau_{xy} \tag{1.8}$$

and introduction of Airy's stress function according to

$$\sigma_x = \frac{\partial^2\phi}{\partial y^2}, \quad \sigma_y = \frac{\partial^2\phi}{\partial x^2}, \quad \tau_{xy} = -\frac{\partial^2\phi}{\partial x \partial y} \tag{1.9-11}$$

the compatibility equation takes the form

$$\frac{1}{E_y}\cdot\frac{\partial^4\phi}{\partial x^4} + \frac{2}{E_{xy}}\cdot\frac{\partial^4\phi}{\partial x^2 \partial y^2} + \frac{1}{E_x}\cdot\frac{\partial^4\phi}{\partial y^4} = 0 \tag{1.12}$$

where

$$\frac{2}{E_{xy}} = \frac{1}{G_{xy}} - \frac{\nu_{xy}}{E_x} - \frac{\nu_{yx}}{E_y} \tag{1.13}$$

The boundary conditions

$$\sigma_y(x,0) = q_0 \cos\frac{2\pi x}{\lambda}, \quad \tau_{xy}(x,0) \equiv 0 \tag{1.14-15}$$

indicate that the stress function may be expressed as

$$\phi(x,y) = F(y) \cdot q_0 \cos\frac{2\pi x}{\lambda} \tag{1.16}$$

which transforms eq. (1.12) into

$$\frac{1}{E_x} \cdot F^{IV}(y) - \left(\frac{2\pi}{\lambda}\right)^2 \cdot \frac{2}{E_{xy}} \cdot F''(y) + \left(\frac{2\pi}{\lambda}\right)^4 \cdot \frac{1}{E_y} \cdot F(y) = 0 \tag{1.17}$$

with solutions of the form

$$F \propto \exp\left(\frac{2\pi c y}{\lambda}\right) \tag{1.18}$$

where

$$c = \pm\sqrt{\frac{E_x}{E_{xy}} \pm \sqrt{\left(\frac{E_x}{E_{xy}}\right)^2 - \frac{E_x}{E_y}}} \tag{1.19}$$

Hence the complete solution may be stated as

$$\phi(x,y) = \cos\frac{2\pi x}{\lambda} \cdot \sum_{i=1}^{4} A_i \exp\left(\frac{2\pi c_i y}{\lambda}\right) \tag{1.20}$$

The boundary conditions (14), (15) and the conditions at infinity ($y \to \infty$) determine the four constants A_1, ..., A_4.

The following expression is then found for the stress σ_x at the loaded edge

$$\sigma_x(x,0) = \sqrt{\frac{E_x}{E_y}} \cdot q_0 \cos\frac{2\pi x}{\lambda} = \sqrt{\frac{E_x}{E_y}} \cdot \sigma_y(x,0) \tag{1.21}$$

The deformation of the loaded edge may then also be determined, and is found to be

$$\Delta(x) = -v(x,0) = \frac{q_0 \lambda}{\pi E_*} \cdot \cos\frac{2\pi x}{\lambda} \tag{1.22}$$

where

$$E_* = \frac{2\sqrt{E_x E_y}}{\sqrt{E_x / E_{xy} + \sqrt{(E_x / E_{xy})^2 - E_x / E_y}} + \sqrt{E_x / E_{xy} - \sqrt{(E_x / E_{xy})^2 - E_x / E_y}}} \tag{1.23}$$

For an *isotropic* half plane, where $E_x = E_y = E_{xy} = E$ the edge deformation takes the form

$$\Delta(x) = \frac{q_0 \lambda}{\pi E} \cdot \cos\frac{2\pi x}{\lambda} \tag{1.24}$$

Replacing now the harmonic edge load by an arbitrary, periodic edge load, which is represented by the Fourier series

$$q(x) = \sum_{n=1}^{\infty} q_n \cdot \cos\frac{2n\pi x}{\lambda} \tag{1.25}$$

the corresponding edge deformation becomes

$$\Delta(x) = \frac{\lambda}{\pi E_*} \cdot \sum_{n=1}^{\infty} \frac{q_n}{n} \cdot \cos\frac{2n\pi x}{\lambda} \tag{1.26}$$

For an *isotropic* half plane the edge deformation takes the form

$$\Delta(x) = \frac{\lambda}{\pi E} \cdot \sum_{n=1}^{\infty} \frac{q_n}{n} \cdot \cos\frac{2n\pi x}{\lambda} \tag{1.27}$$

Corresponding expressions may be stated for the case of tangential edge loading on the half plane:

$$t(x) = \sum_{n=1}^{\infty} \tau_n \cdot \sin\frac{2n\pi x}{\lambda} \tag{1.28}$$

Comparison of the expressions (1.26) and (1.27) shows that results pertaining to the isotropic half plane may be transformed to results for the orthotropic half plane simply by replacing Young's modulus E by the quantity E_* defined by eq. (1.23). In this manner problems of contact involving FRP components may be solved by transforming results available for corresponding isotropic components.

References

[1] Williams, M.L., "The Stresses Around a Fault or Crack in Dissimilar Media", *Bulletin of the Seismological Society of America*, Vol. 49, 1959, pp. 199-204.

[2] Rice, J.R. & Sih, G.C., "Plane Problems of Cracks in Dissimilar Media", *Trans. ASME*, 1965, pp. 418-423.

[3] Williams, M.L., "Stress Singularities, Adhesion, and Fracture", *Proc. 5th U.S. National Congress of Applied Mechanics* 1966, pp. 451-464.

[4] Bogy, D.B., "Edge-Bonded Dissimilar Orthogonal Elastic Wedges Under Normal and Shear Loading", *Trans. ASME*, 1968, pp. 460-466.

[5] Dundurs, J., Discussion remarks to [4], Trans, ASME, 1969, pp. 650-652.

[6] Bogy, D.B., & Wang, K.C., "Stress Singularities at Interface Corners in Bonded Disimilar Isotropic Elastic Materials", *Int. J. Solids and Structures*, Vol.7, 1971, pp. 993-1005.

[7] Bogy, D.B., "Two Edge-Bonded Elastic Wedges of Different Materials and Wedge Angles under Surface Tractions", *J. Appl. Mech.*, 1971, pp. 377-386.

[8] Hein, V.L. & Erdogan, F., "Stress Singularities in a Two-Material Wedge", *Int. J. Fracture Mechanics*, Vol. 7, 1871, pp. 317-330.

[9] Mulville, D.R., "Characteristics of Crack Propagation at the Interface Between Two Dissimilar Media", *Naval Research Laboratory Report 7839*, 1975.

[10] Smelser, R.E. & Gurtin, M.E., "On the J-Integral for Bi-Material Bodies", *Int. J. Fracture*, Vol. 13, 1977, pp. 382-384.

2. JOINTS

A direct comparison of the strength of different types of joints for woodwork is shown in Fig. 2.1. The four joints have the same carrying capacity, viz. 15 kN, for a purely tensile load on the structure. The specifications were the following:

A - adhesive joint, 5 in. overlap
B - bolted joint with four ½ in. bolts, 9 in. overlap
C - bolted joint with one split ring connector and one ½ in. bolt, 11 in. overlap
D - nailed joint with 24 nails in drilled holes, 12 ¼ in. overlap

The adhesive joint (A) is obviously superior to the bolted (B, C) and nailed joint (D). The latter three may be disassembled, even though the nailed joint will lose some carrying capacity on reassembly. Nailing is used only with wood attached to wood, and will not be considered further.

References

Adams, R.D. & Wake, W.C., *Structural Adhesive Joints in Engineering*, Elsevier Applied Science Publishers, London & New York 1984, 309 pages.

Hart-Smith, L.J., "Joints", in *Composites, Enginered Materials Handbook, Vol. 1*, ASM International, Metals Park, OH. USA 1987, pp. 479-495.

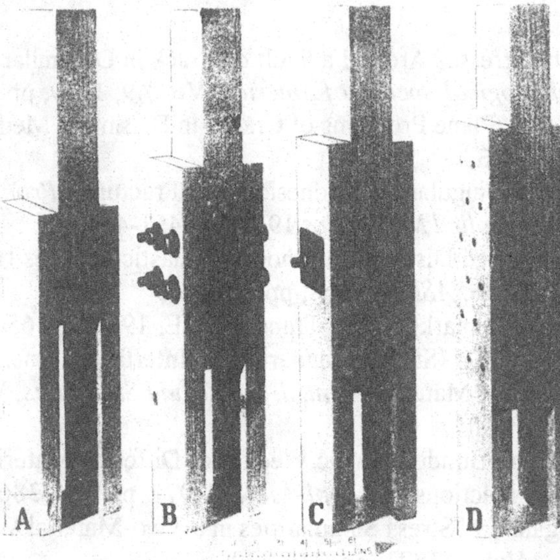

Fig 2.1. Joints of equivalent load capacity with various fastenings. (from R.D. Adams & W.C. Wake, *Structural Adhesive Joints in Engineering*, Elsevier Applied Science Publishers, 1984, p. 4).

2.1 Bolted Joints

Joints may be arranged as single or double lap (Figs. 2:2a,b). The symmetry of the latter case results in a more favourable stress field. Obviously the strength of a bolted joint will increase due to tightening of the bolts, such that load transfer occurs to an increasing amount by frictional forces between the adjoining surfaces. When an FRP component is bolted to a steel component in a lap joint, the FRP component will normally be given a larger cross section (Fig. 2:2c).

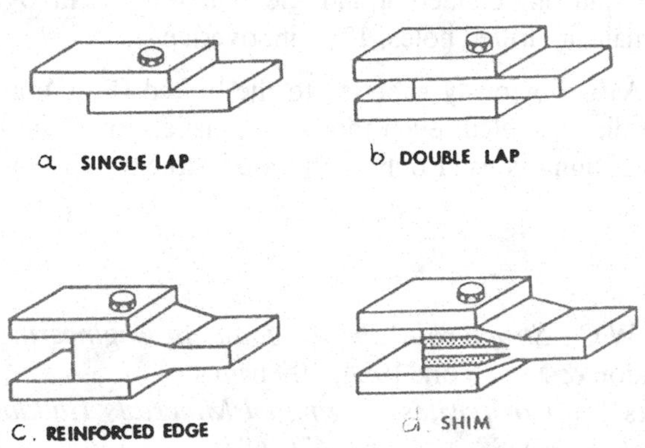

Fig. 2:2. Bolted joints. (from R.M. Jones, *Mechanics of Composite Materials*, McGraw-Hill Book Company, 1975, p. 233).

A bolted lap joint under tension may fail in one or more of four different principal modes, cf. Fig. 2:3

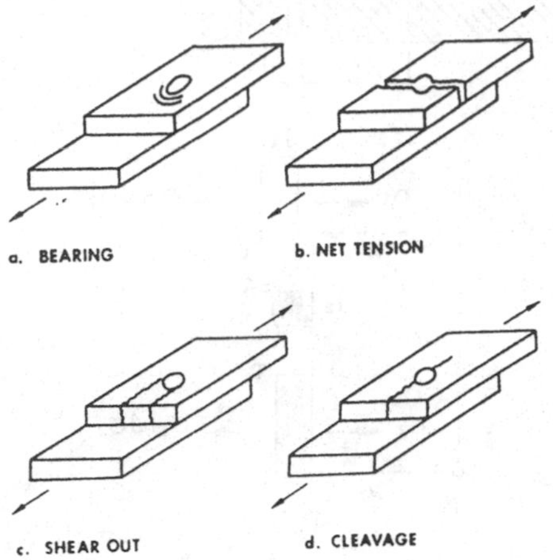

Fig. 2:3. Different failure mechanisms in bolted Joints. (from R.M. Jones, *loc.cit*, p. 234). (a) Excessive elongation of the bolt hole. (b) Tensile failure at the reduced cross section through the bolt hole. (c) Shearing out, causing plug failure. (d) Cleavage failure (often perpendicular to fibre reinforcement).

Finally, the bolt itself may fail due to shearing. Stress relaxation may occur in bolted joints involving FRP members, and the bolts may then have to be retightened at intervals.

Reference

Vinson, J.R. & Sierakowski, R.L., *The behavior of structures composed of composite materials*, Martinus Nijhoff Publishers,Dordrecht 1986, pp. 239-282.

2.2 Adhesive joints

Because of its demonstrated superiority, adhesive joining of structural components, e.g. FRP to steel, will be studied here in some detail. A simple mechanical model due to Volkersen (1938), cf. Fig. 2:4, will first be analyzed, and practical implications will then be discussed.

Two laminæ, denoted 1 and 2, are joined by an adhesive layer, denoted 3. A tensile load in the lamina induces shearing stresses in the adhesive layer. The main assumptions in the Volkersen model are that the laminæ are deformed only in tension, whereas the adhesive layer is deformed only in shearing.

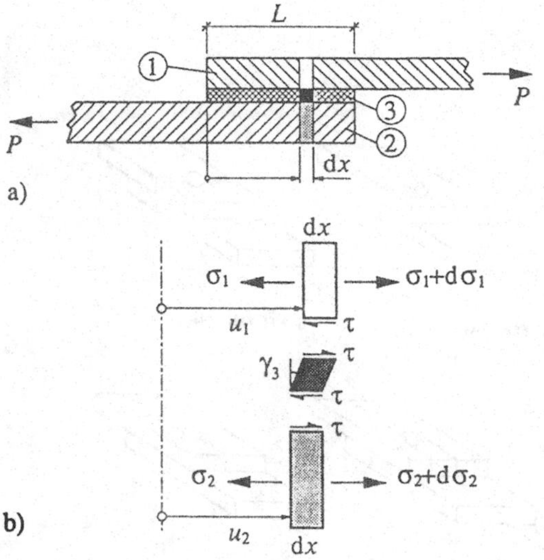

Fig. 2:4. Adhesive lap joint. (from J. Hult & H. Bjarnehed, *Styvhet och styrka: Grundläggande kompositmekanik* [in Swedish], Studentlitteratur, Lund 1993, p. 146).

With the notation of Fig. 2:4 there follows the equilibrium conditions for the two laminæ elements

$$\frac{d\sigma_1}{dx} = \frac{\tau}{t_1}, \quad \frac{d\sigma_2}{dx} = -\frac{\tau}{t_2} \tag{2.1-2}$$

With the constitutive equations for the laminæ and the adhesive layer

$$\frac{du_1}{dx} = \frac{\sigma_1}{E_1}, \quad \frac{du_2}{dx} = \frac{\sigma_2}{E_2}, \quad \frac{u_1 - u_2}{t_3} = \frac{\tau}{G_3} \tag{2.3-5}$$

then follows the differential equation

$$\frac{d^2}{dx^2} = \left(\frac{\sigma_1}{E_1} - \frac{\sigma_2}{E_2}\right) - \lambda^2 \cdot \left(\frac{\sigma_1}{E_1} - \frac{\sigma_2}{E_2}\right) = 0 \tag{2.6}$$

where

$$\lambda^2 = \left(\frac{1}{t_1 E_1} + \frac{1}{t_2 E_2}\right) \cdot \frac{G_3}{t_3} \tag{2.7}$$

The general solution

$$\frac{\sigma_1}{E_1}-\frac{\sigma_2}{E_2}=A\cosh\lambda x+B\sinh\lambda x \tag{2.8}$$

combines with the boundary conditions

$$\sigma_1(0)=0,\quad \sigma_2(0)=\frac{P}{bt_2},\quad \sigma_1(L)=\frac{P}{bt_1},\quad \sigma_2(L)=0 \tag{2.9-12}$$

to give

$$A=-\frac{P}{bt_2E_2},\quad B=\frac{P}{b}\cdot\frac{1/t_1E_1+\cosh(\lambda L/t_2E_2)}{\sinh\lambda L} \tag{2.13-14}$$

Hence the shear stress distribution in the adhesive layer

$$\tau(x)=\sqrt{\frac{t_1t_2E_1E_2}{t_1E_1+t_2E_2}\cdot\frac{G_3}{t_3}}\cdot\frac{P}{b}\cdot\left[-\frac{1}{t_2E_2}\cdot\sinh\lambda x+\left(\frac{1}{t_2E_2}+\frac{\cosh\lambda L}{t_2E_2}\right)\cdot\frac{\cosh\lambda x}{\sinh\lambda L}\right] \tag{2.15}$$

which is shown in Fig. 2:5. The largest shear stresses occur at the ends of the adhesive layer. If $t_1=t_2=t$ and $E_1=E_2=E$, there results a symmetrical stress distribution, where the maximum shear stress equals

$$\tau_{max}=\sqrt{\frac{G_3}{2Ett_3}}\cdot\frac{P}{b}\cdot\coth\sqrt{\frac{G_3L^2}{2Ett_3}} \tag{2.16}$$

If, in addition, the adhesive layer is so long, that $L^2 >> 2E\,t\,t_3\,/\,G_3$, there results a simpler expression for the maximum shear stress

$$\tau_{max}=\sqrt{\frac{G_3}{2Ett_3}}\cdot\frac{P}{b} \tag{2.17}$$

which is independent of the adhesive length L.

An elementary analysis, considering only the shear deformation in the adhesive but assuming the laminæ to be rigid, gives a constant shear stress in the adhesive and linearly varying tensile stresses in the laminæ (Fig. 2:6a). The Volkersen model gives the stress fields shown in Fig. 2:6b.

The average shear stress in the adhesive region is

$$\tau_{ave}=\frac{P}{bL} \tag{2.18}$$

and hence the ratio between maximum and average shear stress is found to be much larger than unity.

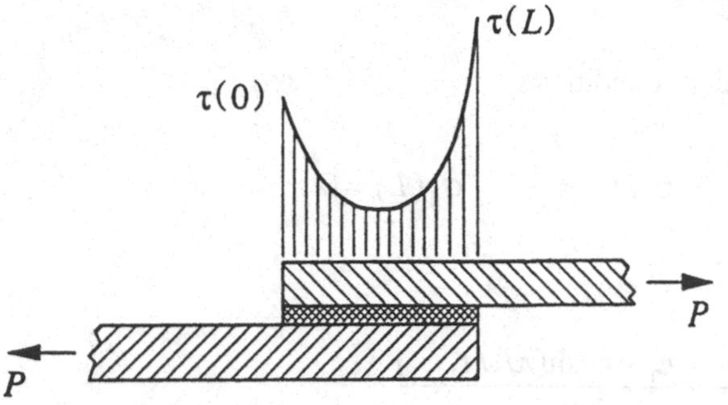

Fig. 2:5. Shear stress distribution in adhesive lap joint under tension. (from J. Hult & H. Bjarnehed, loc.cit., p. 148).

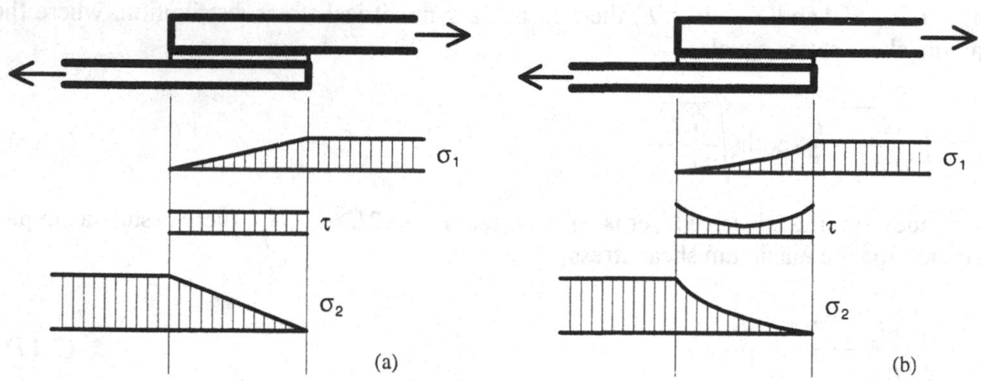

Fig. 2:6. Tensile and shear stress distributions in adhesive lap joint under tension. (a) Rigid members, (b) elastic members.

The analysis may be improved by including also the shearing deformation occurring in the laminæ. An extension has later been performed, which predicts also the stresses which are perpendicular to the tensile direction.

The theory of elasticity predicts a stress singularity at the ends of the adhesive region. The fact that the Volkersen model does not predict such a singularity shows its limitation. It is, nevertheless, an important tool in comparing different adhesive bondings.

Some standard types of adhesive joints are shown in Fig. 2:7. The purpose of the scarf, bevel, and step configurations is to diminish the stress concentrations at the ends of the adhesive region.

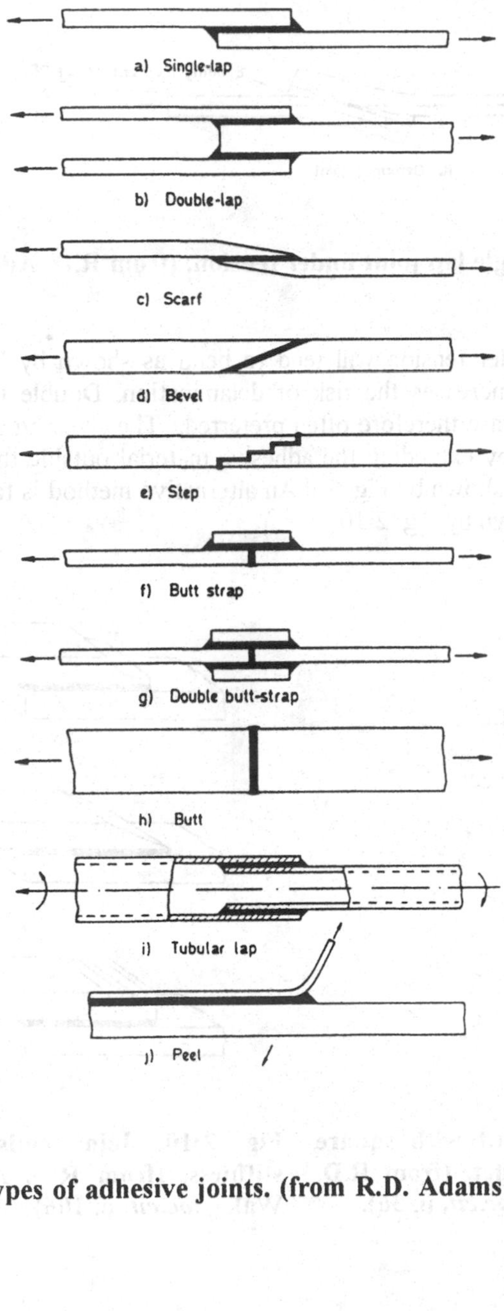

Fig. 2:7. Different types of adhesive joints. (from R.D. Adams & W.C. Wake, *loc.cit*, p. 15).

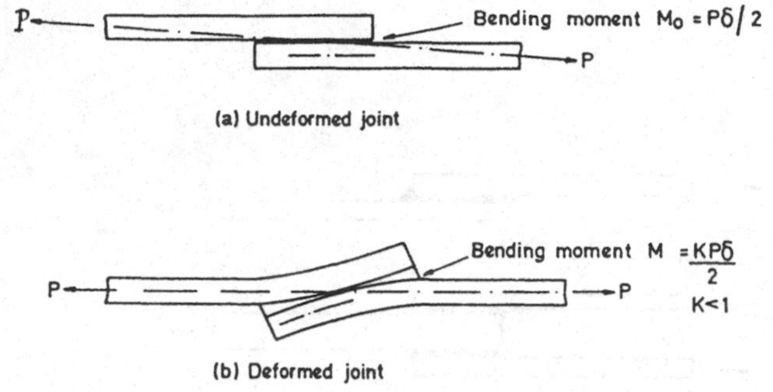

Fig. 2:8. Bending in single lap joint under tension. (from R.D. Adams & W.C. Wake, *loc.cit*, p. 16).

A single lap joint under tension will tend to bend as shown by Fig. 2:8. This causes additional stresses, and increases the risk of delamination. Double lap joints, where the bending effect is smaller, are therefore often preferred,. The shearing stresses at the lamina ends may be diminished by extending the adhesive material outside the lamina ends in the form of a "spew" fillet as shown by Fig. 2:9.An alternative method is to reduce the stiffness of the lamina ends as shown by Fig. 2:10.

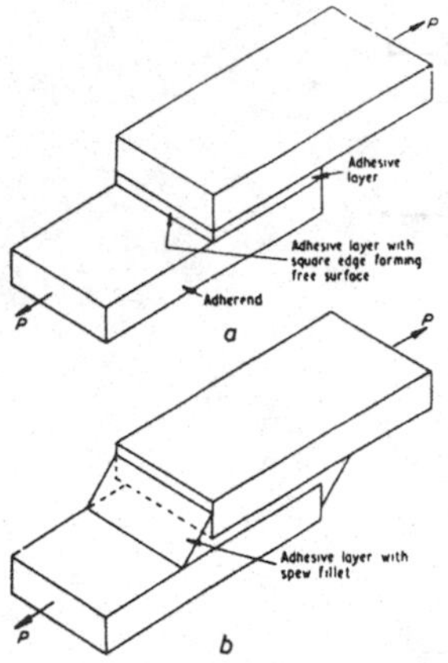

Fig. 2:9. Single lap joint with square edge and with spew fillet. (from R.D. Adams & W.C. Wake, *loc.cit*, p. 36).

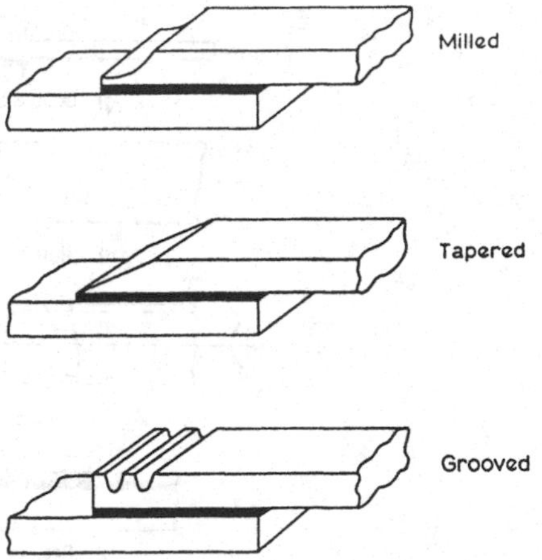

Fig. 2:10. Joint ends with reduced stiffness. (from R.D. Adams & W.C. Wake, *loc.cit*, p. 106).

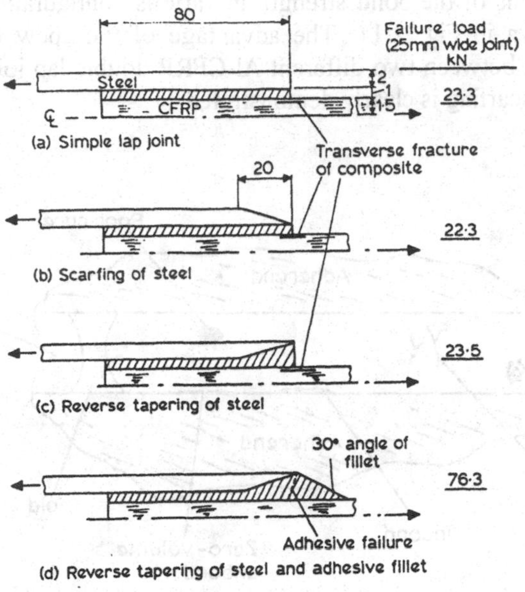

Fig. 2:11. Lap joint with steel bonded to CFRP. (from R.D. Adams & W.C. Wake, *loc.cit*, p. 107).

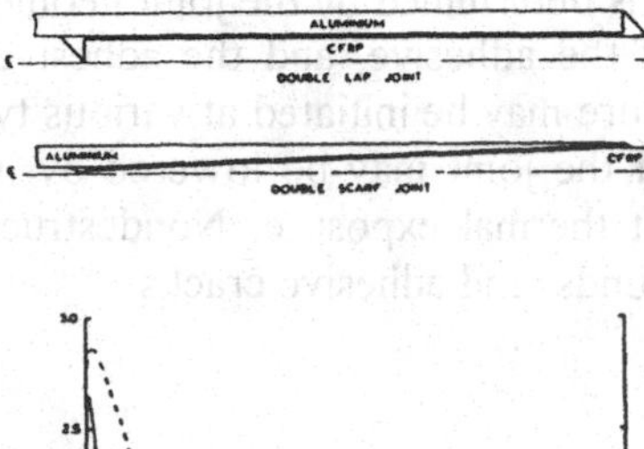

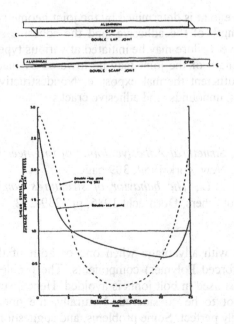

Fig. 2:12. Adhesive shear stress distribution in aluminium-CFRP double-scarf joint. (from R.D. Adams & W.C. Wake, *loc.cit*, p. 87).

Numerical comparisons of the bond strength in various configurations of a steel-CFRP adhesive joint are shown in Fig. 2:11. The advantage of the spew fillet is very clearly indicated. A comparison between two different Al-CFRP double lap joints is shown in Fig. 2:12. The advantage of scarfing is clearly demonstrated.

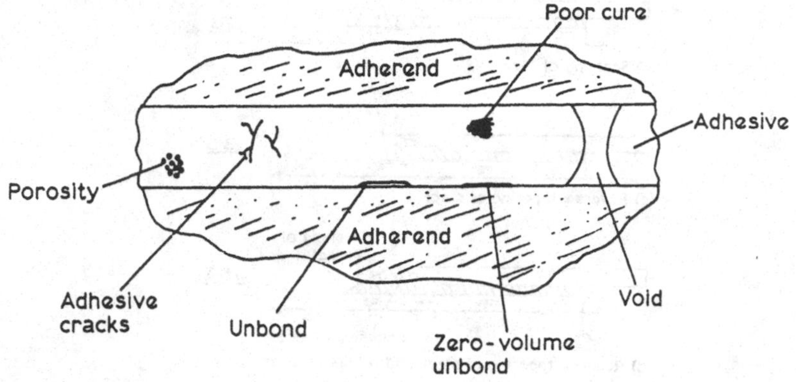

Fig. 2:13. Typical defects in an adhesive bond-line. (from R.D. Adams & W.C. Wake, *loc.cit*, p. 135).

The strength of an adhesive joint is determined by the joint geometry, the strength of the adherends, the cohesive strength of the adhesive, and the adhesive strength of the bond between adherend and adhesive. Failure may be initiated at various types of defects, such as shown in Fig. 2:13. The strength of the joint may be lowered by incorrect mixing of the adhesive, poor curing or insufficient thermal exposure. Nondestructive, ultrasonic testing may be used to discover voids, unbounds, and adhesive cracks.

References

Adams, R.D. & Wake, W.C., *Structural Adhesive Joints in Engineering*, Elsevier Applied Science Publishers, London & New York 1984, 309 pages.

Vinson, J.R. & Sierakowski. R.L., *The behavior of structures composed of composite materials*, Martinus Nijhoff Publishers, Dordrecht 1986, pp. 239-262.

2.3 Clamping joints

Clamping joints are used with advantage when one or both of the members to join consists of FRP (Fibre Reinforced Polymer) components. The problem of manufacturing FRP components with holes as used in bolt jonts is avoided. Hence, stress concentration in regions of such holes has not to be considered. Naturally, the operating function of a clamping joint is not completely perfect. Some problems, and suggestions on their solutions, arising in a steel clamping joint of a FRP leaf spring are considered below.

CHAPTER 9
PART II

STEEL CLAMPING JOINT OF FRP LEAF SPRING

H.L. Bjarnehed
Prosolvia Konsult AB, Göteborg, Sweden

ABSTRACT

The present lecture note considers the hybrid structure of an FRP leaf spring clamping joint. The hybrid structure consists of up to three structural components, *viz.* two steel clamps, an FRP leaf spring, and an intermediate polymer layer, all with different material properties. Some problems in the joint region of the leaf spring due to high contact stresses are discussed, such as delamination and wear. The contact state in the clamping joint region is modelled by the use of FEM and an analytical model involving contact loadings on an orthotropic half-plane edge. Contact stress distributions between a rigid flat punch and the half-plane edge under contact conditions of no friction, complete bonding, and limiting friction are presented. Further, contact stress distributions at the interface between an intermediate layer, a cushion, and a half-plane where the cushion is subjected to an indentation of a rounded rigid punch or a flexible beam is studied. An optimum cushion thickness profile giving uniform contact pressure distribution is presented.

INTRODUCTION

Hybrid structures subjected to contact loading

Problems concerning mechanical contact between elastic bodies are common in design work. Such states occur in regions where forces are transmitted from one component to another as in regions of joints, in bearings, in the contact region of a railway track and wheel, *etc.* Some of the more serious problems which might occur are cracks, wear, plastic collapse, and fatigue. Very high stresses at the interface of contact members are often the underlying cause of these problems. Cracks initiate and propagate in regions of high tensile and shear stresses. Wear occurs between structural components in contact due to high contact shear stresses. Concentrations of contact stresses may cause local plastic deformation of the material, which in dynamical cases has a very strong negative effect on fatigue life.

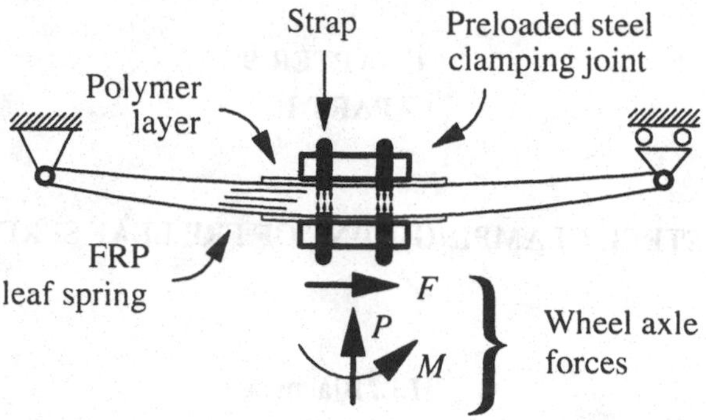

Fig. 2:14. Schematic picture of a truck FRP leaf spring provided with a clamping joint.

A special group of hybrid structures subjected to contact loading is mechanical joints involving composite components. A case in point is the use of a preloaded steel clamping joint for an FRP leaf spring in trucks. Fig. 2:14 shows an FRP leaf spring in a truck provided with two steel blocks transferring the rear wheel axle forces to the spring. The two steel units are clamped onto the leaf spring by the tightening of straps. The aim of the two clamps is two create normal forces at the contact surfaces, through which large enough friction may arise to prevent sliding at the contact surfaces, when the truck is driven on rough roads.

Stress concentration in clamping joint

The contact stresses arising due to the preload and axle forces on the clamps may be quite high near the ends of the steel clamps. Delamination of fibres from matrix in the clamped region due to high shear stresses is a problem to be considered additional to those mentioned above. To illustrate the shear stress concentration in the clamped region of an FRP leaf spring some results from Ref. [1] are presented in Fig. 2:15. The FRP spring and the steel clamps are modelled as rectangular shaped structures in the FE-code SOLVIA by 2-D finite elements. Conditions of complete bonding are applied between the clamps and the spring. A single centrally placed vertical force P is acting on the simply supported spring.

The magnitude of the force is chosen so as to produce the shear stress τ_{xy}=10 [MPa] at the centre of a spring cross section uninfluenced by the clamped region. Fig. 2:15 shows that the shear stress peak occurs some distance inside the clamped region and with a stress concentration factor larger than 5 for this particular configuration.

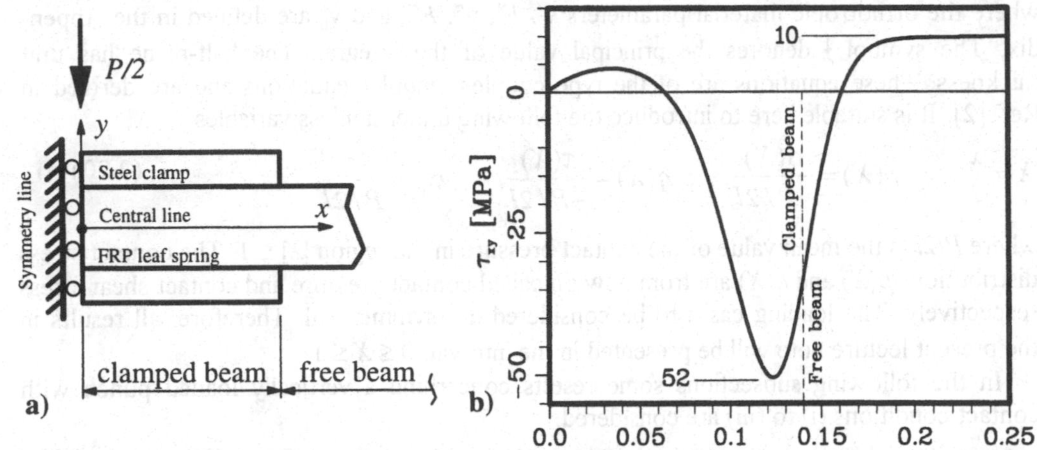

Fig. 2:15. a) Clamped region of FRP leaf spring, b) shear stress τ_{xy} along the central line. The dotted line indicates the transition between free and clamped spring.

CONTACT CONDITIONS

Contact model

Due to the fact that the spring is made of a fibre based composite it is highly sensitive to wear. Slip will always occur at the interface between two components in contact except for a complete bonded interface as in a adhesive joint. In Ref. [2] and [3] different frictional contact conditions in the clamp-spring interface are investigated, *viz*.

(i) no friction,
(ii) complete bonding, and
(iii) limiting friction.

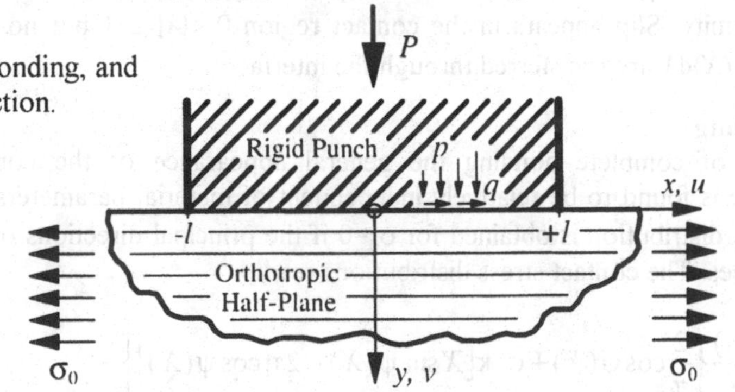

Fig. 2:16. The contact model of the FRP leaf spring clamping joint.

The contact is modelled as shown in Fig. 2:16, *i.e.* a rigid punch (steel clamp) on stressed orthotropic half-plane (spring in bending). The half-plane principal directions of elasticity coincide with the co-ordinates axes *x*, *y*, and *z*.

The governing equations relating the surface stresses (σ,τ) to the surface displacements (u,v) in the half-plane region $|x| \leq l$ read

$$\int_{-l}^{l} \frac{\tau(t)}{x-t}\,dt = -\nu'\sigma(x) + E'\frac{du}{dx} - \nu_0\sigma_0 \tag{2.19a}$$

$$\int_{-l}^{l} \frac{\tau(t)}{x-t}\,dt = \nu''\tau(x) + E''\frac{dv}{dx} \tag{2.19b}$$

where the orthotropic material parameters ν', E', ν'', E'', and ν_0 are defined in the Appendix. The symbol $⨍$ denotes the principal value of the integral. The half-plane has unit thickness. These equations are of the type coupled singular equations and are derived in Ref. [2]. It is suitable here to introduce the following dimensionless variables

$$X=\frac{x}{l},\qquad p(X)=\frac{\sigma(X)}{-P/2l},\qquad q(X)=\frac{\tau(X)}{-P/2l},\qquad \tilde{\sigma}_0=\frac{\sigma_0}{P/2l} \tag{2.20a-d}$$

where $P/2l$ is the mean value of the contact pressure in the region $|X|\leq 1$. The contact stress distributions $p(X)$ and $q(X)$ are from now on called contact pressure and contact shear stress respectively. The loading cases to be considered are symmetrical. Therefore, all results in the present lecture note will be presented in the interval $0\leq X\leq 1$.

In the following subsections some results concerning a vertically loaded punch with contact conditions (i) to (iii) are considered.

No friction

In case of no friction the same material independent contact stress distributions are obtained as for a rigid punch indenting an isotropic half-plane, *viz.*

$$p(X)=\frac{2/\pi}{\sqrt{1-X^2}} \tag{2.21a}$$

$$q(X)=0 \tag{2.21b}$$

It appears from eq. (2.21a) that the contact pressure $p(X)$ approaches infinite values as $|X|$ approaches unity. Slip appears in the contact region $0<|X|\leq 1$ but no wear induced friction forces $\int q(X)\mathrm{d}X$ are transferred through the interface.

Complete bonding

In the case of complete bonding the general appearance of the contact pressure distribution $p(X)$ is found to be relatively independent of material parameters. An identical contact pressure distribution is obtained for $\sigma_0=0$ if the principal directions of elasticity are rotated 90 degrees. The contact stress distributions read

$$p(X)=\frac{\cosh(\pi\eta)}{\sqrt{1-X^2}}\left\{\frac{2}{\pi}\cos\psi(X)+\tilde{\sigma}_0\kappa[X\sin\psi(X)-2\eta\cos\psi(X)]\right\} \tag{2.22a}$$

$$q(X)=\rho\frac{\cosh(\pi\eta)}{\sqrt{1-X^2}}\left\{\frac{2}{\pi}\sin\psi(X)-\tilde{\sigma}_0\kappa[X\cos\psi(X)+2\eta\sin\psi(X)]\right\} \tag{2.22b}$$

where

$$\psi(X)=\eta\ln\left(\frac{1+X}{1-X}\right) \tag{2.23}$$

and η, κ, and ρ are material parameters defined in the Appendix.

It appears from (2.22) and (2.23) that the contact stresses approach infinite values and that their oscillating behaviour ($\sin\psi(X)$, $\cos\psi(X)$) grows stronger as $|X|$ approaches unity. No slip occurs in the contact region $|X|\leq 1$ and hence no wear is present.

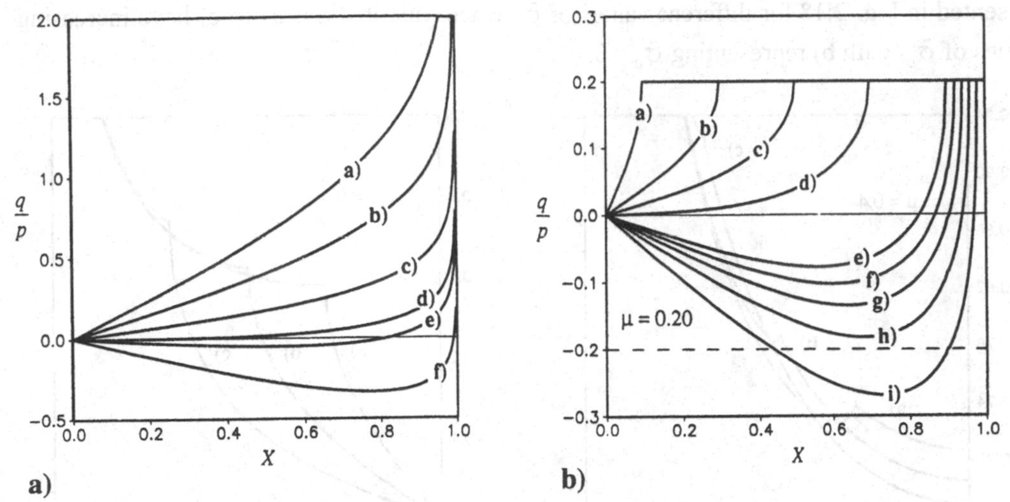

a) b)

Fig. 2:17. Distribution of *q*(*X*)/*p*(*X*) for the case of a) complete bonding, *p*(*X*) and *q*(*X*) according to eq. (2.22) and b) limiting friction with μ=0.2 (from H.L. Bjarnehed [2]).

Fig. 2:17a shows distributions of the ratio q/p for $P > 0$ and different values of $\tilde{\sigma}_0$, which gives a hint where slip can be expected. The ratio gives the necessary magnitude of the coefficient of Coloumb friction μ to prevent slip. Distributions a) to f) have increasing values of $\tilde{\sigma}_0$ where c) represents $\tilde{\sigma}_0 = 0$. The conclusions of Fig. 2:17a are:

(1) For values of the ratio $\tilde{\sigma}_0 < \tilde{\sigma}^*$, where $\tilde{\sigma}^* > 0$ depends on material parameters and coefficient of friction, slip occurs at the outer ends of contact.

(2) For values $\tilde{\sigma}_0 > \tilde{\sigma}^*$ more than two slip regions can be expected. This is due to the fact that the ratio q/p has an extreme value in the region $0 < |X| < 1$.

Limiting friction

In the case of limiting friction a closed form solution for the contact stress distributions, as shown in eq. (2.21) and (2.22), does not exist. A numerical solution with Coloumb friction applied in the contact region $|X| \leq 1$ and predicted slip regions at the outer ends of contact is given by Ref. [2]. This solution is naturally only valid for $P>0$. The boundary conditions for this case read

$$\text{No slip region, } |X| \leq C: \quad \frac{du}{dX} = \frac{dv}{dX} = 0, \qquad p(X) \leq \mu|q(X)| \tag{2.24a}$$

$$\text{Slip region, } C < |X| \leq 1: \quad \frac{dv}{dX} = 0, \qquad p(X) = \mu|q(X)| \tag{2.24b}$$

As it appears in (2.24b) wear is present in the slip region. Contact stress distributions are presented in Fig. 2:18 for different values of $\tilde{\sigma}_0$ where distributions a) to c) have increasing values of $\tilde{\sigma}_0$, with b) representing $\tilde{\sigma}_0 = 0$.

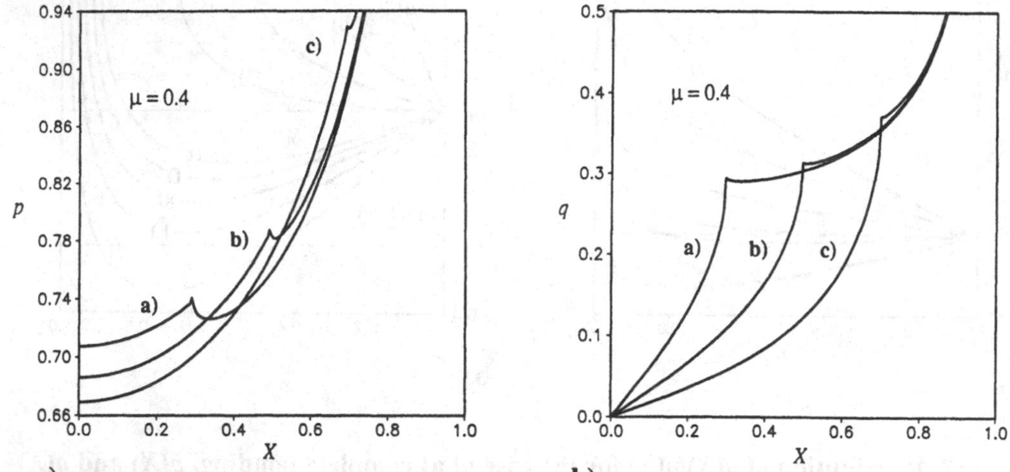

a) b)

Fig. 2:18. Distributions of a) contact pressure *p*(*X*) and b) contact shear stress *q*(*X*) for the case of limiting friction μ=0.4 (from H.L. Bjarnehed [2]).

The slope discontinuities in the distributions indicate the transition between regions with slip and no slip at $|X|=C$. The general appearance of the contact pressure distribution is found to be relatively independent of the coefficient of friction in the contact region.

Fig. 2:17b shows distributions of the ratio q/p for different values of $\tilde{\sigma}_0$ for a case with $\mu = 0.2$. Distributions a) through i) have increasing values of $\tilde{\sigma}_0 > 0$, where h) represents $\tilde{\sigma}_0 = 1.47$. The distribution i) is not a valid solution of the problem but it indicates, as discussed in subsection *Complete bonding*, the possibility of additional slip regions occurring in the contact region $0 < |X| \leq 1$.

INTERMEDIATE CUSHION

Introduction

As discussed in the above subsection *Introduction* it is of main interest to reduce the high contact stresses. Three different methods may be used to diminish the contact pressure non uniformity, *viz*.

(a) Rounding off the sharp corners of the punch.
(b) Introducing a cushion between the punch and the half-plane.
(c) Making the punch more flexible at its ends.

The clamping joint of the FRP leaf spring is often provided with an intermediate layer, a cushion, made of a softer polymeric material. This will make the contact stress distributions more uniform, thus decreasing the severity of the contact stresses. However, the main purpose of the polymer layer is to protect the FRP leaf spring from wear. Basic understanding of the mechanics of such load transfer is essential in much design work involving FRP.

Optimum cushion thickness

The two methods (a) and (b) may also be combined, such as shown in Fig. 2:19. The present configuration of the clamping joint is a hybrid structure including three components, *i.e.* punch, cushion, and half-plane, with different material properties.

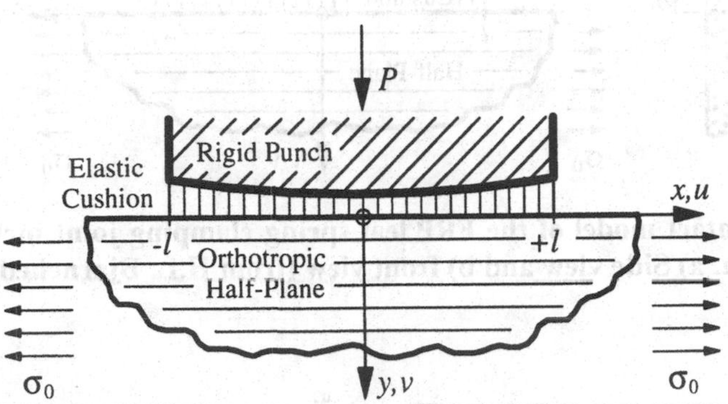

Fig. 2:19. The contact model of the FRP leaf spring clamping joint including a cushion.

The cushion is modeled as a system of an infinite number of linearly elastic springs. It can be considered as an elastic continuum with both normal (E_c) and shear (G_c) stiffness but without any interaction between individual springs. Therefore, the following strain-stress relations are assumed for the cushion

$$\gamma_c(X) = \frac{u(X)}{h(X)} = \frac{\tau(X)}{G_c} \tag{2.25a}$$

$$\varepsilon_c(X) = \frac{v(X) - \Delta_y}{h(X)} = \frac{\sigma(X)}{E_c} \tag{2.25b}$$

where Δ_y denotes the vertical displacement of the rigid punch. An optimum thickness profile $h(X)$ of the cushion has been derived for frictionless contact, which results in a uniform contact pressure $p(X)=P/2l$=constant. The optimum thickness profile reads

$$h^{\text{opt}}(X) = l \cdot \frac{E_c}{E''} \cdot \left[(1+X)\ln(1+X) + (1-X)\ln(1-X)\right] + h_0 \tag{2.26}$$

The minimum cushion thickness at X=0, *i.e.* $h_0=h^{\text{opt}}(X=0)$, is to be chosen by the user. A uniform cushion thickness profile $h(X)$ = constant gives a contact pressure distribution, which is intermediate between the limiting ones of no cushion, $p(X)$ according to eq. (2.21), and an optimal cushion, $p(X)=P/2l=\text{constant}$.

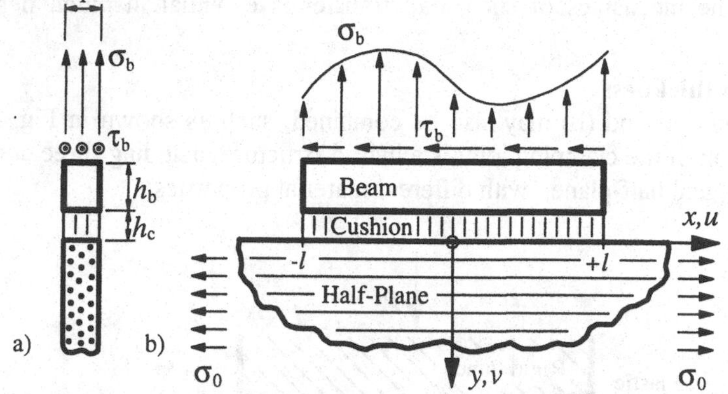

Fig. 2:20. The contact model of the FRP leaf spring clamping joint including flexible beam and cushion. a) Side view and b) front view (from H.L. Bjarnehed [4]).

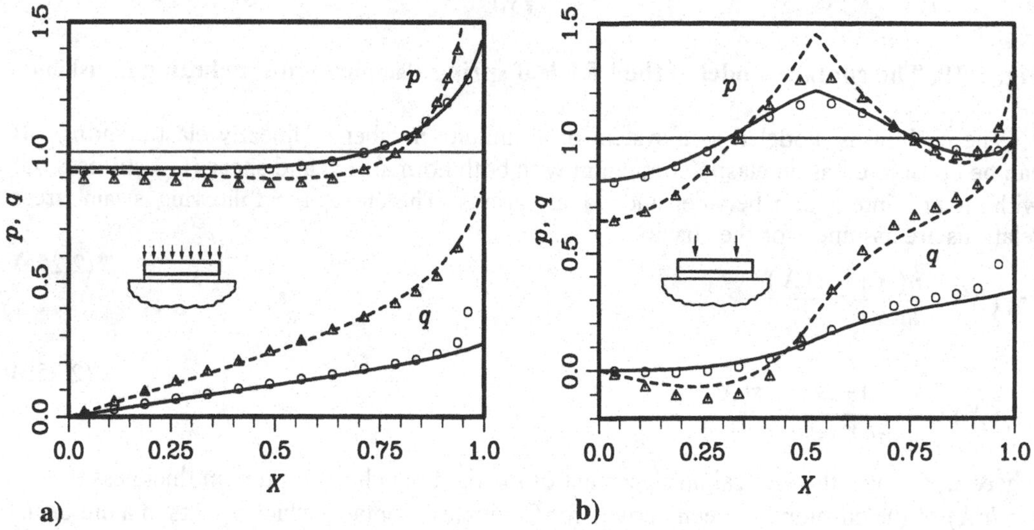

Fig. 2:21. Contact stress distributions for the case of a flexible punch. a) Uniform pressure load and b) two point forces. A solid line represents a more softer cushion than the dotted line. The markers correspond to a FEM solution of the present problem where the contact members are modeled with 2-D plane elements (from H.L. Bjarnehed [4]).

Flexible punch

All contact models of the clamping joint presented so far have a non flexible formulation of the steel clamp (rigid punch formulation). In Ref. [4] a Timoshenko beam is used to model a flexible steel clamp as shown in Fig. 2:20. The beam is loaded with both normal σ_b and tangential τ_b load distributions. The cushion is formulated as described above and has a uniform thickness profile, *i.e.* $h(X)$=constant.

Consider a configuration with a cushion completely bonded to the beam and half-plane. Fig. 2:21 shows contact stress distributions for a vertically loaded beam. The same configuration in the contact model has been used in Figs. 2:21a and 2:21b. Due to the flexibility of the beam the distributions in Figs. 2:21a and 2:21b display a relatively distinct difference in shape. The location of the point force at $|X|$ =0.525 is mirrored in the contact pressure distribution of Fig. 2:21b.

REFERENCES

[1] Bjarnehed, H.L. and Neumeister J.M.: FEM-analysis of shear stress concentrations in a steel-composite clamping joint, Report CTH-1, Volvo Truck Corp., Product Development Div., Chalmers Univ. of Tech., Div. of Solid Mech., Göteborg 1989.

[2] Bjarnehed, H.L.: Rigid punch on stressed orthotropic half-plane with partial slip, ASME J. Appl. Mech., 1 (1991), 128-133, Erratum, ibid, 2 (1991), 565.

[3] Bjarnehed, H.L.: Multiply loaded rigid punch on stressed orthotropic half-plane via a thin elastic layer, ASME J. Appl. Mech.,. (1992), S115-S122.

[4] Bjarnehed, H.L.: Multiply loaded Timoshenko beam on stressed orthotropic half-plane via a thin elastic layer, ASME J. Appl. Mech., (1993), 541-547.

APPENDIX

Material parameters

The material parameters ν', E', ν'', E'', and ν_0 in the governing equations of the half-plane (2:19) all depend on the orthotropic properties of the half-plane. However, eq:s (2:19) are applicable also for an isotropic half-plane. It is only a matter of difference in the material parameters. These parameters can, through their definition, be degenerated to material parameters valid for an isotropic half-plane. They read

$$\nu' = \pi\frac{\alpha\beta c_{11} + c_{12}}{(\alpha+\beta)c_{11}}, \qquad E' = \frac{\pi}{(\alpha+\beta)c_{11}}, \qquad \nu_0 = \frac{\pi}{\alpha+\beta}, \qquad \frac{\nu'}{\nu''} = \frac{E'}{E''} = \alpha\beta \qquad \text{(A1a-e)}$$

where

$$\alpha, \beta = \sqrt{\frac{c_{66}+2c_{12}}{2c_{11}} \pm \sqrt{\left(\frac{c_{66}+2c_{12}}{2c_{11}}\right)^2 - \frac{c_{22}}{c_{11}}}}, \qquad \alpha\cdot\beta = \sqrt{\frac{c_{22}}{c_{11}}} \qquad \text{(A2a-c)}$$

and c_{11}, c_{12}, c_{22}, and c_{66} are the elastic constants in the generalised Hooke's law for an orthoptropic medium, *i.e.*

$$\varepsilon_x = c_{11}\sigma_x + c_{12}\sigma_y \tag{A3a}$$

$$\varepsilon_y = c_{12}\sigma_x + c_{22}\sigma_y \tag{A3b}$$

$$\gamma_{xy} = c_{66}\tau_{xy} \tag{A3c}$$

The material parameters in the solution of contact stress distributions for complete bonding contact presented in eq. (2.22) read

$$\eta = \frac{1}{2\pi}\ln\left(\frac{\pi+\sqrt{\nu'\nu''}}{\pi-\sqrt{\nu'\nu''}}\right), \qquad \kappa = \frac{1/\sqrt{\alpha\beta}}{\alpha+\beta}, \qquad \rho = \sqrt{\alpha\beta} \tag{A4a-c}$$

CHAPTER 10

CURING OF EPOXY MATRIX COMPOSITES

A.C. Loos

Virginia Polytechnic Institute, Blacksburg , VA, USA

and

G.S. Springer

Stanford University, Stanford , CA, USA

ABSTRACT

Models were developed which describe the curing process of composites constructed from continuous fiber-reinforced, thermosetting resin matrix prepreg materials. On the basis of the models, a computer code was developed, which for flat-plate composites cured by a specific cure cycle, provides the temperature distribution, the degree of cure of the resin, the resin viscosity inside the composite, the void sizes, the temperatures and pressures inside voids, and the residual stress distribution after the cure. In addition, the computer code can be used to determine the amount of resin flow out of the composite and the resin content of the composite and the bleeder.

* Virginia Polytechnic Institute, Blacksburg, Virginia, 24061

1. INTRODUCTION

Composite parts and structures constructed from continuous fiber reinforced thermosetting resin matrix prepreg materials are manufactured by arranging the uncured fiber resin mixture into the desired shape and then curing the material. The curing process is accomplished by exposing the material to elevated temperatures and pressures for a predetermined length of time. The elevated temperatures applied during the cure provide the heat required for initiating and maintaining the chemical reactions in the resin which cause the desired changes in the molecular structure. The applied pressure provides the force needed to squeeze excess resin out of the material, to consolidate individual plies, and to compress vapor bubbles.

The elevated temperatures and pressures to which the material is subjected are referred to as the cure temperature and the cure pressure. The magnitudes and durations of the temperatures and pressures applied during the curing process (denoted as the cure cycle) significantly affect the performance of the finished product. Therefore, the cure cycle must be selected carefully for each application. Some major considerations in selecting the proper cure cycle for a given composite material are:

a) the temperature inside the material must not exceed a preset maximum value at any time during cure,

b) at the end of cure, all the excess resin is squeezed out from every ply of the composite and the resin distribution is uniform,

c) the material is cured uniformly and completely,

d) the cured composite has the lowest possible void content, and

e) the curing process is achieved in the shortest amount of time.

At the present time, the cure cycle is generally selected empirically by curing small specimens and by evaluating the "quality" of the specimens after cure. Such empirical methods have several drawbacks; a) an extensive experimental program is usually required to determine the proper cure cycle for a given material, b) a cure cycle found to be satisfactory for a given material under one set of conditions may not apply under a different set of conditions, and c) they do not ensure that the composite was cured completely in the shortest amount of time.

The shortcomings of empirical approaches could be overcome by use of analytical models. Models applicable to different aspects of the curing process have been proposed by Springer and Loos [1-3]. In this paper, the different models are extended and combined into a comprehensive model which relates the cure cycle to the thermal, chemical and physical processes occurring in continuous fiber-reinforced composites during cure, and which then can be used to establish the most appropriate cure cycle in any given application.

2. MODEL

In this section, a model is described which yields the following parameters during cure:

a) the temperature inside the composite as a function of position and time;

b) the pressure inside the composite as a function of position and time;

c) the degree of cure of the resin as a function of position and time;

d) the resin viscosity as a function of position and time;

e) the number of compacted prepreg plies as a function of time;

f) the amount of resin in the bleeder as a function of time;

g) the thickness and the mass of the composite as a function of time;

h) the void sizes, and the pressures and temperatures inside voids as functions of void locations and time; and

i) the residual stresses in each ply after cure.

A model providing the above-mentioned information is developed below in four parts. The first part of the model, referred to as the "thermo-chemical model", yields the temperature, the degree of cure and the viscosity. The second part ("flow model") gives the pressure, the resin flow out of the composite, and the resin content of the composite and the bleeder. The third part ("void model") gives the void size and the temperature and pressure inside the void. The fourth part ("stress model") yields the residual stresses. Details of the models are presented subsequently. First, a brief description of the problem is given.

2.1 Problem Statement

We consider a fiber-reinforced epoxy-matrix composite of initial thickness L_i constructed from unidirectional continuous fiber "prepreg" tape (Figure 1). An absorbent material (referred to as a "bleeder") is placed on one side (or on both sides) of the composite. The thickness of the bleeder is L_b. The composite-bleeder system is placed on a metal tool plate ready for processing. A sheet of non-porous teflon release cloth is placed between the composite and the tool plate, and a sheet of porous teflon release cloth is placed between the composite and the bleeder to prevent sticking. A metal plate is placed on top of the bleeder, and an air breather is added when curing is done in an autoclave. Restraints (called "dams") are also mounted around the prepreg to prevent lateral motion and to minimize resin flow parallel to the tool plate and through the edges. Finally, a plastic sheet ("vacuum bag") is placed around

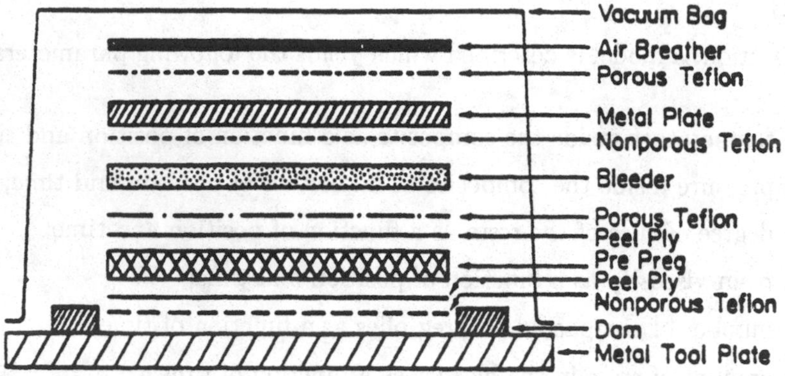

Figure 1 Schematic of the prepreg lay-up.

the entire assembly when vacuum is applied during the cure. Here, we are concerned only with the composite-bleeder system illustrated in Figure 2, because the additional components (vacuum bag, air breather, teflon sheets, etc.) have no direct effect on the model.

Initially time (time<0), the resin is uncured and the bleeder contains no resin. Starting at time $t = 0$, the composite-bleeder system is exposed to a known temperature T_o. The cure temperature T_o may be the same or may be different on the two sides of the composite-bleeder system. At some time $t_p (t_p \geq 0)$, a known pressure P_o is applied to the system. Both the cure temperature T_o and the cure pressure P_o may vary with time in an arbitrary manner. The objective is to determine the parameters listed in points $a - i$ previously.

In formulating the model, resin is allowed to flow in the directions both perpendicular and parallel to the plane of the composite. Resin flow in the plane of the composite is allowed only in the direction parallel to the fibers. In order to emphasize the concepts and the solution methods, the properties in the plane of the composite are taken to be constant. However, the model and the method are general and can readily be extended to complex geometries. It is also noted that the analysis is presented for composites made of continuous fiber, unidirectional tape. The model can also be applied to composites made of woven fabric.

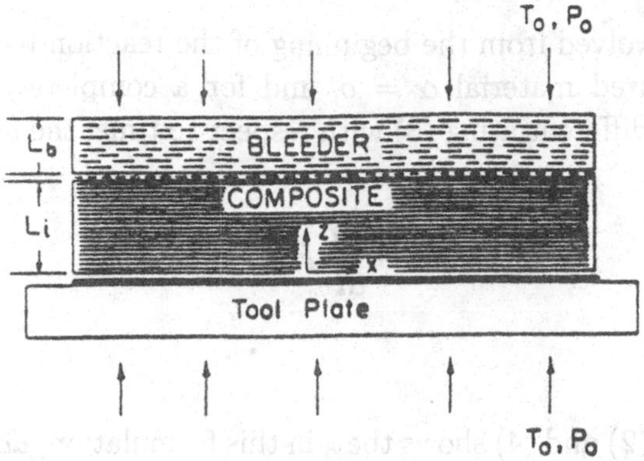

Figure 2 Geometry of the composite-bleeder system.

2.2 Thermo-Chemical Model

The temperature distribution, the degree of cure of the resin, and the resin viscosity inside the composite depend on the rate at which heat is transmitted from the environment into the material. The temperature inside the composite can be calculated using the law of conservation of energy together with an appropriate expression for the cure kinetics. By neglecting energy transfer by convection, the energy equation may be expressed as

$$\rho C \frac{\partial T}{\partial t} = \frac{\partial}{\partial z}\left(K \frac{\partial T}{\partial z}\right) + \varrho \dot{H} \tag{1}$$

where ϱ and C are the density and specific heat of the composite, K is the thermal conductivity in the direction perpendicular to the plane of the composite, and T is the temperature. $\dot{H}$ is the rate of heat generated by chemical reactions and is defined in the following manner

$$\dot{H} = RH_R \tag{2}$$

H_R is the total or ultimate heat of reaction during cure and R is the reaction or cure rate. The degree of cure of the resin (denoted as the degree of cure α) is defined as

$$\alpha \equiv \frac{H(t)}{H_R} \tag{3}$$

$H(t)$ is the heat evolved from the beginning of the reaction to some intermediate time, t. For an uncured material $\alpha = o$, and for a completely cured material α approaches unity. By differentiating (3) with respect to time, the following expression is obtained

$$\dot{H} = \frac{d\alpha}{dt} H_R \tag{4}$$

A comparison of Eqs. (2) and (4) shows that, in this formulation, $d\alpha/dt$ is the reaction or cure rate. If diffusion of chemical species is neglected, the degree of cure at each point inside the material can be calculated once the cure rate is known in the following way

$$\alpha = \int_0^t \left(\frac{d\alpha}{dt}\right) dt \tag{5}$$

In order to complete the model, the dependence of the cure rate on the temperature and on the degree of cure must be known. This dependence may be expressed symbolically as

$$\frac{d\alpha}{dt} = f(T, \alpha) \tag{6}$$

The functional relationship in Eq. (6), along with the value of the heat of reaction H_R for the prepreg material under consideration, can be determined experimentally by the procedures described in ref. [4].

The density ϱ, specific heat C, heat of reaction H_R, and thermal conductivity K depend on the instantaneous, local resin and fiber contents of each ply. The properties ϱ, C, and H_R can be calculated by the rule of mixtures [5]. The thermal conductivity of the prepreg can be calculated by the approximate formula developed by Springer and Tsai [6].

Solutions to Eqs. (1) and (4)-(6) can be obtained once the initial and boundary conditions are specified. The initial conditions require that the temperature and degree of cure inside the composite be given before the start of the cure (time<0). The boundary conditions require that the temperatures on the top and bottom surfaces of the composite be known as a function of time during cure (time$\geq$ 0). Accordingly, the initial and boundary conditions corresponding to Eqs. (1) and (4)-(6) are

Initial conditions:

$$\left.\begin{array}{l} T = T_i(z) \\ \alpha = 0 \end{array}\right\} \begin{array}{l} 0 < z < L \\ t < 0 \end{array} \qquad (7)$$

T_i is the initial temperature in the composite

Boundary conditions

$$\left.\begin{array}{l} T = T_L(t) \text{ at } z = 0 \\ T = T_u(t) \text{ at } z = L \end{array}\right\} t \geq 0 \qquad (8)$$

where T_u and T_L are the temperatures on the top and bottom surfaces of the composite, respectively (Figure 3).

Solutions to Eqs. (1) and (4)-(8) yield the temperature T, the cure rate $d\alpha/dt$, and the degree of cure α as functions of position and time inside the composite.

Once these parameters are known, the resin viscosity can be calculated, provided a suitable expression relating resin viscosity to the temperature and degree of cure is available. If the resin viscosity is assumed to be independent of shear rate, then the relationship between viscosity, temperature, and degree of cure can be represented in the form

$$\mu = g(T, \alpha) \qquad (9)$$

The manner in which the relationship between viscosity, temperature, and degree of cure can be established is described in ref [4].

2.3 Resin Flow Model

At some time $t_p (t_p \geq 0)$, pressure is applied to the composite-bleeder system (Figure 3). As a result of this pressure, resin flows from the composite into the bleeders. Resin flow in the direction parallel to the plane of the composite can be

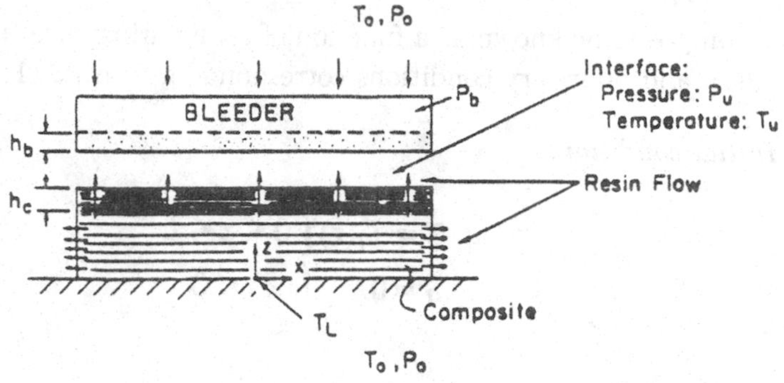

Figure 3 Resin flow model.

neglected if a) both the width and the length of the composite are large compared to the thickness L, and b) if restraints are placed around the sides of the composite. This situation is generally encountered in practice. When modelling the curing process of systems where the aforementioned conditions are met, only resin flow normal to the tool plate need to be taken into account. However, under some conditions, resin flow along the fibers cannot be prevented. This situation may occur when the length of the composite is similar to the thickness. Under these circumstances, resin flow both normal and parallel to the tool plate takes place simultaneously. The model must then consider resin flow in both directions. In the model that is developed below, resin flow both normal and parallel to the tool plate is taken into account.

Before the resin flow model is established, the behavior of the prepreg plies during the squeezing action (cure pressure application) is examined.

The resin flow process normal to the tool plate was demonstrated by Springer [2] to occur by the following mechanism. As pressure is applied, the first (top) ply ($n_s = 1$, Figure 4) moves toward the second ply ($n_s = 2$), while resin is squeezed out from the space between the plies. The resin seeps through the fiber bundles of the first ply. When the fibers in the first ply get close to the fibers in the second ply, the two plies move together toward the third ply, squeezing the resin out of the space between the second and the third ply. This sequence of events is repeated for the subsequent plies. Thus, the interaction of the fibers proceeds down the prepreg in a wavelike manner (Figure 4).

Note that there are essentially two regions in the composite. In region 1, the plies are squeezed together and contain no excess resin, while in region 2 the plies have not moved and have the original resin content. Some compacting of the fibers within the individual plies may also occur but, as a first approximation, this effect is neglected here.

It is noted that there is a pressure drop only across those plies through which resin flow takes place. The pressure is constant (and equal to the applied pressure P_o) across the remaining layers of prepreg.

When there is resin flow in both the normal and parallel directions, resin is squeezed out from between every ply continuously, as long as there is excess resin between adjacent plies. In this case, the thickness between different plies vary and change with time, as illustrated in Figure 5.

Although the resin flow in the normal and parallel directions are related, to facilitate the calculations in the model, the two phenomena are decoupled. Hence, separate models are described below for the resin flow in directions normal and parallel to the tool plate.

The model developed predicts changes in the dimensions of the composite only due to changes in the resin content. Shrinkage due to changes in the molecular structure of the resin during cure is not considered.

2.4 Resin Flow Normal to the Tool Plate

Owing to the complex geometry, the equations describing the resin flow through the composite normal to the tool plate (z direction) and into the bleeder cannot be established exactly. Nevertheless, an approximate formulation of the problem is feasible by treating the resin flow through both the composite and bleeder as flow through porous media. Such an approach was proposed by Bartlett [7] for studying resin flow through glass fabric prepreg, and by Loos and Springer [1-3] for resin flow in continuous fiber prepregs. In the model, inertia forces are considered to be negligible compared to viscous forces. Then, at any instant of time, the resin velocities in the prepreg and in the bleeder may be represented by Darcy's law.

$$V = -\frac{S}{\mu}\frac{dP}{dz} \tag{10}$$

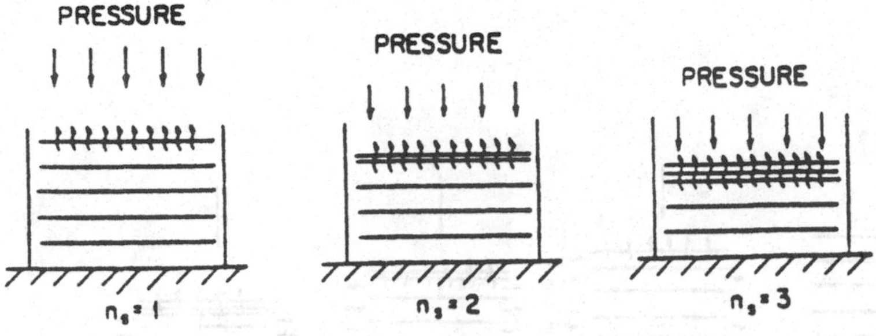

Figure 4 Illustration of the resin flow process normal to the tool plate.

where S is the apparent permeability, μ is the viscosity, and dP/dz is the pressure gradient. The law of conservation of mass (together with Eq. 10) gives the following expression for the rate of change of mass M in the composite

$$\frac{dM}{dt} = -\varrho_r A_z V_z = -\varrho_r A_z S_c \frac{P_c - P_u}{\int_0^{h_c} \mu dz} \tag{11}$$

where ϱ_r is the resin density, A_z is the cross sectional area perpendicular to the z axis, h_c is the thickness of the compacted plies, i.e., the thickness of the layer through which resin flow takes place (Figure 3). P_u is the pressure at the interface between the composite and the bleeder. The subscript c refers to conditions in the composite at position h_c. Accordingly, P_c is the pressure at h_c and is the same as the applied pressure ($P_c = P_o$) [2].

At any instant of time, the resin flow rate through the composite is equal to the resin flow into the bleeder

$$\varrho_r A_z V_z = \varrho_r A_z V_b \tag{12}$$

The temperature, and hence the viscosity, of the resin inside the bleeder is assumed to be independent of position (but not of time). Thus, Eqs. (10) and (12) yield

$$\varrho_r A_z V_z = \varrho_r A_z \frac{S_b}{\mu_b} \frac{P_u - P_b}{h_b} \tag{13}$$

where h_b is the instantaneous depth of resin in the bleeder. The subscript b refers to conditions in the bleeder. In developing the above expressions, the pressure drop across the porous teflon sheet between the bleeder and the composite was neglected.

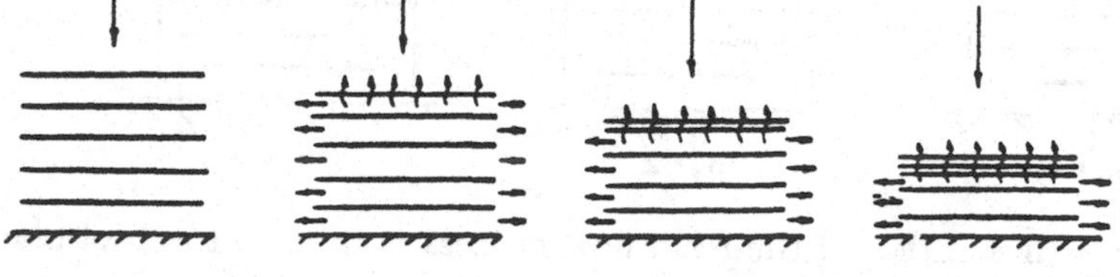

Figure 5 Simultaneous resin flow both normal and parallel to the tool plate.

Noting that the mass of the fibers in the composite remains constant, Eqs. (11)-(13) may be rearranged to yield the following expression for the rate of change of resin in the composite

$$\frac{dM_r}{dt} = \frac{-\varrho_r A_z S_c}{\int_0^{h_c} \mu dz} \left(\frac{P_o - P_b}{1 + G(t)} \right) \tag{14}$$

The parameter $G(t)$ is defined as

$$G(t) \equiv \frac{S_c}{S_b'} \frac{\mu_b h_b}{\int_0^{h_c} \mu dz} \tag{15}$$

M_r is the mass of resin in the composite at any instant of time. The mass of resin that leaves the composite and enters the bleeder in time t is

$$M_T = \int_0^t \frac{dM_r}{dt} dt \tag{16}$$

The instantaneous resin depth in the bleeder is related to the mass of resin that enters the bleeder by the expression

$$h_b = \frac{1}{\varrho_r \phi_b A_z} \int_0^t \frac{dM_r}{dt} dt \tag{17}$$

where ϕ_b is the porosity of the bleeder and represents the volume (per unit volume) which can be filled by resin. The thickness of the compacted plies is

$$h_c = n_s h_1 \tag{18}$$

where h_1 is the thickness of one compacted prepreg ply and n_s is the number of compacted prepreg plies. The value of n_s varies with time, depending on the amount of resin that has been squeezed out of the composite.

Equations (11)-(18) are the relationships needed for calculating the resin flow normal to the tool plate.

2.5 Resin Flow Parallel to the Tool Plate

In principle, in the plane of the composite, resin may flow along the fibers and in the direction perpendicular to the fibers. In practice, resin flow perpendicular to the fibers is small because of a) the resistance created by the fibers and b) the restraints placed around the edges of the composite. If such restraints were not provided, fiber spreading ("wash-out") would occur, resulting in a non-uniform distribution of fibers in the composite. Therefore, in this section, only resin flow along the fibers is considered.

It is assumed that resin flow along the fibers and parallel to the tool plate can be characterized as viscous flow between two parallel plates separated by a distance d_n ("channel flow", Figure 6). The distance d_n, separating the plates is small compared to the thickness of the composite ($d_n < L$). The variation in resin properties across and along the channel are taken to be constant. The pressure drop between the center of any given channel and the channel exit ($P_H \cong -P_L$, Figure 6) can then be expressed as [8]

$$\frac{2(P_H - P_L)}{\varrho_r (V_x^2)_n} = \lambda \frac{X_L}{d_n} \tag{19}$$

where $(V_x)_n$ is the average resin velocity in the channel, X_L is the channel length. The subscript n refers to the channel located between the n and $n-1$ prepreg plies (i.e., beneath the fiber bundles of prepreg ply n). The thickness of nth channel is calculated by assuming that a) there is one channel per ply and b) all the excess resin is contained in the channel. Accordingly, the thickness of the channel is given by the following expression:

$$d_n = \frac{M_n}{\varrho_n A_z} - \frac{M_{com}}{\varrho_{com} A_z} \tag{20}$$

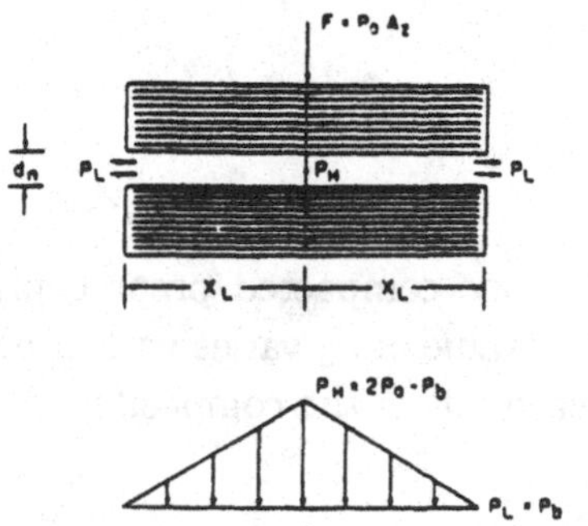

Figure 6 Geometry of the resin flow model parallel to the tool plate.

The mass M_n and density ϱ_n of prepreg ply n can be calculated by the rule of mixtures [5]. M_{com} and ϱ_{com} are the mass and the density of a compacted prepreg ply, respectively. The technique used to determine M_{com} is described in the next section, along with an appropriate expression for determining ϱ_{com}. For laminar flow between parallel plates, λ is defined as

$$\lambda \equiv \frac{(1/B)\,\mu_n}{\varrho_r (V_x)_n d_n} \tag{21}$$

where μ_n is the viscosity of the resin in the channel. Substitution of Eq. (21) into Eq. (20) yields the following expression for the average velocity in the channel

$$(V_x)_n = B\,\frac{d_n^2}{\mu_n}\,\frac{(P_H - P_L)}{X_L} \tag{22}$$

where B is a constant which must be determined experimentally. The resin mass flow rate is

$$(\dot{m}_{rx})_n = \varrho_r A_x (V_x)_n \tag{23}$$

where A_x is the cross sectional area defined as the product of the channel width W and thickness d_n.

The law of conservation of mass, together with Eqs. (22)-(23), gives the following expression for the rate of change of mass in the nth prepreg ply

$$\frac{d(M_r)_n}{dt} = -2(\dot{m}_{rx})_n = -\,2B\,\frac{d_n^3}{\mu}\,\varrho_r W\,\frac{(P_H - P_L)}{X_L} \tag{24}$$

The amount of resin leaving the nth prepreg ply in time t is

$$(M_E)_n = \int_0^t \frac{d(M_r)_n}{dt}\,dt \tag{25}$$

The total amount of resin flow parallel to the tool plate can be determined by summing Eq. (25) over all plies containing excess resin

$$M_E = \sum_{n=1}^{N-n_s} (M_E)_n \tag{26}$$

where N is the total number of prepreg plies.

The pressure at the centerline of the channel P_H can be estimated from the force balance applied along the boundaries of the channel. Assuming that the pressure gradient in the x direction is linear, and that the centerline pressure P_H is the same in each ply, the pressure distribution of each channel may be expressed as

$$P = \left(\frac{P_L - P_H}{X_L}\right) x + P_H \tag{27}$$

where P_L is the pressure at the exit of the channel and is assumed to be equal to the pressure of the environment surrounding the composite P_b. A force balance applied along the channel surface gives (Figure 6)

$$F = \int_A P dA = 2W \int_0^{X_L} P dx \tag{28}$$

F is the applied force which can be related to the cure pressure P_o as

$$F = P_o A_z = 2 P_o W X_L \tag{29}$$

Equations (27)-(29) yield the centerline pressure

$$P_H = 2P_o - P_b \tag{30}$$

Equations (19)-(30) are the relationships needed for calculating the resin flow in the direction along the fibers.

2.6 Total Resin Flow

The total resin flow out of the composite in time t is the sum of the resin flows both normal and parallel to the tool plate. The law of conservation of mass gives the following expression for the total rate of change of mass M in the composite

$$\frac{dM}{dt} = -\left[\dot{m}_{rz} + 2\sum_{n=1}^{N-n_t} (m_{rx})_n\right] \tag{31}$$

where m_{rz} and m_{rx} are the resin mass flow rates normal to the tool plate (z direction) and parallel to the tool plate (x direction), respectively.

The total mass of the composite at time t is

$$M = M_i - M_T - M_E \tag{32}$$

where M_T and M_E are defined by Eqs. (16) and (26) and M_i is the initial mass of the composite. The composite thickness at time t is

$$L = \frac{M}{\varrho 2 X_L W} \tag{33}$$

where ϱ is the density of the composite.

2.7 Void Model

Void nuclei may be formed either by mechanical means (e.g., air or gas bubble entrapment, broken fibers) or by homogeneous or heterogeneous nucleation [9]. Once a void is established, its size may change due to three effects: a) changes in vapor mass inside the void caused by vapor transfer through the void-prepreg interface, b) changes in pressure inside the void due to changes in temperature and pressure in the prepreg, and c) thermal expansion (or shrinking) due to temperature gradients in the resin. The model described below takes into account the first two of these effects, namely vapor transfer and changes in temperature and pressure.

A spherical nucleus of diameter d_i is assumed to be at a given location in the prepreg. The nucleus is filled with water vapor resulting from the humid air surrounding the prepreg during lay-up. The partial pressure of the water vapor in the nucleus PP_{wi} is related to the relative humidity ϕ_a by the expression

$$PP_{wi} = \phi_a P_{wga} \tag{34}$$

where P_{wga} is the saturation pressure of the water vapor at the ambient temperature. From the known values of the initial partial pressure PP_{wi} and the initial nucleus volume, the initial mass m_{wi} and the initial concentration $(c_{vw})_i$ of the water vapor in the nucleus can be determined.

During the cure, the volume of the void changes because a) water and other types of molecules are transported across the void-prepreg interface, and b) the cure pressure increases the pressure at the location of the void. For a spherical void of diameter d, the total pressure inside the void P_v is related to the pressure in the prepreg surrounding the void p by the expression

$$P_v - P = \frac{4\sigma}{d} \tag{35}$$

σ is the surface tension between the resin and the void. P_v is the total pressure inside the void and is the sum of the partial pressures of the air and the different types of vapors present in the void. In the model outlined below, it is assumed that only water-vapor is transported through the void-prepreg interface. However, this assumption does not affect the formulation of the model. If necessary, other types of vapors can readily be included in the calculations, as described by Springer [1].

The pressure inside the void is

$$P_v = PP_w + PP_{air} \tag{36}$$

PP_w and PP_{air} are the partial pressures of the water vapor and the air in the void. The partial pressure is a known function of the temperature, mass, and the void diameter

$$\begin{aligned} PP_{air} &= f(T, m_{air}, d) \\ PP_w &= f(T, m_w, d) \end{aligned} \tag{37}$$

Thus, if the pressure in the prepreg around the void, the temperature inside the void (taken to be the same as the temperature of the prepreg at the void location), and the mass of vapor in the void are known, the partial pressure, the total pressure, and the void diameter can be calculated from Eqs. (35)-(37). The temperature and the pressure are given by the thermo-chemical-resin-flow models. The air mass in the void is taken to be constant. Thus, it remains to evaluate the mass of water vapor in the void as a function of time. The change in water vapor mass may be calculated by assuming that the vapor molecules are transported through the prepreg by Fickian diffusion. Fick's law gives

$$\frac{dc}{dt} = D\left(\frac{\partial^2 c}{\partial r^2} + \frac{2}{r}\frac{\partial c}{\partial r}\right) \tag{38}$$

c is the water vapor concentration at a radial coordinate r, with $r = 0$ at the center of the void. D is the diffusivity of the water vapor through the resin in the r direction.

Initially, ($t < 0$) the vapor is taken to be distributed uniformly in the prepreg at the known concentration c

$$c = c_i \text{ at } r \geq d_i/2 \quad t < 0 \tag{39}$$

At times $t \geq 0$ the vapor concentration at the void-prepreg interface must be specified. By denoting the concentration at the prepreg surface by the subscript m, we write

$$\left.\begin{aligned} c &= c_m \text{ at } r = d/2 \\ c &= c_i \quad at\ r \to \infty \end{aligned}\right\} t \geq 0 \tag{40}$$

The second of the above expressions reflects the fact that the concentration remains unchanged at a distance far from the void. The surface concentration is related to the maximum saturation level M_m in the prepreg by the expression

$$c_m = \varrho M_m \tag{41}$$

The value of M_m can be determined experimentally for each vapor-resin system [10].

Solutions of Eqs. (38)-(41) yield the vapor concentration as a function of position and time $c = f(r,t)$. The mass of vapor transported in time t through the surface of the void is

$$m_T = -\int_0^t \pi d^2 D \left(\frac{\partial c}{\partial r}\right)_{r=d/2} dt \tag{42}$$

The mass of vapor in the void at time t is

$$m = m_i - m_T \tag{43}$$

The initial mass of water vapor in the void is known, as was discussed previously.

Solutions to Eqs. (34)-(43) give the void size and the pressure inside the void as functions of time, for a void of known location and initial size.

2.8 Stress Model

For a symmetric laminate, the residual stress in any given ply is [11]

$$\sigma_i = Q_{ij}(e_{oj} - e_j) \tag{44}$$

Q_{ij} is the modulus as defined by Tsai and Hahn [11]. The strain e_j is

$$e_j = \alpha_j (T - T_a) \tag{45}$$

where α_j is the thermal expansion coefficient, T_a is the ambient temperature, and T is the temperature in the ply at the end of the cure given by the thermo-chemical model. The laminate curing strain is

$$e_{oj} = a_{ij} \int_0^L Q_{ij} e_j dz \tag{46}$$

a_{ij} is the inplane compliance of a symmetric laminate as defined by Tsai and Hahn.

Solution of Eqs. (44)-(46) yield the residual stress in each ply.

3. METHOD OF SOLUTION

Solutions to the thermo-chemical, flow, void, and stress models must be obtained by numerical methods. A computer code (designated as "CURE") suitable for generating solutions was developed, and is available from the Department of Aeronautics and Astronautics, Stanford University.

Solutions of the model (and the corresponding computer code) require that the parameters listed in Tables 1 and 2 be specified. The parameters pertaining to the geometry, along with the initial and boundary conditions are specified by the user of the prepreg. The properties of the prepreg, the fiber, the resin, and the bleeder cloth are either specified by the manufacturer or can often be found in the published literature. Items (11) through (14) in Table 1 and item 8 in Table 2 are generally unknown. In the following, a brief description is given of the methods which can be used to determine these parameters.

The compacted prepreg ply thickness and the compacted prepreg ply resin content can be determined by constructing a thin (4 to 16 ply) composite panel. The panel is cured employing a cure cycle that will ensure that all the excess resin is squeezed out of every ply in the composite (i.e., all the plies are consolidated $n_s = N$). The total mass of the composite M is measured after cure. The resin content of one compacted prepreg ply $(M_r)_{com}$ is related to the composite mass by the expression

$$(M_r)_{com} = \frac{M}{N} - M_f \tag{47}$$

M_f is the fiber mass of one prepreg ply, and N is the total number of plies in the composite. The compacted prepreg ply thickness h_1 is

$$h_1 = \frac{M/N}{\varrho_{com} A_z} \tag{48}$$

where ϱ_{com} is compacted ply density which can be derived from the rule of mixtures [5].

The apparent permeability of the prepreg normal to the fibers S_c can be determined by curing a thin (4 to 8 ply) composite specimen for a predetermined length of time. During the cure, the resin squeezed out through the plane of the composite normal to the tool plate is collected in the bleeder placed on the top of the composite. After the cure is terminated, the amount of resin in the bleeder (i.e., the resin flow into the bleeder) is determined by measuring the difference between the original bleeder weight (mass) and the final weight (mass) of the resin-soaked bleeder. An initial value for the apparent permeability is estimated, and the resin flow normal to the fibers is calculated using the flow model. The value of the permeability is adjusted and the calculations are repeated until the calculated and measured resin flows match.

The flow coefficient of the prepreg parallel to the fibers (B) can be estimated from the following procedure. A thick composite (approximately 30-60 ply thick) is cured for a predetermined length of time. Resin squeezed out from between the individual plies (parallel to the fibers) is collected by bleeders placed around the edge of the composite. The resin flow through the edges is determined by measuring the difference between the original weight of the "edge" bleeders and the final weight of the resin-soaked bleeders. Assuming a value for the flow coefficient, the resin flow parallel to the fibers is calculated using the flow model. The value of the flow coefficient is adjusted and the calculations are repeated until the measured resin flow matches the calculated resin flow.

The surface tension at the void resin interface may be approximated by the surface tension of water.

The contents of this chapter are reprinted from the JOURNAL OF COMPOSITE MATERIALS, Vol. 17, pp. 135-152, March 1983, by permission of Technomic Publishing Company.

Table 1. Input Parameters Required for Solutions of the Thermo-Chemical and Resin Flow Models.

A. Geometry
  1) Length of the composite
  2) Width of the composite
  3) Number of plies in the composite

B. Initial and Boundary Conditions
  4) Initial temperature distribution in the composite
  5) Initial degree of cure of the resin in the composite
  6) Cure temperature as a function of time
  7) Cure pressure as a function of time
  8) Pressure in the bleeder

C. Prepreg Properties
  9) Initial thickness of one ply
  10) Initial resin mass fraction
  11) Resin content of one compacted ply
  12) Compacted ply thickness
  13) Apparent permeability of the prepreg normal to the plane of the composite
  14) Flow coefficient of the prepreg parallel to the fibers

D. Resin Properties
  15) Density
  16) Specific heat
  17) Thermal conductivity
  18) Heat of reaction
  19) Relationship between the cure rate, temperature, and degree of cure
  20) Relationship between the viscosity, temperature, and degree of cure

E. Fiber Properties
  21) Density
  22) Specific heat
  23) Thermal conductivity

F. Bleeder Properties
  24) Apparent permeability
  25) Porosity

Table 2. Input Parameters Requires in the Computer Code for Calculating Void Sizes. (These parameters are in addition to those in Table 1.)

Void Model

A. Initial and Boundary Conditions
   1) Initial void size
   2) Initial void location
   3) Initial water vapor concentration in prepreg
   4) Ambient relative humidity
   5) Ambient temperature

B. Resin Properties
   6) Expression relating the relative humidity to the maximum saturation level of water vapor in the resin
   7) Diffusivity of water vapor through the resin
   8) Surface tension at the resin-void interface

Stress Model

A. Properties of the Composite
   1) Longitudinal and transverse Young's moduli
   2) Longitudinal and transverse Poisson's ratios
   3) Longitudinal and transverse shear moduli
   4) Longitudinal and transverse thermal expansion coefficients

B. Stacking Sequence
   5) Orientation of each ply

C. Environment
   6) Ambient temperature

REFERENCES

1.) Springer, G.S., "A Model of the Curing Process of Epoxy Matrix Composites", in *Progress in Science and Engineering of Composites*, (T. Hayashi and K. Kawaka, eds.) Japan Society of Composite Materials, pp. 23-35 (1982).

2.) Springer, G.S., "Rėsin Flow During the Cure of Fiber Reinforced Composites", *Journal of Composite Materials*, Vol. 16, pp. 400-40, (1982).

3.) Loos, A.C. and Springer, G.S., "Calculation of Cure Process Variables During Cure of Graphite-Epoxy Composites", in Composite Materials, Quality Assurance and Processing, (C.E. Browning, ed.) ASTM STP 797, pp. 110-118, (1983).

4.) Lee, W.I., Loos, A.C. and Springer, G.S., "Heat of Reaction, Degree of Cure, and Viscosity of Hercules 3501-6 Resin", *Journal of Composite Materials*, Vol. 16, pp. 510-520, (1982).

5.) Loos, A.C. and Springer, G.S., "Curing of Graphite/Epoxy Composites", Air Force Materials Laboratory Report AFWAL-TR-83-4040 Wright Aeronautical Laboratories, Wright Patterson Air Force Base, Dayton, OH. (1983).

6.) Springer, G.S. and Tsai, S.W., "Thermal Conductivities of Unidirectional Materials", *Journal of Composite Materials*, Vol. 1, pp. 166-173 (1967).

7.) Bartlett, C.J., "Use of the Parallel Plate Plastometer to Characterize Glass-Reinforced Resins: I. Flow Model", *SPE Technical Papers*, Vol. 24, pp. 638-640 (1978).

8.) White, F.M., "Viscous Fluid Flow", Mc-Graw Hill, New York (1974) pp. 336-337.

9.) Kardos, J.L., Dudukovic, J.P., McKague, E.L. and Lehman, M.W., "Void Formation and Transport during Composite Laminate Processing", in *Composite Materials, Quality Assurance and Processing* (C.E. Browning, ed.) ASTM STP 797, pp. 96-109, (1983).

10.) Springer, G.S., "Environmental Effects on Composite Materials", Technomic Publishing Co. (1981).

11.) Tsai, S.W. and Hahn, H.T., "Introduction to Composite Materials", Technomic Publishing Co. (1980).

CHAPTER 11

ENVIRONMENTAL EFFECTS

G.S. Springer
Stanford University, Stanford , CA, USA

ABSTRACT

A model is presented which describes the reponse of organic matrix composites to hot and moist environments. Equations are presented for calculating the temperature and moisture distributions, the total moisture content, and the hygrothermal deformations of composite laminates. Methods for measuring the relevant material properties are described. A technique for accelerated moisture conditioning is discussed.

I. INTRODUCTION

When an organic matrix composite is exposed to humid air or to a liquid, both the moisture content and the temperature of the material may change with time. These changes, in turn, affect the thermal and mechanical properties, resulting in a decrease in performance. Therefore, to utilize the full potential of composite materials their response to environmental exposure must be known. Specifically, answers to the following problems are sought.

A composite material is exposed to an environment in which the temperature and the moisture level vary with time in a prescribed manner. It is required to find the following parameters:

(a) The temperature inside the material as a function of position and time

(b) The moisture concentration inside the material as a function of position and time

(c) The total amount (mass) of moisture inside the material as a function of time

(d) The moisture and temperature induced ("hygrothermal") stresses inside the material as a function of time

(e) The dimensional changes of the material as a function of time

(f) Changes in the "performance" of the material as a function of time.

Here the word "performance" is used in a broad sense to denote any mechanical, chemical, thermal, or electrical property of interest, such as strength, modulus, fatigue life, glass transition temperature, thermal and electrical conductivities.

Generally, the problem is attacked in three steps, as illustrated in Figure 1. First, the temperature distribution and the moisture content inside the material are calculated. Second, from the known temperature and moisture distributions the hygrothermal deformations and stresses are calculated. Third, the changes in performance due to temperature and moisture are determined. The procedures employed in each of these steps are summarized in this chapter.

2. TEMPERATURE AND MOISTURE DISTRIBUTIONS

The temperature and moisture distributions inside the composite can readily be calculated when moisture penetrates into the material by "Fickian" diffusion. Such diffusion is assumed to take place when the following conditions are met.

(a) Heat transfer through the material is by conduction only and can be described by Fourier's law.

(b) The moisture diffusion can be described by a concentration-dependent form of Fick's law.

(c) The temperature inside the material approaches equilibrium much faster than the moisture concentration, and, hence, the energy (Fourier) and mass transfer (Fick) equations are decoupled.

(d) The thermal conductivity and mass diffusivity depend only on temperature and are independent of the moisture concentrations or the stress levels inside the material.

Below, solutions are presented for the temperature and moisture distributions in single and multilayered composites under Fickian diffusion.

2.1 Single-Layer Composite-Constant Environmental Conditions

The following problem is considered (Figure 2):

(a) The composite is a single-layer plate in which the moisture content and temperature vary only in the direction normal to the face of the plate.

(b) The ambient temperature and ambient moisture content are constant and are the same on both sides of the plate.

(c) The temperature inside the material approaches equilibrium much faster than the normal concentration, and hence, the temperature inside the material can be taken to be the same as the ambient temperature.

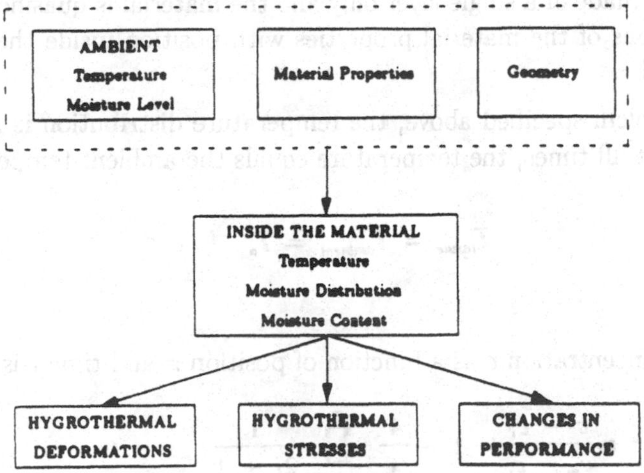

Figure 1. Procedure for assessing environmental effects.

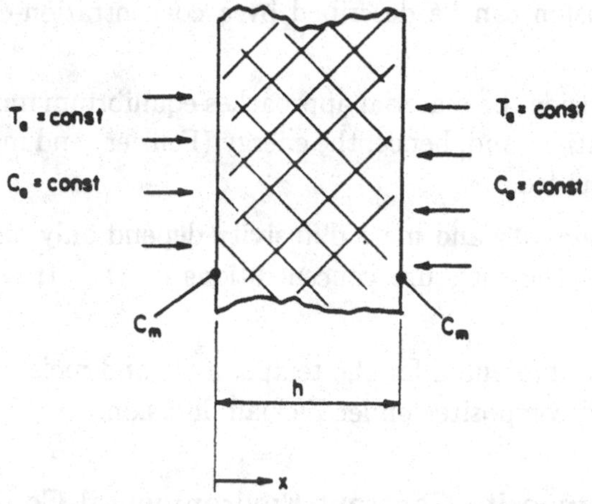

Figure 2. Single-layer composite - constant environmental conditions.

(d) Initially, the temperature and moisture distributions are uniform inside the material.

(e) The thermal conductivity and the mass diffusivity depend only on temperature and are independent of the moisture concentration and the stress levels inside the material.

(f) The plate is made of a single layer only and the material is quasihomogeneous, so that variations of the material properties with position inside the material may be neglected.

For the problem specified above, the temperature distribution is uniform across the plate and, at all times, the temperature equals the ambient temperature:

$$T_{inside} = T_{ambient} = T_a \tag{1}$$

The moisture concentration c as a function of position x and time t is [1]:

$$c^* = \frac{c - c_i}{c_m - c_i} = 1 - \frac{4}{\pi} \sum_{j=0}^{\infty} \frac{1}{2j+1} \sin\frac{(2j+1)\pi x}{h} \exp\left[-\frac{(2j+1)^2\pi^2 Dt}{h^2} \right] \tag{2}$$

where c_i is the uniform, initial moisture concentration inside the material, c_m is the maximum moisture concentration that is reached in the material for a given ambient condition, h is the plate's thickness, and D is the mass diffusivity in the direction normal to the plate.

The total mass of moisture inside the plate is:

$$m = A \int_0^h c \, dx \tag{3}$$

where A is the exposed surface area.

Equations (2) and (3) yield the following expression for the total mass:

$$G \equiv \frac{m - m_i}{m_m - m_i} = 1 - \frac{8}{\pi^2} \sum_{j=0}^{\infty} \frac{\exp\left[-(2j+1)^2 \pi^2 \left(\frac{Dt}{h^2}\right)\right]}{(2j+1)^2} \tag{4}$$

where m_i is the initial mass of the material (that is, the mass prior to exposure to the moist environment) and m_m is the mass of moisture in the material when the material is fully saturated, in equilibrium with its environment:

$$m_m = (A)(c_m)(h) \tag{5}$$

The parameter G may conveniently be approximated by:

$$G = 1 - \exp\left[-7.3\left(\frac{Dt}{h^2}\right)^{0.75}\right] \tag{6}$$

This expression is simpler to use than Equation (4).

A parameter of practical interest is the percent weight gain, defined as:

$$M = \frac{\text{wt. of moist matrl.} - \text{wt. of dry matrl.}}{\text{weight of dry matrl.}} \times 100 \tag{7}$$

$$= \frac{W - W_{dry}}{W_{dry}} \times 100 = \frac{m}{m_{dry}} \times 100$$

Equations (4) and (7) give:

$$M = G(M_m - M_i) + M_i \tag{8}$$

The subscripts refer to the uniform initial and fully saturated conditions, respectively. Accordingly, we have

$$M_m = \frac{m_m}{m_{dry}} \times 100 = \frac{c_m}{\varrho} \times 100 \qquad (9)$$

$$M_i = \frac{m_i}{m_{dry}} \times 100 = \frac{c_i}{\varrho} \times 100 \qquad (10)$$

(9)
(10)

where ϱ is the density of the dry material.

Note that the moisture concentration and the total moisture content can be calculated by the above expressions even when one side of the plate is insulated so that moisture enters through one face only. In this case the thickness h must be replaced by twice the thickness $2h$ in Equations (2), (4) and (6).

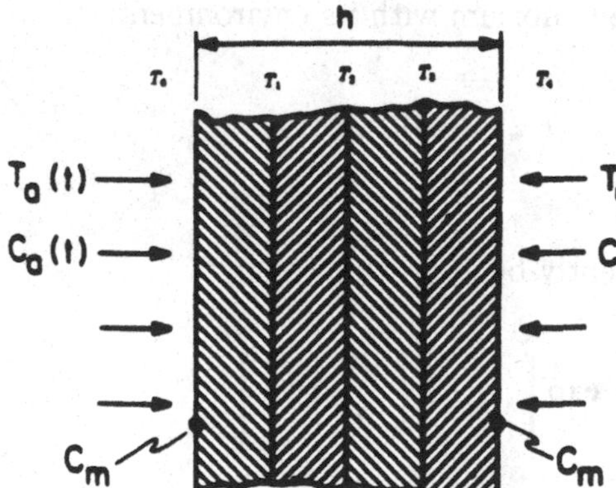

Figure 3. Multilayered composite - time varying environmental conditions.

2.2 Multilayered Composites - Time Varying Environmental Conditions

The following problem is considered (Figure 3):

(1) The temperature and moisture content inside the the material vary only in the direction normal to the face of the plate.

(2) The temperature inside the material equilibrates much faster than the moisture concentration, and hence at each instant of time the temperature distribution inside the material correspondents to the instantaneous ambient temperature.

(3) The material properties depend only on temperature and are independent of moisture concentration and stress level.

(4) The environmental conditions (temperature, moisture level) vary in arbitrary but known manner.

The temperature distribution inside the plate is [2]:

$$
\begin{aligned}
T_0 - T_1 &= qR_1 \\
T_1 - T_2 &= qR_2 \\
T_2 - T_3 &= qR_3 \\
T_3 - T_4 &= qR_4
\end{aligned}
\tag{11}
$$

where

$$
R_j = \frac{K_j}{h_j}; \quad q = \frac{T_0 - T_4}{\sum_{j=1}^{4} R_j} \tag{12}
$$

T_0 through T_4 are the surface temperatures; K_j is the thermal conductivity of the j-th layer (in the direction normal to the face of the plate). Although the above expressions are written for a plate consisting of four layers, the results can readily be extended to plates consisting of arbitrary number of layers.

The moisture concentration and the moisture content must be obtained by numerical methods. A computer code (designated as W8GAIN) was developed for performing the calculations. The program listing and a sample input-output can be found in Reference [1].

The required input parameters for the program are:

(a) The ambient temperature and relative humidity as a function of time on both sides of the plate

(b) The initial moisture concentration and initial temperature distribution inside the plate; these concentrations need not be uniform

(c) The density of each layer

(d) The maximum moisture content of each layer as a function of ambient conditions

(e) The thermal conductivity and mass diffusivity as a function of temperature for each layer

The outputs of the program are:

(a) Moisture concentration as a function of position and time in each layer

(b) Weight (mass) change of each layer as a function of time

(c) Total weight (mass) change as a function of time

2.3 Non-Fickian Diffusion

The aforementioned calculation procedures can be used only if the moisture diffusion is by a Fickian process. The conditions under which Fickian diffusion exists must be determined by tests. Generally, Fickian diffusion takes place at low temperatures and for materials exposed to humid air. Deviations from Fickian diffusion occur at elevated temperatures and for materials immersed in liquids. It is noteworthy that Fickian diffusion is a reasonable approximation for many materials, including graphite-epoxy composites.

3. TEST PROCEDURES FOR DETERMINING D AND M_m

The following test procedure may be used to determine the diffusivity D and the maximum moisture content M_m:

(a) A test specimen is fabricated in the form of thin plate.

(b) The specimen is completely dried in an oven and the dry weight W_{dry} is measured.

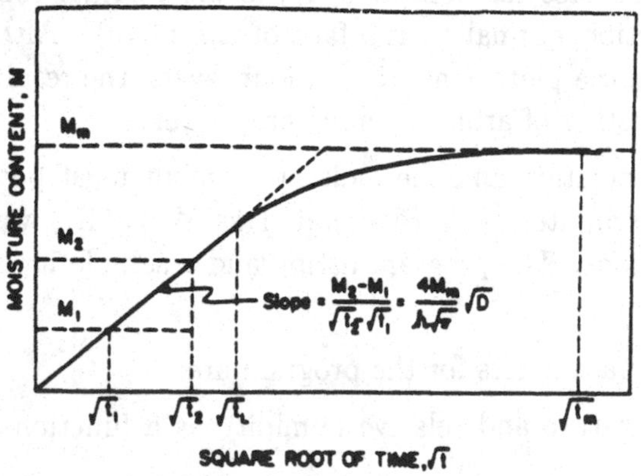

Figure 4. Illustration of the change of moisture content with the square root of time for Fickian diffusion. For $t < t_L$ the slope is constant.

(c) The specimen is placed in a constant temperature; constant moisture level environment and the weight is recorded as a function of time.

(d) The moisture content M [see Equation (7)] is plotted versus the square root of time (Figure 4)

In case of Fickian diffusion, after a long period of time the M versus $\sqrt{t}$ curve approaches asymptotically the maximum moisture content M_m. The initial slope of the curve is proportional to the diffusivity:

$$D = \pi \left(\frac{h}{4M_m} \right)^2 \left(\frac{M_2 - M_1}{\sqrt{t_2} - \sqrt{t_1}} \right)^2 \tag{13}$$

Alternately, the value of D can be found using Equation (8). An arbitrary value of D is selected and M is calculated by Equations (8) and (6) for different times. The calculated results are compared with the data. The procedure is repeated with different D values until a "best fit" to the data is obtained.

The error introduced in the value of D due to diffusion through the edges can be estimated by the procedure given in Reference [1].

4. MAXIMUM MOISTURE CONTENT

The value of M_m is nearly constant when the material is submerged in a liquid. For materials exposed to humid air, M_m can be related to the relative humidity ϕ by the expression:

$$M_m = a(\phi)^b \tag{14}$$

where a and b are constants. Values of a and b are given in Tables 1-3.

5. MASS DIFFUSIVITY

When the diffusion is Fickian and D is a function of temperature only, the diffusivity is related to temperature by the Arrhenius relationship:

$$D = D_0 \exp\left[-\frac{C}{T} \right] \tag{15}$$

where D_0 and C are constants, and T is the absolute temperature. Values of the constants are given in Tables 4-6.

For fiber reinforced composites the diffusion coefficients in the direction parallel and perpendicular to the fibers (D_{11}andD_{22}) may be estimated by the expressions ($\nu_f < 0.785$) [1]:

$$D_{11} = (1 - v_f)D_r + v_f D_f \tag{16}$$

$$D_{22} = (1 - 2\sqrt{v_f/\pi}\,)\,D_r + \frac{D_r}{B_D}\left[\pi - \frac{4}{\sqrt{1-(B_D^2 v_f/\pi)}}\tan^{-1}\frac{\sqrt{1-(B_D^2 v_f/\pi)}}{1 + B_D\sqrt{v_f/\pi}}\right] \tag{17}$$

$$B_D \equiv 2\left(\frac{D_r}{D_f} - 1\right) \tag{18}$$

where D_r is the diffusivity of the resin, D_f is the diffusivity of the fiber and ν_f is the volume fraction of the . These expressions become invalid if moisture propagates along fiber-resin interfaces or through cracks and voids.

Table 1. Maximum moisture contents of selected graphite-epoxy composites immersed in liquid [1].

	Maximum Moisture Content, M_m (%)		
Liquid	T300/1034	AS/3501-5	T300/5208
Distilled water	1.70	1.90	1.50
Saturated salt water	1.25	1.40	1.12
No. 2 diesel fuel	0.50	0.55	0.45
Jet A fuel	0.45	0.52	0.40
Aviation oil	0.65	0.65	0.60

Table 2. Maximum moisture contents of selected graphite-epoxy composites and epoxy resins exposed to humid air, $M_m = a\phi^b$ when $b = 1$; $M_m = a(\phi/100)^b \times 100$ when $b \neq 1$ [1].

Material	a	b
T300/1034	0.017	1
AS/3501-5	0.019	1
T300/5208	0.015	1
934 (resin) 3501-5 (resin) 5208 (resin)	0.060	1.4 ϕ < 60% 1.8 ϕ < 60%

Table 3. The apparent maximum moisture contents of selected polyester E-glass and vinylester E-glass composites (percent) [1].

Substance	Temperature (°C)	SMC-R25	VE SMC-R50	SMC-R50
Humid air, 50%	23	0.17	0.13	0.10
	93	0.10	0.10	0.22
Humid air, 100%	23	1.00	0.63	1.35
	93	0.30	0.40	0.56
Distilled water	23	3.60	—	—
	50	3.50	—	—
Salt water	23	0.85	0.50	1.25
	93	2.90	0.75	1.20
No. 2 diesel fuel	23	0.29	0.19	0.45
	93	2.80	0.45	1.00
Lubricating oil	23	0.25	0.20	0.30
	93	0.60	0.10	0.25
Antifreeze	23	0.45	0.30	0.65
	93	4.25	3.50	2.25
Indolene	23	3.50	0.25	0.60
	93	4.50	5.00	4.25

Table 4. Transverse diffusivities of selected graphite-epoxy composites immersed in liquids [1].

	T300/1034		AS/3501-5		T300/5208	
Liquid	D_o	C	D_o	C	D_o	C
Distilled water	16.3	6211	768	7218	132	6750
Saturated salt water	5.85	6020	5.38	6472	6.23	5912

The transverse diffusivity is $D_{22} = D_o \exp(-C/T)$ where D_o is in mm^2s^{-1} and C is in K

Table 5. Transverse diffusivities of selected graphite-epoxy composites and epoxy resins exposed to humid air [1].

Material	$D_o(mm^2s^{-1})$	C(K)
T300/1034	2.28	5554
AS/3501-5	6.51	5722
T300/5208	0.57	4993
934 (resin)	4.85	5113
3501-5 (resin)	16.10	5690
5208 (resin)	4.19	5448

$D_{22} = D_o \exp(-C/T)$

Generally, the diffusivity of the fiber is small compared to the diffusivity of the matrix ($D_f \ll D_r$) and we may write ($\nu_f < 0.785$):

$$D_{11} = (1 - v_f)D_r \tag{19}$$

$$D_{22} = (1 - \sqrt{v_f/\pi})D_r \tag{20}$$

In a direction making α degrees with the orientation of the fibers the diffusivity is

$$D_\alpha = D_{11}\cos^2\alpha + D_{22}\sin^2\alpha \tag{21}$$

6. THERMAL CONDUCTIVITY

The thermal conductivity of a composite may be determined by tests or may be approximated from the known fiber and matrix properties ($\nu_f < 0.785$) [1]:

$$K_{11} = (1 - v_f)K_r + v_fK_f \tag{22}$$

$$K_{22} = (1 - 2\sqrt{v_f/\pi})\, K_r + \frac{K_r}{B_K}\left[\pi - \frac{4}{\sqrt{1 - (B_K^2 v_f/\pi)}}\tan^{-1}\frac{\sqrt{1 - (B_K^2 v_f/\pi)}}{1 + B_K\sqrt{v_f/\pi}}\right] \tag{23}$$

$$B_K = 2\left(\frac{K_r}{K_f} - 1\right) \tag{24}$$

where K_{11} and K_{22} are the thermal conductivities in a direction along the fiber ($\alpha = 0^o$) and perpendicular to the fiber ($\alpha = 90^o$). K_r and K_f are the thermal conductivities of the resin and the fiber, respectively. In the α direction:

$$K_\alpha = K_{11}\cos^2\alpha + K_{22}\sin^2\alpha \tag{25}$$

Other approximations of the thermal conductivity may be found in the relevant literature. [3].

Table 6. Apparent transverse diffusivities of selected polyester E-glass and vinylester E-glass composites [1].

Substance	Temperature (°C)	SMC-R25	VE SMC-R50	SMC-R50
Humid air, 50%	23	10.0	10.0	30.0
	93	50.0	50.0	30.0
Humid air, 100%	23	10.0	5.0	9.0
	93	50.0	50.0	50.0
Salt water	23	10.0	5.0	15.0
	93	5.0	30.0	80.0
No. 2 diesel fuel	23	6.0	5.0	5.0
	93	6.0	10.0	5.0
Lubricating oil	23	10.0	10.0	10.0
	93	10.0	10.0	10.0
Antifreeze	23	50.0	30.0	20.0
	93	5.0	0.8	10.0
Indolene	23	1.0	10.0	10.0
	93	40.0	1.0	3.0

Values given are $D_{22} \times 10^7$ mm²/s.

7. ACCELERATED MOISTURE CONDITIONING

The mechanical, thermal, and chemical properties of organic matrix composites change during environmental exposure. To determine the magnitudes of these changes the usual procedure is to expose the material to a moist environment until the moisture level inside the material reaches the required value. The material is then subjected to the appropriate tests to measure the changes in properties caused by the absorbed moisture. Unfortunately, moisture conditioning of the material may last months or years. Under most circumstances such long conditioning times are unacceptable. In the following a procedure is described which reduces the time required to moisturize the material during environmental conditioning.

7.1 Problem Statement

We consider a plate of thickness h made of a fiber reinforced organic matrix composite. The moisture concentration c and the temperature T inside the plate is taken to vary only in the direction normal to the face of the plate. Initially (at time $t < 0$), the temperature T_i and the moisture concentration c_i are known at every point inside the plate. At time $t = 0$ the plate is exposed to humid air. Both the ambient temperature T_a and the ambient humidity ϕ_a may vary with time. As time progresses the temperature and moisture distributions inside the plate and the total moisture content of the plate change.

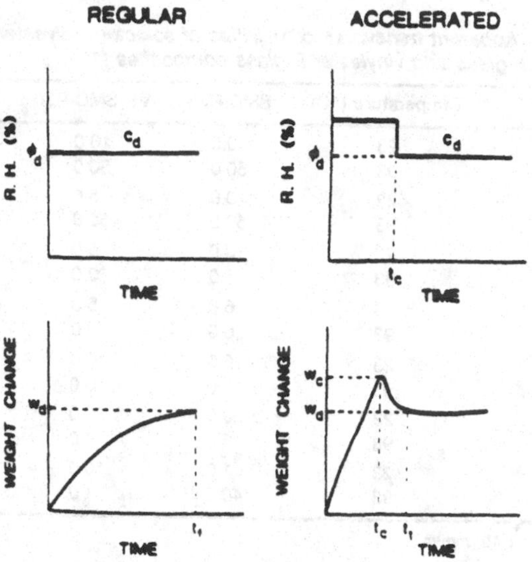

Figure 5. Illustration of the relative humidity employed in regular and accelerated tests, and the corresponding weight change of the material.

After some time t_f the temperature, moisture concentration, and the total moisture content reach the values T_d, c_d and M_d.

The objective is to establish the environmental conditions (temperature, relative humidity) which yield the same T_d, c_d, and M_d values as the actual ambient, but yield these in a shorter time, i.e., in a time t, which is less than t_f ($t_t < t_f$).

Here, a procedure for accelerated environmental conditioning is presented for the case when:

(1) The moisture distribution to be reached is uniform across the plate.

(2) At any instant of time, the temperature inside the plate corresponds to the ambient temperature.

For this problem the temperature distribution is uniform across the plate and, at all times, the temperature equals the ambient temperature:

$$T_{inside} = T_{ambient} = T_a \tag{26}$$

The required uniform moisture concentration c_d inside the plate can be established by several different environments. Commonly, the plate is exposed to humid air in which the humidity is constant and corresponds to the required value of c_d ("regular" method, see Figure 5). The time to reach c_d by the regular method is t_f.

The procedure for establishing c_d in a time t_t which is less than t_f consists of two major steps:

(1) The plate is exposed to air at 100 percent relative humidity for a period of time t_c (Figure 5).

(2) At time t_c, the relative humidity of the ambient air is reduced to the value which corresponds to c_d, and the plate is exposed to this ambient until time t_t. The relationship between c_d and ϕ_d is [see Equations (3), (7), and (14)]:

$$c_d = \varrho M_d/100 = \varrho a(\phi_d)^b/100 \tag{27}$$

It is necessary to select the "changeover time" t_c and the "conditioning time" t_t. These times must be selected to satisfy the following two conditions:

(1) At the end of the exposure (time $t = t_t$) at every point inside the plate the moisture concentration c agrees (within a prescribed limit) with the desired uniform moisture concentration c_d.

(2) The desired uniform moisture concentration c_d is reached in the shortest possible time t_t.

7.2 Selection of the "Changeover" and "Conditioning" Times

The changeover time and the conditioning time can be determined by the solution of the partial differential equations describing moisture diffusion through the material [4]. The results can conveniently be summarized in charts. To accomplish this we introduce a dimensionless moisture concentration and dimensionless time:

$$c^* = c/c_{100} \tag{28}$$

$$t^* = Dt/h^2 \tag{29}$$

where, as before h is the thickness of the plate and c_{100} is the maximum uniform moisture concentration which is reached in the material after exposure to air at 100 percent relative humidity.

The dimensionless "changeover time" t_c^* versus the dimensionless desired moisture concentration c_d^* is given in Figure 6. The dimensionless conditioning time t_t^* required to ensure that the moisture concentration is everywhere within a prescribed limit of c_d^*, is presented in Figure 7. The results in Figure 7 are for the three cases when the moisture concentrations everywhere are within 99, 95 or 90 percent of the desired, constant value of c_d^*.

Once the dimensionless times t_c^* andt_t^* are known (Figures 6 and 7) the actual times t_c and t_t can be calculated by Equation (29). These calculations require a knowledge of the diffusivity D. Unfortunately, there is always an uncertainty in the value of D; 100 percent variation in D is quite common [1]. Any error in D also manifests itself in the values of t_c and t_t. More accurate values of t_c and t_t can be obtained by placing a test coupon (often referred to as the "witness coupon") into the environmental chamber and by monitoring the weight change of this coupon. The change in humidity from $\phi = 100$ percent to $\phi = \phi_d$ is made when the weight change (percent) of the coupon M reaches the value M_c.

$$M_c = (M_{100})(M_c^*) \tag{30}$$

where M_{100} is the maximum moisture content corresponding to 100 percent relative humidity [Equation (14)].

$$M_{100} = a(100)^b \tag{31}$$

M_c^* is the moisture content in the material at the changeover time $t^* = t_c^*$. The changeover time t_C^* is given by Figure 6. At time t_c^* the moisture concentration c_C^* as well as the moisture content M^* can be calculated [5]. The calculated values of M_c^* are summarized in Figure 8. The value of M_c at which the changeover from $\phi =$ 100(31). The weight of the witness coupon also indicates the end of the conditioning. The conditioning is complete when the weight of the coupon becomes constant.

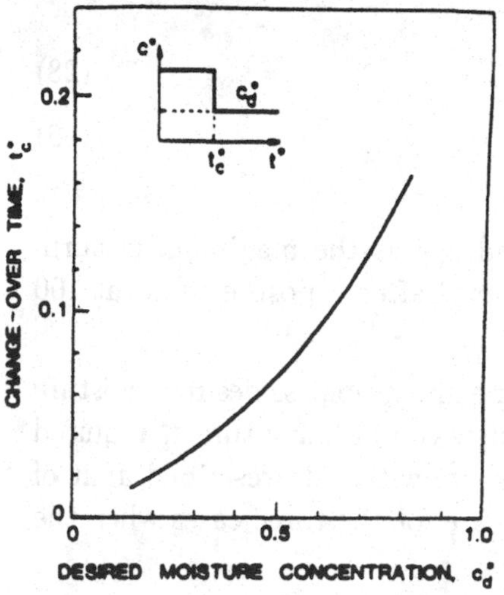

Figure 6. The dimensionless changeover time t_c^* versus the dimensionless desired moisture concentration c_d^*.

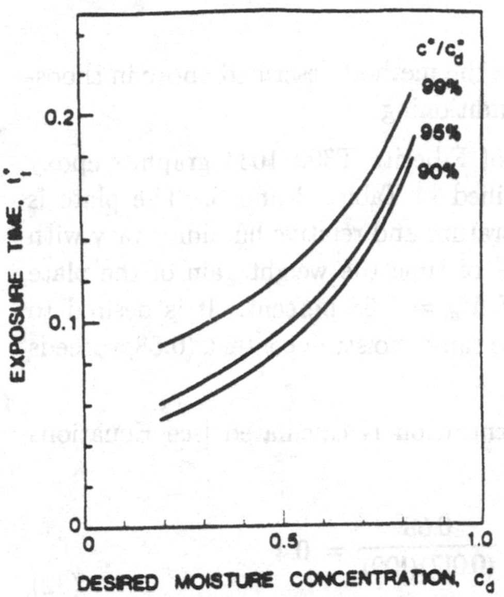

Figure 7. The dimensionless exposure time t_t^* versus the dimensionless desired moisture concentration c_d^*. The three curves correspond to the conditions where the moisture concentration agrees within 90, 95 and 99 percent of the desired moisture concentration at every point inside the composite.

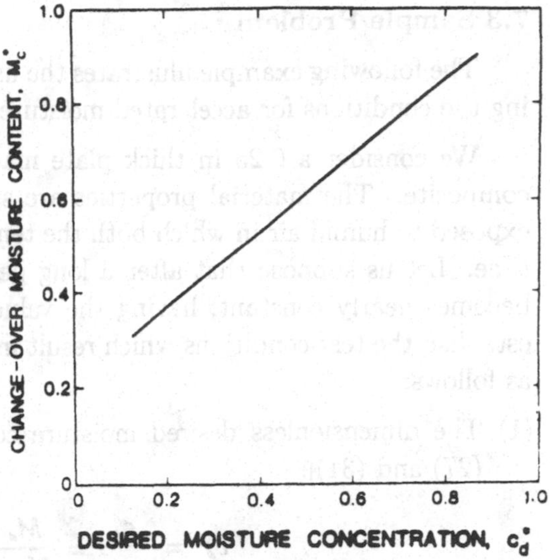

Figure 8. The dimensionless changeover moisture content M_c^* versus the dimensionless moisture concentration c_d^*.

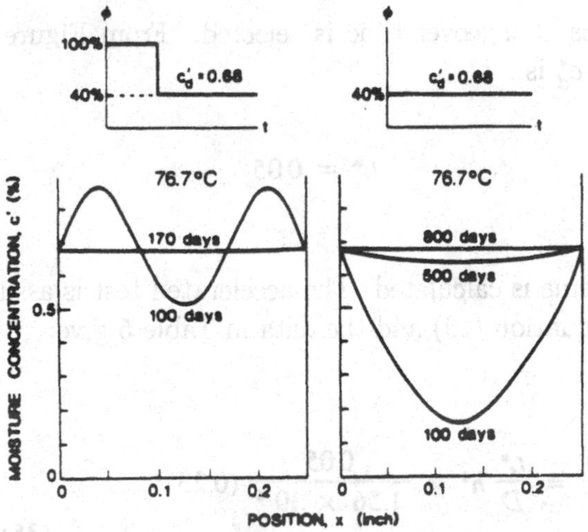

Figure 9. The variation of moisture with time in a 1/4" thick Fiberite T 300/934 graphite-epoxy composite exposed to two different ambients. Note that both ambients result in the uniform moisture concentration of 0.68 percent ($c' = c/\varrho x100$).

7.3 Sample Problem

The following example illustrates the use of the method described above in choosing the conditions for accelerated moisture conditioning.

We consider a 0.25 in thick plate made of Fiberite T300/1034 graphite-epoxy composite. The material properties are specified in Tables 2 and 5. The plate is exposed to humid air in which both the temperature and relative humidity vary with time. Let us suppose that after a long period of time the weight gain of the plate becomes nearly constant, having the value of $M_d = 0.68$ percent. It is desired to establish the test conditions which result in the same moisture content (0.68proceeds as follows:

(1) The dimensionless desired moisture concentration is calculated [see Equations (27) and (31)]:

$$c_d^* = \frac{c_d}{c_{100}} = \frac{M_d}{M_{100}} = \frac{0.68}{(0.017)(100)} = 0.4 \tag{32}$$

(2) The relative humidity corresponding to M_d is calculated:

$$\phi_d = (M_d)^{1/b}/a = \frac{0.68}{0.017} = 40 \text{ percent} \tag{33}$$

(3) The dimensionless changeover time is selected. From Figure 6 the value of t_c^* corresponding to c_d^* is:

$$t_c^* = 0.05 \tag{34}$$

(4) The changeover time is calculated. The accelerated test is assumed to take place at170°F. Then Equation (29) and the data in Table 5 give:

$$t_c = \frac{t_c^*}{D} h^2 = \frac{0.05}{1.56 \times 10^{-6}} (0.25)^2 \tag{35}$$

(35)

$$= 2000 \text{ hrs} = 84 \text{ days}$$

(5) The dimensionless conditioning time is determined. From Figure 7 (corresponding to the 99 percent level), we obtain:

$$t_t^* = 0.104 \tag{36}$$

(6) The actual conditioning time is calculated:

$$t_t = \frac{t_t^*}{D} h^2 = \frac{0.104}{1.56 \times 10^{-6}} (0.25)^2 \tag{37}$$

$$= 4200 \text{ hrs} = 170 \text{ days}$$

If a "witness coupon" were to be placed in the chamber, steps 3-6 would be as follows:

(3a) The dimensionless changeover weight M_c^* is determined from Figure 8:

$$M_c^* = 0.43 \tag{38}$$

(4a) The actual changeover weight M_c is calculated. Equations (30) and (31) give

$$M_c = (M_{100})(M_c^*) = (1.7)(0.43) = 0.73 \text{ percent} \tag{39}$$

(5a-6a) The final conditioning time is determined from the observed weight change of the witness coupon.

According to this example the plate is to be kept in humid air at 100 percent relative humidity for 84 days. The relative humidity is to be changed then to 40 percent. After 170 days the moisture concentration will be within 99 percent of the desired concentration throughout the plate.

The desired moisture concentration could also be achieved by exposing the plate to humid air at 40 percent relative humidity and 170° F. As can be seen from Figure 9, under this condition the desired moisture concentration would be reached in 800 days. This is a long time compared to the 180 days needed to reach c_d by the present method.

8. GLASS TRANSITION TEMPERATURE

Fickian diffusion is more likely to occur in rubbery polymers than in glassy polymers. Transition from glassy to a rubbery state occurs at the glass transition temperature. Hence the glass transition temperature T_g is an important parameter in the moisture transfer process.

The absorbed moisture may change (generally decrease) the glass transition temperature, thereby affecting the diffusion behavior of the material. The Bueche-Kelley theory provides the following estimate of T_g:

$$T_g = \frac{\beta_r(1 - v_f)T_{gr} + \beta_f v_f T_{gf}}{\beta_r(1 - v_f) + \beta_f v_f} \qquad (40)$$

where β_r and β_f and ν_r and ν_f are the moisture expansion coefficients and the volume fractions of the resin and the fiber. Typical values of the glass transition temperature are presented in Figures 10 and 11.

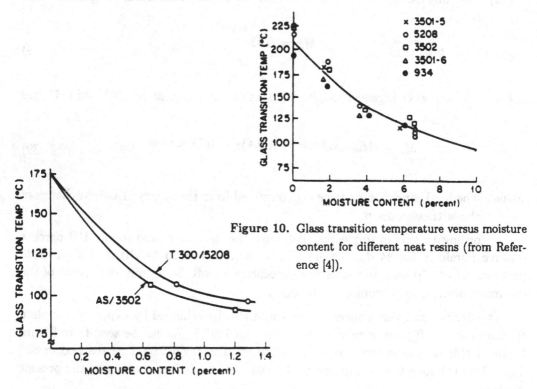

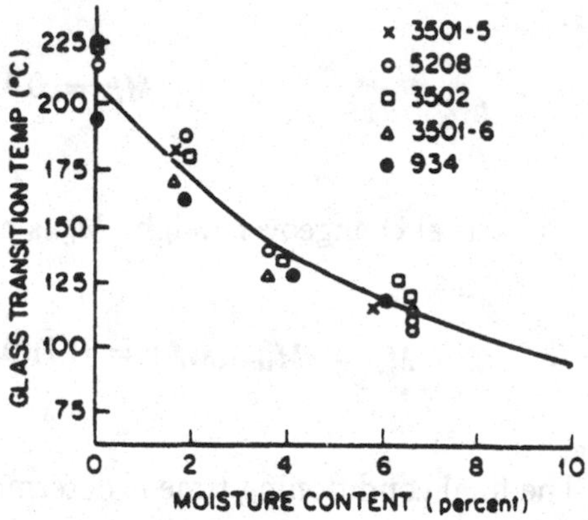

Figure 10. Glass transition temperature versus moisture content for different neat resins (from Reference [4]).

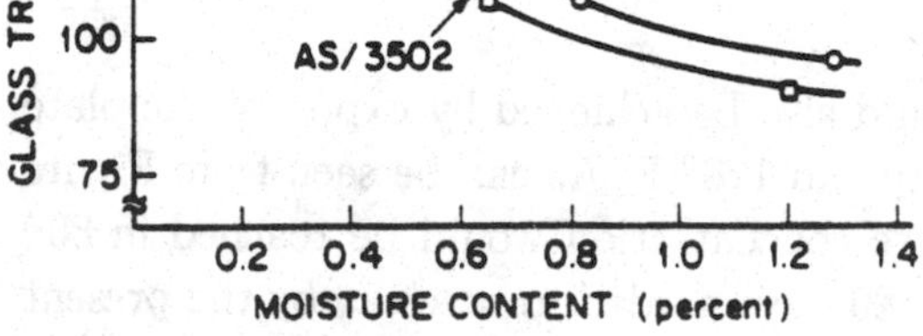

Figure 11. Glass transition temperature versus moisture content for two types of graphite-epoxy composites (from Reference [4]).

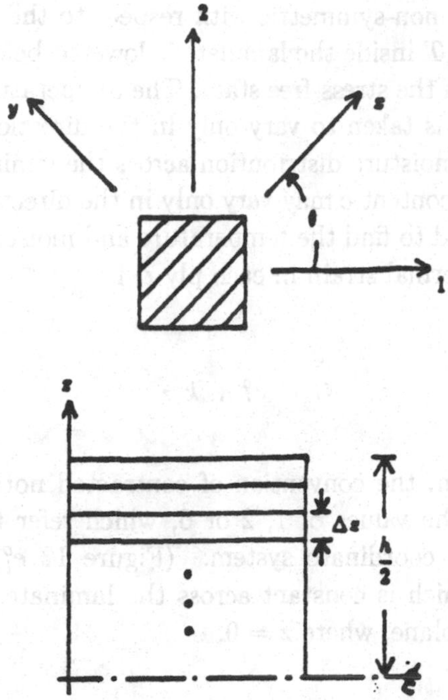

Figure 12. Coordinate system used in calculating hygrothermal deformations.

9. HYGROTHERMAL DEFORMATIONS

Changes in temperature and in moisture content may introduce significant changes in dimension and in shape which must be taken into account when designing composite structures. Hygrothermal deformation of plates can be calculated by the procedures described below.

We consider a plate of length l, width w, and thickness h made of N plies of unidirectional fiber reinforced plies. The layup of the laminate is arbitrary and may be either symmetric or non-symmetric with respect to the center plane of the laminate. The temperature T inside the laminate is lowered below the temperature T_a at which the laminate is in the stress-free state. The temperature distribution inside the laminate is known but is taken to vary only in the direction normal to the surface. There is also a known moisture distribution across the laminate. Similar to the temperature, the moisture content c may vary only in the direction normal to the surface of the plate. It is desired to find the temperature and moisture induced strains in the laminate. The hygrothermal strain in each ply ϵ_i is

$$\epsilon_i = e_i^0 + k_i z \tag{41}$$

In writing this equation, the convention of contracted notation was adopted. The subscript i may have the values of 1, 2 or 6, which refer to the normal and shear components in the 1-2 coordinate system. (Figure 12 e_i^o) is the in-plane overall hygrothermal strain which is constant across the laminate. k_i is the hygrothermal curvature of the centerplane, where $z = 0$.

Table 7. Summary of equations for calculating hygrothermal deformations.

	$D'_{ij} = D_{ij} - (B_{ik} a_{kl} B_{lj})^{-1}$
	$A'_{ij} = a_{ij} + a_{ik} B_{kl} D'_{lm} B_{mn} a_{nj}$
$\epsilon = e^o + kz$	$B'_{ij} = -a_{ik} B_{kl} D'_{lj}$
$\epsilon_i^o = A'_{ij} N_j + B'_{ij} M_j$	$a_{ij} = A_{ij}^{-1}$
$k_i = B'_{ij} N_j + D'_{ij} M_j$	$A_{ij} = \int_{-h/2}^{h/2} Q_{ij}\,dz$
$N_i = \int_{-h/2}^{h/2} Q_{ik} e_k\,dz$	$B_{ij} = \int_{-h/2}^{h/2} Q_{ij} z\,dz$
$M_i = \int_{-h/2}^{h/2} Q_{ik} e_k z\,dz$	$D_{ij} = \int_{-h/2}^{h/2} Q_{ij} z^2\,dz$
$e_k = \alpha_k \Delta T + \beta_k c$	$\alpha_1 = \cos^2\theta\alpha_x + \sin^2\theta\alpha_y$
	$\alpha_2 = \sin^2\theta\alpha_x + \cos^2\theta\alpha_y$
	$\alpha_6 = 2\cos\theta\sin\theta\alpha_x - 2\cos\theta\sin\theta\alpha_y$

Transformation of the moisture expansion coefficient β_k from ply to laminate coordinate system follows the same rule as the transformation of α_k

Expressions for Q_{ij} are given in Table 8

The overall in-plane hygrothermal strain e_i^o and hygrothermal curvature k_i are given by

$$e_i^o = A'_{ij}N_j + B'_{ij}M_j \tag{42}$$

$$k_i = B'_{ij}N_j + D'_{ij}M_j \tag{43}$$

The coefficients A'_{ij}, B'_{ij} and D'_{ij} depend on the modulus Q_{ij} and on the laminate configuration (see Table 7). N_j is the in-plane hygrothermal stress resultant (j = 1,2,6):

$$N_j = \int_{-h/2}^{h/2} Q_{jk}e_k dz \tag{44}$$

and M_j is the hygrothermal moment:

$$M_j = \int_{-h/2}^{h/2} Q_{jk}e_k z dz \tag{45}$$

e_k represents the off-axis strain which would occur in each ply if the ply were unrestrained (k = 1,2,6):

$$e_k = \alpha_k \Delta T + \beta_k c \tag{46}$$

α_k and β_k are the off-axis thermal and moisture expansion (swelling) coefficients of each ply, respectively α_k and β_k can be calculated from the known on-axis values of these coefficients by the equations given in Table 7. ΔT is the difference between the lamina (ply) temperature T and the lowest temperature at which the laminate is still in a stress-free state T_a ($\Delta T = T - T_a$). c is the moisture concentration.

For specified temperature and moisture distributions, Equations (41) to (46) provide the overall in-plane hygrothermal strain e_i^o and curvature k_i. The calculations must be performed by numerical methods. A computer code, HSTRESS, is available which can be used to perform the calculations.

10. IN-PLANE HYGROTHERMAL DEFORMATIONS OF SYMMETRIC LAMINATES

Changes in curvature do not take place, and only in-plane deformations occur in plates consisting of symmetric lay-ups in which the temperature and moisture distributions are uniform across the thickness. The in-plane hygrothermal deformation (strain) then may be written as [6]:

$$\epsilon_i^o = e_i^o = \alpha_i^o \Delta T + \beta_i^o c \tag{47}$$

ΔT is the difference between the uniform laminate temperature T and the lowest temperature T_a in which the laminate is in a stress-free state ($\Delta T = T - T_a$). c is the uniform moisture concentration in the laminate. α_i^o and β_i^o are the laminate thermal and moisture expansion coefficients, respectively. For given moisture and temperature distributions the deformations (strains) can easily be determined once α_i^o and β_i^o are known.

Analytic expressions for calculating α_i^o and β_i^o are given in Table 7. Results for typical graphite-epoxy, boron-epoxy, glass-epoxy, and Kevlar-epoxy laminates are given in Tables 10-13. The material systems and material properties used in the calculations are described in Tables 8 and 9.

The laminate thermal and moisture expansion coefficients given in Tables 10, 12, and 13 may be applied to laminates not included in these tables by applying the relationships summarized in Table 14.

11. HYGROTHERMAL STRESSES

Once the hygrothermal strains ϵ_j in each ply are known [see Equations (41) and (47)], the corresponding hygrothermal stresses may be calculated by the expression:

$$\sigma_i = Q_{ij}(\epsilon_j - e_j) \tag{48}$$

Q_{ij} is the off-axis ply modulus given in Table 7; ϵ_j is hygrothermal strain in the "unrestrained" ply [Equation (46)].

Table 8. Summary of equations for calculating in-plane hygrothermal deformations of symmetric laminates with uniform temperature and moisture distributions.

$\epsilon_i^e = \alpha_i^e \Delta T + \beta_i^e \Delta c$	$\alpha_1 = p + q\cos 2\theta$	$p = 1/2(\alpha_x + \alpha_y)$	$V_1 = 1/h\ \Sigma_{m=1}^{N} \cos 2\theta_m (\Delta z)_m$
$\alpha_1^e = (r + s)/\lvert A_{ij}\rvert$	$\alpha_2 = p - q\cos 2\theta$	$q = 1/2(\alpha_x - \alpha_y)$	$V_2 = 1/h\ \Sigma_{m=1}^{N} \cos 4\theta_m (\Delta z)_m$
$\alpha_2^e = (r - s)/\lvert A_{ij}\rvert$	$\alpha_6 = 2q\sin 2\theta$		$V_3 = 1/h\ \Sigma_{m=1}^{N} \sin 2\theta_m (\Delta z)_m$
$\alpha_6^e = t/\lvert A_{ij}\rvert$	$U_1 = 1/8[3Q_{xx} + 3Q_{yy} + 2Q_{xy} + 4Q_{ss}]$		$V_4 = 1/h\ \Sigma_{m=1}^{N} \sin 4\theta_m (\Delta z)_m$
$r = 1/2[RT + SU_2(V_1W_1 + V_3W_2)]$	$U_2 = 1/2[Q_{xx} - Q_{yy}]$		$Q_{11} = U_1 + U_2\cos 2\theta + U_3\cos 4\theta$
$s = 1/2[qYW_1]$	$U_3 = 1/8[Q_{xx} + Q_{yy} - 2Q_{xy} - 4Q_{ss}]$		$Q_{12} = U_4 - U_3\cos 4\theta$
$t = qYW_2$	$U_4 = 1/8[Q_{xx} + Q_{yy} + 6Q_{xy} - 4Q_{ss}]$		$Q_{22} = U_1 - U_2\cos 2\theta + U_3\cos 4\theta$
$\lvert A_{ij}\rvert = 1/2[T(U_1 + U_4) + U_2^2(V_1W_1 + V_3W_2)]$	$U_5 = 1/8[Q_{xx} + Q_{yy} - 2Q_{xy} + 4Q_{ss}]$		$Q_{16} = 1/2U_2\sin 2\theta + U_3\sin 4\theta$
$R = p(U_1 + U_4) + qU_2$	$Q_{xx} = E_x/(1 - \nu_x^2 E_y/E_x)$		$Q_{26} = 1/2U_2\sin 2\theta - U_3\sin 4\theta$
$S = pU_2 + q(U_1 + 2U_3 - U_4)$	$Q_{xy} = \nu_x E_y/(1 - \nu_x^2 E_y/E_x)$		$Q_{66} = U_5 - U_3\cos 4\theta$
$T = (U_4 - U_1)^2 - 4U_3^2(V_2^2 + V_4^2)$	$Q_{yy} = E_y/(1 - \nu_x^2 E_y/E_x)$		
$W_1 = 2U_3(V_3V_4 + V_1V_2) + V_1(U_4 - U_1)$	$Q_{ss} = E_s$		
$W_2 = 2U_3(V_1V_4 + V_3V_2) + V_3(U_4 - U_1)$			
$Y = U_4^2 - 2U_3U_4 - 2U_1U_3 + U_3^2 - U_1^2$			

The same expressions apply to the laminate moisture coefficients β^e (Replace α with β in the above expressions.)

Table 9. The material systems used in calculating in-plane thermal and moisture expansion coefficients of symmetric laminates.

I.	T300/5208
II	T300/976
III.	AS/3501
IV	B(4)/5505
V	Scotchply 1002
VI.	Kevlar 49/Epoxy

Table 10. Properties of the material systems used in calculating in-plane thermal and moisture expansion coefficients of symmetric laminates.

Material*	I	II	III	IV	V	VI
E_x	26.3	20.2	20.0	29.6	5.6	11 0
E_y	1 49	0 5	1 3	2 68	1 2	0.8
E_s	1 04	1 0	1 03	0 81	0 6	0 33
ν_x	0.28	0.09	0.30	0.23	0 26	0 34
α_x	0.01	0.3	−0 17	3 4	4 8	−2 2
α_y	12.5	16 0	15 6	16.8	12 3	43 9
β_x	0.0					
β_y	0 44					
Δz	6	6	6	6	6	6

E_x Longitudinal Young's modulus, × 10^6 psi
E_y Transverse Young's modulus × 10^6 psi
E_s Longitudinal shear modulus, × 10^6 psi
ν_x. Longitudinal Poisson's ratio
α_x. Longitudinal thermal expansion coefficient, × 10^{-6} °F^{-1}
α_y Transverse thermal expansion coefficient, × 10^{-6} °F^{-1}
β_x Longitudinal moisture expansion (swelling) coefficient
β_y Transverse moisture expansion (swelling) coefficient
Δz. Thickness of one composite ply, × 10^{-3} in

*The numbers designate the material systems. See Table 8

Table 11. The in-plane laminate thermal expansion coefficients of graphite-epoxy laminates.

Material	I			II			III		
	α_1^a	α_2^a	α_6^a	α_1^a	α_2^a	α_6^a	α_1^a	α_2^a	α_6^a
[0/90]	0 847	0 847	0 0	0 712	0 712	0 0	1 04	1 04	0 0
[0/0]	0 011	12.5	0.0	0 30	16 0	0.0	−0 17	15 6	0.0
[+30/−30]	−1 32	6 31	0 0	−0 440	3 94	0 0	−1 47	7 24	0 0
[+45/−45]	0.847	0 847	0.0	0 712	0 712	0 0	1 04	1 04	0.0
[0/0/0/90]	0 398	1 98	0.0	0 460	1 42	0 0	0.398	2 63	0 0
[0/+30/−30/90]	0 454	1 46	0 0	0 491	1 07	0.0	0 496	1 88	0 0
[0/+45/−45/90]	0.847	0 847	0.0	0 712	0.712	0 0	1 04	1 04	0 0

a) The units of α_i^a are (× 10^{-6} °F^{-1})
b) The roman numerals designate the materials systems described in Tables 8 and 9

Table 12. The in-plane laminate thermal expansion coefficients of boron-epoxy, glass-epoxy, and Kevlar-epoxy laminates.

	IV			V			VI		
Material	α_1^o	α_2^o	α_6^o	α_1^o	α_2^o	α_6^o	α_1^o	α_2^o	α_6^o
[0/90]	4.71	4.71	0.0	6.31	6.31	0.0	1.77	1.77	0.0
[0/0]	3.40	16.8	0.0	4.8	12.3	0.0	−2.2	43.9	0.0
[+30/−30]	0.751	14.0	0.0	4.54	9.57	0.0	−9.47	28.9	0.0
[+45/−45]	4.71	4.71	0.0	6.31	6.31	0.0	1.77	1.77	0.0
[0/0/0/90]	3.97	6.44	0.0	5.46	7.89	0.0	−0.314	6.79	0.0
[0/+30/−30/90]	4.06	5.70	0.0	5.69	7.15	0.0	−0.0705	4.57	0.0
[0/+45/−45/90]	4.71	4.71	0.0	6.31	6.31	0.0	1.77	1.77	0.0

a) The units of α_i^o are ($\times 10^{-6}$ °F^{-1})
b) The roman numerals designate the materials systems described in Tables 8 and 9

Table 13. The in-plane laminate moisture expansion coefficients of graphite-epoxy T300/5208 laminates.

	β_1^o	β_2^o	β_6^o
[0/90]	0.029	0.029	0.0
[0/0]	0.0	0.44	0.0
[+30/−30]	−0.047	0.22	0.0
[+45/−45]	0.029	0.029	0.0
[0/0/0/90]	−0.014	−0.069	0.0
[0/+30/−30/90]	−0.016	0.051	0.0
[0/+45/−45/90]	0.029	0.029	0.0

The unit β_i^o are dimensionless

Table 14. The relationships between the laminate thermal expansion coefficients for different types of layups.

$$\alpha_i^o([+45/-45]) = \alpha^o([0/90]) \qquad (i = 1,2,6)$$
$$\alpha_1^o([+60/-60]) = \alpha_2^o([+30/-30])$$
$$\alpha_2^o([+60/-60]) = \alpha_1^o([+60/-60])$$
$$\alpha_6^o([+60/-60]) = 0$$
$$\alpha_1^o([+90/-90]) = \alpha_2^o([+0/-0])$$
$$\alpha_2^o([+90/-90]) = \alpha_1^o([+0/-0])$$
$$\alpha_6^o([+90/-90]) = 0 \, ,$$
$$\alpha_1^o[(0/\pm 60/90]) = \alpha_2^o([0/\pm 30/90])$$
$$\alpha_2^o[(0/\pm 60/90]) = \alpha_1^o([0/\pm 30/90])$$
$$\alpha_6^o([0/\pm 60/90]) = 0$$
$$\alpha_1^o[(0/\pm 90/90]) = \alpha_2^o([0/\pm 0/90])$$
$$\alpha_2^o[(0/\pm 90/90]) = \alpha_1^o([0/\pm 0/90])$$
$$\alpha_6^o([0/\pm 90/90]) = 0$$

The same expressions apply to the laminate moisture expansion coefficients β_i^o (Replace α_i^o with β_i^o in the above expressions.)

REFERENCES

1.) Springer, G. S., ed., *Environmental Effects on Composite Materials*, Vol. 1, Lancaster, Pennsylvania: Technomic Publishing Company, Inc., (1981).

2.) Kreith, F., *Principles of Heat Transfer*. International Textbook (1966).

3.) Springer, G. S., ed., *Environmental Effects on Composite Materials*, Vol. 2, Lancaster, Pennsylvania: Technomic Publishing Company, Inc., (1984).

4.) Springer, G. S., "Moisture Absorption in Fibre-Resin Composites", in *Developments in Reinforced Plastics*, G. Pritchard, ed., *Applied Science*, pp. 43-65 (1982).

5.) Ciriscioli, P. R., W. I. Lee, D. G. Peterson, G. S. Springer and J. M. Tang, "Accelerated Environmental Testing of Composites", *Environmental Effects on Composites Materials*, Vol. 3. George S. Springer, ed. Lancaster, Pennsylvania: Technomic Publishing Company, Inc., (1987).

6.) Doxsee, L. E., W. I. Lee, G. S. Springer and S. S. Chang, "Temperature and Moisture Induced Deformations in Composite Sandwich Panels", *Journal of Reinforced Plastics and Composites*, 4:326-353 (1985).

7.) Dexter, H. B. and D. J. Baker, "Worldwide Flight and Ground-Based Exposure of Composite Materials", ACEE Composite Structures Technology Conference, Seattle, Washington, pp. 17-49 (August 1984).

8.) Coggeshal, R. L., "Environmental Exposure Effects on Composite Materials for Commercial Aircraft", NASA Contractor Report 177929 (November 1985).

The contents of this chapter are reprinted from the REFERENCE BOOK FOR COMPOSITE TECHNOLOGY (S. M. Lee, ed.) 2, pp. 17-30,1981, by permission of Technomic Publishing Company.

Printed in the United States
By Bookmasters